Veterinary Protozoan and Hemoparasite Vaccines

Editor

I. G. Wright

Senior Principal Research Scientist
Division of Tropical Animal Production
CSIRO
Brisbane, Queensland, Australia

CRC Press, Inc.
Boca Raton, Florida

Library of Congress Cataloging-in-Publication Data

Veterinary protozoan and hemoparasite vaccines.
  Bibliography: p.
  Includes index.
  1. Protozoan vaccines.   2. Rickettsial vaccines.
3. Veterinary protozoology—Research.   I. Wright, I. G.
QR189.5.P76V47   1989   636.089′6936   88-22121
ISBN 0-8493-4757-2

Direct all inquiries to CRC Press, Inc., 2000 Corporate Blvd., N.W., Boca Raton, Florida, 33431.

© 1989 by CRC Press, Inc.
International Standard Book Number 0-8493-4757-2

Library of Congress Card Number 88-22121
Printed in the United States

THE EDITOR

Ian Gordon Wright, Ph.D., D.V.Sc., B.A., is Leader of the Babesiosis Vaccine Project, CSIRO Division of Tropical Animal Production, Indooroopilly, Brisbane, Australia. Dr. Wright graduated in 1965 from the University of Queensland, Brisbane, with a degree in Veterinary Science, obtained an honors degree from the Department of Parasitology, University of Queensland, in 1966, and a Ph.D. from the same department in 1971. In 1983 he was awarded a D.V.Sc. from the same university for his work on *Babesia* infections.

Dr. Wright has almost exclusively studied the pathogenesis and immunity of *Babesia* infections since graduation and in addition studied the pathogenesis of Trypanosome infections at Imperial College, London, in 1975. He has undertaken numerous consultancies for the joint FAO/IAEA Division, Vienna, Austria. These consultancies have been related to the establishment of serodiagnostic and related immunological techniques for the study of *Babesia* infections in developing countries.

Dr. Wright has presented numerous invited papers at international conferences and has published over 90 research papers. His current major research interests include the immunopathophysiology of *Babesia* and related infections and the development of a synthetic vaccine against various *Babesia* species.

CONTRIBUTORS

Patricia C. Augustine, Ph.D.
Research Scientist
Protozoan Diseases Laboratory
Livestock and Poultry Science Institute
Agricultural Research Service
U.S. Department of Agriculture
Beltsville, Maryland

J. D. Bezuidenhout, B.V.Sc.
Deputy Director
Department of Agriculture and Water
 Supply
Veterinary Research Institute,
 Onderstepoort
Pretoria, South Africa

Harry D. Danforth, Ph.D.
Research Scientist
Protozoan Diseases Laboratory
Livestock and Poultry Science Institute
Agricultural Research Service
U.S. Department of Agriculture
Beltsville, Maryland

B. V. Goodger, Ph.D.
Senior Principal Research Scientist
Division of Tropical Animal Production
CSIRO
Indooroopilly, Queensland, Australia

Anthony D. Irvin, Sc.D.
Senior Animal Health Advisor
Overseas Development Administration
London, England

Alan M. Johnson, Ph.D.
Principal Hospital Scientist and Senior
 Lecturer
Department of Clinical Microbiology
Flinders Medical Center
 and
Flinders University of South Australia
Bedford Park, South Australia, Australia

Ibulaimu Kakoma, Ph.D.
Associate Professor
Department of Veterinary Pathobiology
College of Veterinary Medicine
University of Illinois
Urbana, Illinois

Sonia Montenegro-James, Ph.D.
Visiting Assistant Professor and
 Research Immunoparasitologist
Department of Veterinary Pathobiology
College of Veterinary Medicine
University of Illinois
Urbana, Illinois

W. Ivan Morrison, Ph.D.
Program Coordinator Theileriosis
Department of Immuno-Pathology
I.L.R.A.D.
Nairobi, Kenya

Guy H. Palmer, Ph.D.
Associate Professor
Department of Veterinary Microbiology
 and Pathology
College of Veterinary Medicine
Washington State University
Pullman, Washington

Miodrag Ristic, Ph.D.
Professor
Department of Veterinary Pathobiology
University of Illinois
Urbana, Illinois

Stuart Z. Shapiro, M.D., Ph.D.
Assistant Professor
Department of Veterinary Pathobiology
College of Veterinary Medicine
University of Illinois
Urbana, Illinois

S. M. Taylor, Ph.D., B.V.M.S., M.R.C.V.S.
Senior Veterinary Research Officer
Department of Parasitology
Veterinary Research Laboratories
Belfast, Northern Ireland

I. G. Wright, Ph.D.
Principal Research Scientist
Division of Tropical Animal Science
CSIRO
Brisbane, Australia

TABLE OF CONTENTS

Chapter 1

ANAPLASMA VACCINES

Guy H. Palmer

TABLE OF CONTENTS

I. INTRODUCTION

Arthropod-borne hemoparasitic diseases are enzootic in half of the livestock production areas of the world and remain a severe constraint to agricultural development in the tropics.[1] Currently these diseases are controlled primarily by reducing or eliminating transmission using regular application of acaricides to cattle. Paradoxically, efficient arthropod control results in a highly disease-susceptible cattle population and sets the stage for devastating epizootics if breaks in transmission control occur.[2] Increased hemoparasite transmission caused by use of blood-contaminated fomites between cattle, development of acaricide resistance in ticks, and disruption in dipping programs caused by currency or labor shortages are continual threats to animal health in the tropics.[2,3] This is well illustrated by the loss of over 1 million cattle to hemoparasitic diseases in Zimbabwe when highly efficient tick control was disrupted by war in the late 1970s.[2] In contrast, vaccines may be used to protect cattle in regions with either high- or low-transmission levels as well as fully susceptible cattle to be imported into an enzootic area. Vaccines against bovine anaplasmosis have been used successfully to prevent severe morbidity and mortality in susceptible cattle, and research to develop more effective vaccines continues. In this chapter, the development of vaccines against *Anaplasma* infections is reviewed with focus upon recent advances in antigenically defined vaccines.

Anaplasmosis is an arthropod-borne hemoparasitic disease of cattle and wild ruminants caused by the rickettsia *Anaplasma marginale* or the less virulent *A. centrale*.[4] Anaplasmosis occurs worldwide and, along with babesiosis, cowdriosis, theileriosis, and trypanosomiasis, remains the greatest obstacle to meat, milk, and fiber production in tropical, lesser developed nations.[1,2,3] Infection results from inoculation of cattle with either a tick stage of *Anaplasma* (following transstadial or intrastadial replication in the midgut epithelium of a variety of ixodid ticks) or the intraerythrocytic stage (after mechanical transmission on the mouth parts of biting flies or by blood-contaminated fomites, i.e., syringes or dehorning instruments).[5,6] Following transmission there is a prepatent period of 20 to 40 d, during which there is a low but increasing percentage of parasitized erythrocytes.[5,6a] The prepatent period, during which infected cattle are clinically normal, is followed by an acute phase, during which the parasitemia increases dramatically and severe hemolytic anemia occurs.[5] Dramatic weight loss, abortion, and death frequently occur during the acute phase.[7] Cattle recovered from acute disease remain persistently infected with a low-level parasitemia for long periods and serve as a reservoir for transmission of the organism.[8] These cattle are fully resistent to challenge with the homologous isolate; however, they remain susceptible to infection from certain heterologous isolates of *Anaplasma*.[9]

Subtropical and tropical regions enzootic for anaplasmosis correspond to the distribution of the arthropod vectors including 29 species of ticks and numerous hematophagous flies and mosquitos shown experimentally to transmit the disease.[10,11] Several of these vectors also transmit other hemoparasites; as a result, anaplasmosis frequently exists coenzootically with babesiosis, cowdriosis, and theileriosis.[12] Due to this complex of hemoparasitic diseases, the economic impact strictly attributable to anaplasmosis is difficult to estimate in many of the nations with enzootic regions. Within the U.S., where anaplasmosis occurs exclusive of the other major hemoparasitic diseases, the annual loss of 50,000 to 100,000 head of cattle has been estimated at $300 million (1986 $U.S.).[13] Recent epidemiologic data from Texas, with approximately 30% enzootic regions, attributed severe losses to anaplasmosis, including death in 36% of clinical cases, abortion in 24% of clinical cases in pregnant cows, an average weight loss of 86 kg during acute infection, and increased veterinary and management costs of $52 and $30 per head, respectively.[7] Severe production losses attributed to prolonged subclinical anemia in persistently infected cattle, especially in regions where anaplasmosis and babesiosis occur together, have recently been recognized and add dramatically to the already severe constraint upon livestock production in enzootic regions.[14]

In enzootic regions with a high level of transmission an "enzootically stable" livestock population may develop if no control measures are instituted. Calves are uniformly infected at an early age and usually suffer only mild clinical disease, presumably due to partial protection from colostral-derived antibody.[15] The calves upon recovery are protected from severe morbidity and mortality when subsequently challenged. A high level of transmission ensures continual reexposure to boost immunity. Allowing an enzootically stable herd to develop may be considered desirable; however, many regions thought to be stable may actually be relatively unstable as transmission levels vary greatly seasonally and annually depending upon climatic conditions being favorable for arthropod vectors.[16-18] Prolonged levels of low transmission allow some calves to mature without becoming infected and to coexist within a herd with carrier cattle.[18] A return to high levels of transmission following climatic changes may result in disastrous epizootics. In addition, recent epidemiologic data demonstrates that the persistently infected *Anaplasma* carriers that create enzootic stability suffer from prolonged subclinical anemia and poor weight gain, especially in regions endemic for both anaplasmosis and babesiosis.[14] The risk of devastating losses due to instability created by climatic changes and the production losses resulting from chronic anemia may combine to make a goal of enzootic stability undesirable for nations dependent upon efficient livestock production and with the infrastructure to develop anaplasmosis control programs.

Current control measures include (1) control of arthropod transmission, (2) premunization or vaccination of susceptible cattle, and (3) use of antibiotics to treat or prevent *Anaplasma* infections. Commonly a combination of these measures is employed, and the prevailing control programs vary markedly between regions and nations depending upon principal means and seasonality of transmission, presence of other major arthropod-borne hemoparasitic diseases, livestock management systems used, availability and economics of antibiotics, acaricides, and vaccines, and the goals of the program, i.e., eradication vs. control, and the level of control desired.[19] Despite the availability of these programs, the distribution of anaplasmosis has not been reduced, and the severe losses continue on five continents, clear evidence that current control programs are not sufficient.[1,3,12] Control of arthropod transmission by regular use of acaricides is labor intensive, expensive (frequently requiring a significant percentage of foreign capital expenditure by lesser developed nations), and creates a highly susceptible cattle population. This population is vulnerable to disruptions in the control program, development of acaricide resistance in ticks, and any non-arthropod mechanisms of transmission.[2,3,19] Similarly, reliance on antibiotic prophylaxis by continual feeding of tetracyclines results in herds vulnerable to program disruptions. The labor and expense involved to ensure that all cattle receive a minimum daily amount of antibiotic is not amenable to most tropical livestock systems. In spite of the efficacy of tetracyclines in treatment and chemoprophylaxis of anaplasmosis, antibiotic-based control programs have found only limited acceptance in the U.S. and virtually none in other regions. The effect of pending U.S. legislation regarding low-level antibiotic feeding to cattle upon this control method is not presently clear.

The protection engendered by recovery from *Anaplasma* infection clearly indicates that immunoprophylaxis is a realistic and achievable goal. Vaccination provides a means to efficiently and economically protect cattle both within enzootically stable and unstable areas as well as susceptible adult cattle to be imported into these areas. Although premunization (deliberate infection of cattle to induce immunity) has been practiced for 75 years and commercially available killed and live whole organism vaccines have been used for nearly 20 years, the need remains to develop a widely cross-protective, economical vaccine suitable for use in tropical regions.[20]

II. POSTINFECTION IMMUNITY

Cattle recovered from acute infection with *Anaplasma marginale* are protected from severe morbidity and mortality when subsequently challenged with a homologous isolate of *A. marginale*.[9] Parasitemia upon challenge is very low and frequently undetectable. This post infection immunity is not dependent upon the animal remaining persistently infected, as cattle chemotherapeutically cleared of infection retain immunity for at least 8 months.[21] Chemical immunosuppression or splenectomy abrogates this immunity as would be expected; however, the cellular and molecular mechanisms responsible for immunity are not clear.[22-24] Tenable mechanisms include (1) antibody blockade of the erythrocyte binding site on the *A. marginale* initial body surface; (2) direct lysis of initial bodies by antibody or complement; and (3) phagocytosis and intracellular killing of initial bodies or parasitized erythrocytes coated with antibody or complement. To date no significant data to verify or exclude any of these mechanisms in postinfection immunity has been presented. Effector antibody in any of the three mechanisms requires a defined epitope specificity and affinity that need to be identified. Correlative data between the level of protection and antibody titer to crude *A. marginale* antigens from serologic tests have not been helpful in understanding immune mechanisms.[25] Previously a correlation has been made between the level of postinfection immunity and the lymphocyte migration inhibition test.[26-28] Subsequent inability to identify a discrete migration inhibition factor (MIF) has made it difficult to interpret this finding. Several lymphokines, notably the interferons, have some MIF-like activity. This suggests that the previously noted correlation with immunity most likely reflects macrophage and/or T lymphocyte (including helper functions) activation. Complete understanding of the mechanistic basis of postinfection immunity will be significant in designing rational strategies to develop improved vaccines.

The presence of postinfection immunity provides a strong basis for vaccine development. Initial ''vaccines'' attempted to mimic the immunity engendered by natural field exposure while avoiding the severe morbidity and mortality of virulent infections. This method, premunization, has provided a method to partially protect cattle from field infections (both homologous and more virulent heterologous isolates) and remains in use in several countries (advantages and disadvantages of premunization will be discussed in Section IV). Importantly, both experimental and field experience with premunization have provided information regarding cross-species and cross-isolate protection, the likelihood of antigenic variation, and immunity to tick challenge vs. challenge with parasitized erythrocytes. Infection of cattle with *A. centrale* (a less virulent *Anaplasma* species decribed by Theiler and originally confined to Africa) does not prevent challenge infection by *A. marginale,* but decreases the severity of the clinical signs with most A. marginale isolates.[29,30] Other *A. marginale* isolates produce severe morbidity in *A. centrale*-premunized cattle.[31,32] In contrast, *A. marginale*-premunized cattle are uniformly protected against *A. centrale* or homologous *A. marginale* challenge, but clinical signs of variable severity result upon challenge with heterologous isolates of *A. marginale*.[9,33,34] This variation in disease upon heterologous challenge ranges from mild anemia (25% decrease in packed cell volume [PCV]) to severe anemia (75% decrease in PCV) with parasitemias greater than 20%. These experiments indicated for the first time that antigenic differences critical to protection existed between *A. centrale* and *A. marginale* and between different isolates of *A. marginale,* and that effective vaccines must protect against the different organisms. A second potential obstacle to effective immunoprophylaxis is antigenic variation. Although there is no evidence for cyclical emergence of antigenic variants in anaplasmosis, the ability of the parasite to persist in spite of an immune response and the occurrence of recrudescent parasitemias in carrier cattle indicates that a complex host-parasite relationship develops. The role of antigenic variation in this relationship and, therefore, its effect on immunization efficacy is unknown.[35,36] A third consideration

Table 1

REACTIVITY OF ISOLATE-RESTRICTED MONOCLONAL ANTIBODIES TO GEOGRAPHICALLY DISTINCT *A. MARGINALE* ISOLATES[a]

Monoclonal Ab	Ab Reactivity[b]									
	FL	ID	TX	VA	WA-C	WA-O	ISNT	IST	KKb	KKp
Tryp 1E1[c]	−	−	−	−	−	−	−	−	−	−
Ana F22A4	+	−	−	−	−	−	−	−	−	−
Ana 011C2	−	−	−	−	+	+	−	−	+	+
Ana 012B5	−	−	−	−	+	+	−	−	+	+
Ana 023D5	−	+	+	+	+	+	−	+	−	−
Ana 024D5	−	+	+	+	+	+	−	+	−	−
Ana 070A2	+	+	+	+	+	+	+	+	−	+
Ana R19A6	−	−	−	+	−	−	−	−	+	+
Ana R94C1	−	−	−	+	−	−	−	−	+	+
Ana R17A6	−	−	−	+	−	−	−	−	−	−
Ana R83B3	−	−	−	+	−	−	−	−	−	−

[a] Isolate abbreviations: FL (Florida, U.S.), ID (South Idaho, U.S.), TX (North Texas, U.S.), VA (Virginia, U.S.), WA-C (Clarkston, Washington, U.S.), WA-O (Okanogan, Washington, U.S.), ISNT (Israel-non-tailed), IST (Israel-tailed), KKb (Kabete, Kenya), KKp (Kapiti, Kenya).

[b] Reactivity was determined using indirect immunofluoresence on acetone-fixed smears of parasitized erythrocytes. Monoclonal antibodies were unreactive with *Anaplasma ovis, Babesia bigemina, Babesia bovis,* and *Trypanosoma brucei.*

[c] Tryp 1E1 is an anti-*Trypanosoma brucei* monoclonal antibody used as a negative control.

is the relative roles of intraerythrocytic stages and tick stages of *A. marginale* in inducing protective immunity and the stage-specificity of the immune response. These influences upon vaccine development will be addressed in the following section.

III. ANTIGENS RELEVANT TO PROTECTION

A. Isolate-Restricted and Isolate-Common Antigens

Cross-protection experiments in premunized cattle established that antigenic differences among geographically distinct *A. marginale* isolates existed.[9,34] Structural and antigenic differences among isolates have been established by (1) identification of morphologicaly distinct isolates;[37] (2) demonstration of protein structural variation among isolates;[38] and (3) presence of isolate-common and isolate-restricted epitopes.[39-42] Morphologically distinct isolates were identified using phase contrast and fluorescence microscopy initially using Oregon and Florida isolates of *A. marginale.*[37] The Oregon isolate was shown to contain an appendage lacking in the Florida isolate. Numerous isolates of either morphology have been subsequently described within the U.S. and in other countries. The morphologic type remains constant within an infection and between passages in splenectomized and nonsplenectomized cattle.[43-45] Adsorption of polyclonal antisera made to appendaged *A. marginale* with the Florida isolate demonstrated that the appendage bears specific antigens not represented on the initial body itself.[33] That antigenic differences are not limited to the appendage has been shown by cross-adsorption experiments using two-appendaged isolates.[41] In our laboratories, we have used a panel of ten monoclonal antibodies (MAbs) that recognize isolate-restricted epitopes to discriminate among *A. marginale* isolates from the U.S., Israel, and Kenya (Table 1).[40,45] Using this panel, we have shown that both within a morphologic type or between morphologic types, there are multiple epitopes that differ between *A. marginale* isolates.[40,45]

Despite the identification of isolate-restricted epitopes, the majority of antigens are conserved between isolates.[40-42] Resolution of parasite-specific proteins using two-dimensional

IEF⊖ **(a)** ⊕

SDS

Wa.

(b)

Fl.

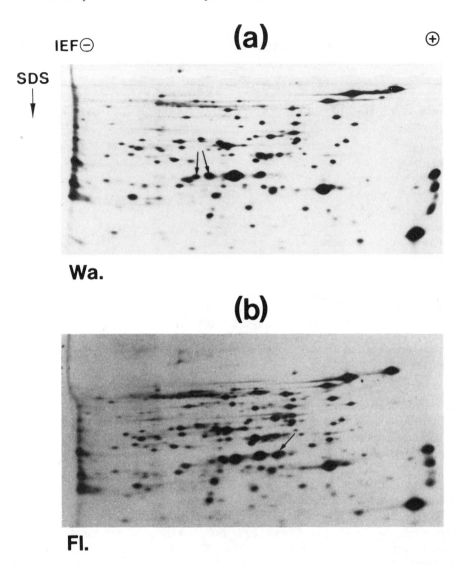

FIGURE 1.   Comparison of ^{35}S-radiolabeled proteins synthesized by the Washington-O and Florida isolates of *A. marginale*.[38] *A. marginale* initial bodies were radiolabeled with ^{35}S-methionine during short-term *in vitro* erythrocyte culture. Barbet et al.[38] previously demonstrated that the radiolabel incorporated exclusively into the *A. marginale* initial body using this procedure. Radiolabeled initial bodies were solubilized and subjected to two-dimensional electrophoresis followed by detection using fluorography.[38] (A) Washington-O isolate and (B) Florida isolate. The majority of the proteins are conserved in molecular weight and isoelectric point between the two isolates; however, three major radiolabeled proteins appeared unique to either the Washington-O or Florida isolate (arrows).[38] Comparison of ^{3}H-radiolabeled proteins (using a mixture of ^{3}H-radiolabeled amino acids not including methionine) from the two isolates confirmed the presence of these three unique proteins.[38]

gel electrophoresis allowed direct structural comparison of two morphologically distinct isolates (Figure 1).[38] The majority of proteins are common to both isolates, as would be expected for organisms of the same species and equally dependent upon intraerythorcytic invasion and replication. However, several major proteins were found to vary in molecular weight and isoelectric point, indicating primary protein structural differences between isolates (Figure 1).[38] Additional work comparing proteins from the Florida and Illinois (an appendaged isolate) identified variant high-molecular-weight antigens that shared common epitopes and one Florida isolate antigen that may bear only isolate-restricted epitopes.[39] The critical question relevant to cross-protective immunity and vaccine development is whether the protection-inducing epitopes are common or variant. Epitopes capable of generating protective immune responses (by direct initial body lysis, receptor blockade, or antibody-mediated phagocytosis) must be surface exposed. We have identified a panel of seven MAbs that recognize highly conserved epitopes on initial body surface proteins (Table 2).[45] The epitopes recognized by the seven MAbs are limited to two of the major initial body surface proteins, a protein with an apparent molecular weight of 105 kDa in the Florida isolate, AmF 105, and a 36-kDa protein, AmF 36.[40,45-47] The MAbs directed against epitopes on AmF 105 (15D2, 22B1, F34C1) recognize 100% of the organisms within parasitized erythrocytes regardless of the isolate.[40,45] This 100% binding is constant throughout the cycle of infection from <1% parasitemia through peak parasitemia with hemolytic crisis.[40] This striking epitope conservation exists despite antigenic, morphologic, and virulence differences among these 13 A. marginale isolates.[38-42,45] In contrast, MAbs against AmF 36 epitopes (F19E2, 050A2, 058A2, and 066A2) bind only 65 to 75% of the organisms in parasitized erythrocytes in all isolates.[40,45] This approximately 70% binding is constant between passages and regardless of the level of parasitemia.[40] The explanation for this limited epitope expression and its influence on cross-protective immunity remains unknown. There are undoubtedly additional conserved epitopes on AmF 105 and AmF 36 as well as on other surface-exposed initial body proteins that should be identified as potentially cross-protective immunogens.

A single, apparently immunodominant epitope on AmF 105 is recognized by MAbs 15D2 and 22B1 (Table 2).[47,48] Both antibodies, produced against a mixed inoculum of Virginia and Washington-O A. marginale, can neutralize 100% of the infectivity of the antigenically, morphologically, and structurally distinct Florida isolate (Table 3).[47] The high degree of conservation of this neutralization-sensitive epitope provided rationale for testing of native AmF 105 bearing this epitope and recombinant or synthetic constructs of this epitope for efficacy as a widely cross-protective immunogen (Section V).

B. Antigenic Variation

The differential reactivity of the various A. marginale isolates with a panel of isolate-restricted MAbs demonstrates that mechanisms to generate antigenic variants are present.[40,45] The limited cross-protective immunity seen with certain isolates in premunization or using a killed whole organism vaccine indicates that some of these antigenic variants are important in protection.[9,34] Critically relevant to development of effective immunoprophylaxis is whether this variation can be generated rapidly enough to avoid a neutralizing host response. While cyclical episodes of high parasitemia analagous to African trypanosomiasis are not seen, the persistence of Anaplasma in an immunologically hostile host and the occurrence of relapse parasitemias may result from emergence of antigenic variants. Systematic investigations of this phenomenon at the molecular level have not been reported for anaplasmosis. We have used panels of isolate-restricted and isolate-common MAbs to test for variation in a limited number of epitopes.[49] Constant patterns of reactivity using 18 MAbs on six U.S. isolates of A. marginale (Florida, Idaho, North Texas, Washington-C, Washington-O, and Virginia) were observed from <1% parasitemia through peak parasitemia with hemolytic crisis.[40] In addition, the antigenic profile of the Florida isolate was invariant after 18 months persistent infection in six calves.[49] These observations were limited to a relatively few epitopes and would not detect variation in other epitopes. Importantly, however, seven of these MAbs recognize surface protein epitopes, and that variation did not occur in these epitopes may indicate that cyclical variation will not be a severe impediment to immunoprophylaxis.[45] The highly cross-isolate-conserved AmF 105 epitope recognized by MAbs 15D2 and 22B1 was among the invariant epitopes and may represent an epitope required for erythrocyte invasion, metabolism, replication, or erythrocyte exit that is essential for A. marginale

Table 2
RECOGNITION OF A. MARGINALE INITIAL BODY SURFACE PROTEIN EPITOPES HIGHLY CONSERVED AMONG ANTIGENICALLY DISTINCT ISOLATES[a]

Monoclonal Ab	Surface protein reactivity	Ab reactivity[b]									
		FL	ID	TX	VA	WA-C	WA-O	ISNT	IST	KKb	KKp
Tryp 1E1[c]	—	–	–	–	–	–	–	–	–	–	–
Ana 15D2	AmF 105	+	+	+	+	+	+	+	+	+	+
Ana 22B1	AmF 105	+	+	+	+	+	+	+	+	+	+
Ana F34C1	AmF 105	+	+	+	+	+	+	+	+	+	+
Ana F19E2	AmF 36	+	+	+	+	+	+	+	+	+	+
Ana 050A2	AmF 36	+	+	+	+	+	+	+	+	+	+
Ana 058A2	AmF 36	+	+	+	+	+	+	+	+	+	+
Ana 066A2	AmF 36	+	+	+	+	+	+	+	+	+	+

[a] Isolate abbreviations are as in Table 1.

[b] Reactivity was determined by indirect immunofluorescence on acetone-fixed smears of parasitized blood smears. All Ana monoclonal antibodies were unreactive with other hemoparasites as in Table 1.

[c] Tryp 1E1 is an anti-*Trypanosoma brucei* monoclonal antibody used as a negative control.

Table 3
NEUTRALIZATION OF *A. MARGINALE*
INITIAL BODY INFECTIVITY BY
MONOCLONAL ANTIBODIES Ana 15D2
AND Ana 22B1[a]

	Number infected/challenged			
A. Monoclonal Ab	10^7	10^8	10^9	10^{10}
Tryp 1E1	3/3	3/3	3/3	7/7
Ana 15D2/22B1	0/4[b]	4/4	4/4	9/10

	Mean number of days between innoculation and 1% parasitemia			
B. Monoclonal Ab	10^7	10^8	10^9	10^{10}
Tryp 1E1	34	30	28	25
Ana 15D2/22B1	>75[b]	37	35	33
p[c]	ND[d]	≤0.01	≤0.01	≤0.01

[a] Graded numbers of initial bodies were purified from Florida isolate-parasitized erythrocytes and mixed with ascitic fluid from mice bearing Tryp 1E1 hybridomas (anti-*Trypanosoma brucei*) or Ana 15D2/22B1 hybridomas (anti-*A. marginale* reactive with a single epitope on AmF 105). The mixture was incubated at 20°C for 45 min and then inoculated intramuscularly into splenectomized calves. Blood samples were collected daily for 75 d postinoculation in order to determine packed cell volume and presence of parasitized erythrocytes.

[b] No parasitized erythrocytes were seen through 75 d postinoculation.

[c] The mean number of days between inoculation and 1% parasitemia were calculated for all infected cattle in each challenge group and compared between Tryp 1E1 and Ana 15D2/22B1 groups using the pooled *t* test.

[d] ND: not determined.

survival. The requirement that *Anaplasma* invade and adapt to intracellular existence undoubtedly limits the variation tolerable in certain surface epitopes. Determination of proteins and epitopes involved in these functions may be a fruitful strategy to identify invariant antigens for incorporation in vaccines. Nonetheless, the persistent nature of *Anaplasma* infection and occurrence of relapse parasitemias indicate a complex host-parasite relationship. Understanding the molecular basis of persistence and relapses will be necessary to develop optimal vaccines that completely prevent *Anaplasma* infection and the accompanying carrier state.

C. Antigenic Differences Between *Anaplasma* Species

The classification of a second *Anaplasma* species, *A. centrale,* was based upon the observation that the organism differed in position in parasitized erythrocytes and that this centrally located organism caused a less severe hemolytic disease.[29,50,51] Theiler noted in 1912 that although *A. marginale* challenge of *A. centrale*-premunized cattle resulted in establishment of an *A. marginale* infection, the clinical disease was less severe than in nonpremunized cattle.[29] (Subsequent work has shown the level of protection to be variable and is reviewed in Section IV.) The partial protection afforded by *A. centrale* premunization suggested that both species-common and species-specific epitopes related to protection were

Table 4
REACTIVITY OF ANTI-*A. MARGINALE*
MONOCLONAL ANTIBODIES WITH *A. CENTRALE*

Panel reactivity w/*A. marginale*	Reactivity w/*A. centrale*[a]
Panel 1: isolate restricted[b]	−
Panel 2: isolate common[c]	−
Panel 3: isolate common, reactive w/AmF 36[d]	+
Panel 4: isolate common, reactive w/AmF 105[e]	+

[a] Reactivity was determined by indirect immunfluorescence on acetone-fixed smears of *A. centrale* paratized erythrocytes.
[b] Includes all ten Ana monoclonal antibodies in Table 1.
[c] Monoclonal antibody Ana F35A1.
[d] Monoclonal antibodies Ana F19E2, Ana 050A2, Ana 058A2, and Ana 066A2.
[e] Monoclonal antibodies Ana 15D2, Ana 22B1, and Ana F34C1.

likely to be present.[29,50-52] Comparisons between capillary-tube agglutination test antigens of *A. marginale* (East African and USDA isolates) and of *A. centrale* demonstrated that the organisms were highly cross-reactive, but that *A. centrale* had significant antigenic differences from the two *A. marginale* isolates.[53] From a taxonomic viewpoint it is unclear whether the speciation is justified on other than a historical basis. The pathophysiologic changes in infection with either are indistinguishable, and the severity varies as greatly between *A. marginale* isolates as between *A. centrale* and *A. marginale*.[9,31,50,54] The validity of the present speciation will most likely be resolved by antigenic, protein structural, and genomic comparisons involving numerous *A. marginale* isolates and *A. centrale*. More important is whether the antigenic differences are relevant to cross-species protection. Cross-protection experiments using *A. centrale*-premunized cattle (experiments which originally supported the speciation) do not differ significantly from similar experiments using various heterologous isolates of *A. marginale*.[9,31,55-57] Briefly, the severity of disease seen upon *A. marginale* challenge of *A. centrale*-premunized cattle does not differ significantly from *A. marginale* challenge of cattle premunized with an *A. marginale* isolate of mild virulence. Conversely, the mild reaction seen upon *A. centrale* challenge of *A. marginale*-premunized cattle does not differ significantly from that seen with mild virulence *A. marginale* challenge. On the basis of these experiments, there is little evidence to indicate the *A. centrale*-*A. marginale* differences are greater than differences among *A. marginale* isolates. Molecular approaches to understanding antigenic differences between the two species have only recently begun. The panels of MAbs produced against *A. marginale* and used to characterize different *A. marginale* isolates have been used to search for species-common epitopes on the Israel isolate of *A. centrale*.[40,45] Four different patterns of reactivity were seen (Table 4): (1) none of the ten *A. marginale* isolate-restricted MAbs reacted with *A. centrale;* (2) a MAb that reacted with all 13 *A. marginale* isolates did not react with *A. centrale* and was tentatively considered species specific; and MAbs directed against either (3) the AmF 36 or (4) the AmF 105 surface proteins of *A. marginale* reacted with *A. centrale*. These latter groups reacted with *A. centrale* in the same pattern as with *A. marginale* — the anti-AmF 105 MAbs bound 100% of the *A. centrale* within parasitized erythrocytes, while the anti-AmF 36 MAbs bound 70%.[45] The lack of *A. centrale* reactivity with the panel of *A. marginale* isolate-restricted MAbs and the putative *A. marginale*-specific MAb may indicate that true species differences occur; however, the relevance of the observation to protection is not clear. The reactivity of the MAbs to AmF 105 and AmF 36 with *A. centrale* is particularly significant because of the surface location of these proteins. The reactivity of the anti-AmF 105 MAbs with *A. centrale* included the antibodies (15D2/22B1) demonstrated to neutralize 100% of *A. mar-*

ginale infectivity. We are currently identifying the *A. centrale* protein that bears this epitope, its surface topography, and its role in protection. The reactivity of MAbs against key epitopes on the AmF 105 *A. marginale* surface protein with *A. centrale* suggests that cross-species as well as cross-isolate protection is feasible. The ability of AmF 105 or other relevant proteins to provide protection against anaplasmosis is addressed in Section V.

D. Antigenic Differences Between Tick Stages and the Intraerythrocytic Stage

Erythrocyte-to-erythrocyte transmission of the initial body stage of *Anaplasma* has been demonstrated *in vitro* and is responsible for the pathologic effects of the infection *in vivo*.[58-60] The infection is initiated by either direct inoculation of parasitized erythrocytes or by ixodid tick transmission of an infective tick stage.[5-6] Because morbidity is mediated through the intraerythrocytic stage (the initial body) and this stage is a common pathway regardless of the mode of transmission, development of vaccines and antigenic studies have focused upon the initial body. This strategy appears sound, as many regions have primary transmission by mechanical inoculation of parasitized erythrocytes rather than by tick transmission.[61,62] Within the U.S., anaplasmosis in the southeast region does not depend upon tick transmission which appears to be significant in the western U.S.[10,63-65] Recent research identified two U.S. isolates (Florida and Illinois) as non-tick transmitted using *Dermacentor andersoni* (Florida and Illinois tested) or *D. variablis* (Illinois only tested).[66] If a significant number of field isolates are not tick transmitted, then immunization against initial body stages would be required to provide acceptable protection to field challenge. Therefore, a goal of developing a tick stage-based vaccine for sole use is not recommended. However, the most effective vaccine may encompass both tick stage and initial body antigens.

While premunization and other initial body-based immunization experiments demonstrated that anti-initial body immunity generated protection against disease, corresponding studies examining immunity against tick stages have been few. Whether anti-tick-stage immunity would provide protection against tick-stage challenge similar to the antisporozoite immunity in hemoprotozoan infections has not been fully investigated. Identifying which of the morphologically distinct tick stages of *A. marginale* is the infective stage is needed to progress on these objectives. Current emphasis is on the salivary gland stage of *Anaplasma* and determination of its infectivity and antigenic structure.[67]

In addition, emphasis has been placed upon identification of tick stage antigens that cross react with initial body antigens.[46] Antibodies produced against *A. marginale* midgut tick stages in *Dermacentor andersoni* were used to immunoprecipitate *A. marginale* initial body proteins (Figure 2).[46] The results demonstrated that multiple initial body proteins in the range of >14 to 200 kDa shared at least one epitope with tick stage *Anaplasma*. The stage-common antigens included epitopes on several of the initial body surface proteins, AmF 105, AmF 86, AmF 61, and AmF 36 (Figure 3).[46] The significance of these shared-surface epitopes is potentially threefold: (1) if the epitopes shared with the initial body stage are present on the surface of the infective tick stage, immunization with initial body surface antigens may protect *directly* against tick stage challenge as well as through the common pathway of the intraerythrocytic stage of *A. marginale* (2) immunization with tick stage antigens may protect against both tick stage and initial body challenge; (3) the shared epitopes, even if not surface exposed on the infective tick stages, would serve to boost anti-initial body immunity in areas where repeated tick challenge occurs and, therefore, extend the duration of immunity. This mechanism will be of significance only if the shared epitopes on the initial body surface proteins are those relevant in protection. Whether these stage-common epitopes include the AmF 105 epitope shown to be capable of inducing neutralizing antibodies is unknown, but is presently being determined using colloidal gold-labeling studies in sections of tick-stage organisms in tick midgut and salivary gland epithelium.[68]

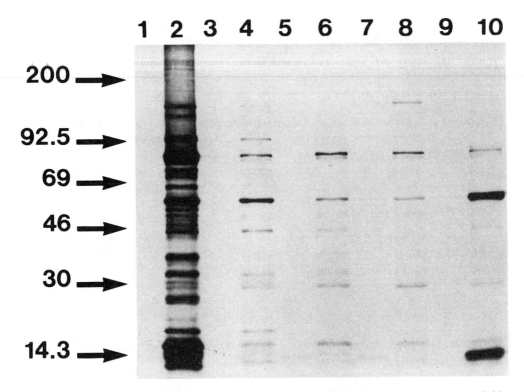

FIGURE 2.   Identification of *A. marginale* antigens common to both the erythrocyte stage (initial body) and tick stages.[46] *A. marginale* initial bodies were radiolabeled with ^{35}S-methionine during short-term *in vitro* erythrocyte culture and immunoprecipitated with sera from cattle infected with parasitized erythrocytes (preinfection sera, Lane 1; postinfection sera, Lane 2) or sera from cattle immunized with killed tick-stage *A. marginale* (preimmunization sera, Lanes 3, 5, 7, and 9; postimmunization sera, Lanes 4, 6, 8, and 10).[38,46] Immunoprecipitated proteins were separated by SDS polyacrylamide gel electrophoresis followed by detection using fluorography. [14]C molecular weight standards are in thousands (arrows at left margin). The precipitation of numerous initial body proteins (in the molecular weight range of <14 to 200 kDa) by cattle sera exposed only to tick-stage *A. marginale* demonstrates that multiple epitopes are common to both vertebrate and invertebrate stages.[46]

IV. CURRENT METHODS OF IMMUNIZATION

Current methods of immunization are principally divided into two strategies: first, premunization using *Anaplasma* parasitized erythrocytes or, second, immunization with killed whole initial body antigens. These methods have been available for years, and specific data regarding their effectiveness under different field conditions as well as experimental use has been reviewed.[5,31,56,69-72] In this chapter, the antigenic basis, the host response, and the primary advantages and disadvantages of the different immunization methods will be discussed primarily as guideposts in developing strategies for improved vaccines.

A. Premunization using *Anaplasma*-Parasitized Erythrocytes

Induction of immunity by premunization requires that the individual develop a patent primary infection.[9] Following resolution of the acute infection, the animal remains persistently infected and develops solid immunity to homologous isolate challenge. To date, this overall strategy remains the most effective available method to induce protective immunity against anaplasmosis. Primary disadvantages are apparent: (1) all inoculated cattle must receive sufficient live organisms to cause a primary infection — this requires either an uninterrupted cold-chain if a standardized dose is used or use of a nonstandardized inoculum from a local carrier animal; (2) the immunity is dependent upon development of an acute

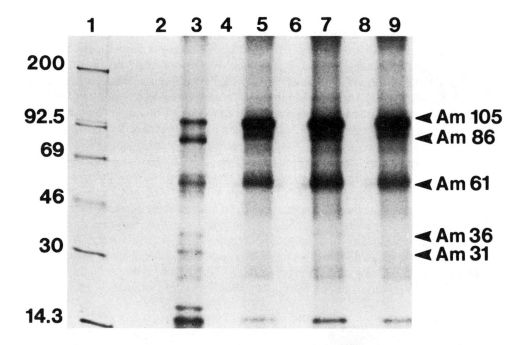

FIGURE 3. Identification of *A. marginale* initial body surface protein epitopes common to tick-stage *A. marginale.*[46] Initial bodies purified from *A. marginale*-parasitized erythrocytes were surface radiolabeled with ^{125}I and immunoprecipitated with sera from cattle immunized with killed tick-stage *A. marginale* (preimmunization sera, Lanes 2, 4, 6, and 8; postimmunization sera, Lanes 3, 5, 7, and 9).[46,104] [14]C molecular weight standards are in thousands (Lane 1). Precipitation of initial body surface proteins, AmF 105, 86, 61, and 36 by sera following tick-stage immunization demonstrates that stage-common epitopes include those on protection-inducing (AmF 105 and AmF 36) and potential protection-inducing (AmF 86 and AmF 61) surface proteins.[46,47,104,114]

infection, yet the severe morbidity and mortality frequently seen in acute infections must be avoided; (3) the premunized animals are persistently infected and are subject to relapse to acute disease during periods of stress; (4) persistently infected cattle may be less productive, especially in areas with poor nutrition and coenzootic with babesiosis; (5) persistently infected cattle maintain the organism in the population for transmission to susceptible cattle; and (6) although homologous immunity appears solid (except during stress-induced relapses), the degree of heterologous immunity varies widely depending on the premunizing isolate.[8,9,14,33,34] The paradox created by the noted disadvantages has limited the overall effectiveness of premunization — the more highly virulent isolates usually induce the most widely cross-isolate protection, but are also most likely to cause morbidity or mortality.[9,54] The virulent isolates also cause the more severe stress-induced relapses, carrier-associated production losses, and disease when transmitted to susceptible cattle. The particular methods of premunization developed share the disadvantages of this paradox to different degrees.

1. Premunization with Virulent A. marginale Followed by Antibiotic Treatment

This method most closely resembles the immunity engendered by natural infection in enzootic areas. Cattle, preferably calves partially protected by colostral antibody, are inoculated with a virulent local isolate and monitored clinically to detect early signs of illness which, if severe, can be treated with imidocarb or tetracyclines.[56,73] Comprehensive trials in Colombia have demonstrated that, given good veterinary supervision, this method is effective, particularly if cattle are reexposed periodically by natural challenge.[74] The primary disadvantage is the reliance upon prompt detection and treatment of animals undergoing severe premunizing infections. This requires skilled animal health personnel and close surveillance not available or amenable to production systems in many tropical regions. The premunizing isolate selected needs to protect against all local isolates or isolates to be brought in by new additions to the herd. The usual source of the inocula is a persistently infected carrier cow. This obviates the need for a prolonged cold-chain, but results in variation in infective doses and, therefore, makes the efficacy of establishing a premunizing infection unreliable. Failure to establish premunition in an individual within a herd of persistently infected cattle will frequently result in severe morbidity or mortality during the subsequent vector season. In addition, direct use of carrier blood as a premunizing inocula carries the risk of transmitting unsuspected infectious agents, including other hemoparasites. Alternatively, inocula can be standardized and maintained in liquid nitrogen at a regional animal health laboratory. While ensuring more consistent premunition, this again requires a prolonged cold-chain frequently unavailable in the tropics.

Improvement of the basic method has been attempted by antibiotic treatment at set intervals in order to reduce the need for continual supervision.[56] Unfortunately, variation in prepatent period among individuals can result in severe morbidity in some, while others are treated with antibiotics prior to developing immunity and remain fully susceptible.[56,73] In summary, the variability in the response to the premunizing inocula and the need for detailed veterinary supervision are severe limitations on the overall effectiveness of this method.

2. Premunization with Minimal Infective Doses of A. marginale

The severity of acute infection and prepatent interval have been shown to be related to the number of organisms inoculated.[47,75] Based upon these findings, several investigators have attempted to use a minimal infective dose that would uniformly infect (premunize) all recipients, but avoid serious morbidity and mortality associated with higher numbers of organisms of the same isolate.[56,74] The obvious advantage would be to avoid the need for veterinary surveillance of premunized cattle. Unfortunately, the dose response varies widely depending upon the isolate of *A. marginale,* the length of storage in liquid nitrogen, route of inoculation, and the age, breed, and nutritional status of the inoculated cattle.[54,74,76-79] Although experimental work showed that the principle was sound, the number of variables make the method unfeasible for field use.[56]

3. Premunization with A. centrale

This method of protecting cattle against anaplasmosis was first demonstrated by Theiler in 1912, when he showed that premunization of cattle with the less virulent *A. centrale* prevented severe morbidity upon *A. marginale* challenge.[29] *A. centrale* premunition has subsequently been used extensively in Israel and Australia and throughout Africa.[80-82] Although conflicting reports are common, the efficacy of this method under ideal conditions has been demonstrated in Israel.[81] Within Israel, a well-coordinated program of *A. centrale* premunization exists. Government veterinarians are responsible for producing and standardizing the inocula, maintaining a liquid nitrogen cold-chain, inoculating the cattle, and ensuring that a premunizing infection occurs without significant morbidity.[81,82] The effectiveness of the method as practiced in Israel certainly reflects upon the available veterinary intrastructure as much as on the particular method. In contrast, very few other subtropical or tropical nations, including regions in the U.S., have livestock production systems and regulated veterinary services amenable to the *A. centrale* program as practiced in Israel.[19] Disadvantages include the cold-chain requirement and the need for veterinary supervision to recognize and treat severe clinical disease following premunition with vaccine isolates of *A. centrale.*[32,56,71,83] Whether these vaccine isolates causing severe morbidity have varied antigenically from the Israel vaccine *A. centrale* isolate is unknown. As well, the efficacy of *A. centrale* premunization has been shown to vary depending upon the *A. marginale* isolate used in challenge.[31,57] Infection, but not severe disease, results from challenge with

Israel isolates of *A. marginale,* while severe morbidity has been reported upon challenge with other *A. marginale* isolates.[31,57] The effect upon production in cattle harboring both the premunizing *A. centrale* infection and *A. marginale* isolates following natural challenge infection has not been described in *Babesia* coenzootic regions. In summary, although the *A. centrale* program in Israel demonstrates the potential of the method, its applicability in most tropical regions is limited, and its use has been reported to be diminishing.[56]

4. Premunization with Attenuated A. marginale

The search for naturally occurring *A. marginale* isolates of low virulence that could be used to premunize cattle without morbidity and protect against challenge with virulent isolates has been largely unrewarding. In contrast, laboratory attenuation of a virulent *A. marginale* isolate (Florida isolate) using irradiation combined with passages through deer and sheep resulted in an isolate meeting these criteria.[69,84] Premunization of young cattle with a standardized dose of attenuated Florida isolate usually results in low parasitemia and mild anemia, but no or minimal clinical disease including effect on weight gain.[54,69,84,85] Premunization of older bulls and cows may cause severe morbidity, abortion, or mortality.[86] In addition, severe anemia accompanied by lethargy and depression has been reported to occur even in yearlings.[86] Upon challenge, cattle develop a low parasitemia, but are protected against severe clinical disease.[55,69,87] The immunity includes protection against several heterologous isolates upon both experimental and field challenge.[55,69,84,88] There is one report of failure to protect against challenge with a Colombian field isolate.[89] Premunization of fully susceptible cattle prior to introduction into an enzootic-susceptible region in Colombia provided solid protection against disease and production losses.[55] In addition, the attenuated *A. marginale* is not tick transmissible and, therefore, allows greater control over its spread to fully susceptible older animals.[90] The attenuated organism is within ovine erythrocytes and does not induce significant titers of isoantibodies in recipients, thus avoiding hemolytic isoerythrolysis in calves ingesting colostral antibodies.[91]

The primary disadvantages of this method of premunization is its reliance on a cold-chain in the tropics. Although reversion to virulence has been reported following 12 passages in splenectomized calves, the significance of this potential problem is not clear.[92] The occurrence of severe premunizing infections and the failure to protect against challenge with certain heterologous isolates can be serious drawbacks to the efficacy of the method, even if infrequent.[86,89] Nevertheless, the standardization of the inoculum and the use of an attenuated *A. marginale* isolate make this method the most effective form of premunization available.

B. Immunization with Killed Whole *A. marginale* Organisms

Immunization with nonliving antigens would be highly desirable as it would avoid the necessity for a prolonged cold-chain in the tropics, the morbidity of premunizing infections, and the introduction of premunizing infections into disease-free but susceptible herds. As well, the risk of transmitting other infectious agents in a cryopreserved inoculum and production losses from the induced carrier status would be abrogated. The challenge in developing a killed vaccine is to induce solid, cross-protective immunity — a goal more readily achieved in most infectious diseases by a live vaccine due to the repeated antigenic stimulus. Presentation of key protection-inducing antigens, avoiding isoantibody formation, and adjuvant selection are all critical criteria in developing an effective vaccine. To date, the available killed vaccines have been limited in their effectiveness by their inability to protect against virulent, heterologous isolates and by the induction of isoantibodies which, upon colostral transfer, cause hemolytic isoerythrolysis in calves.[34,93]

Immunization with the Anaplaz® vaccine (Fort Dodge Laboratories, Fort Dodge, IA) induces partial protection against challenge with heterologous isolates.[34] The severity of the clinical disease is usually reduced compared to nonvaccinated controls; however, significant parasitemia, anemia, and weight loss occur.[34,55] Challenge with virulent isolates, notably the Florida isolate, has been reported to cause severe anemia and high mortality similar to that in unvaccinated cattle.[44] Cattle which have recovered from acute infection remain persistently infected and serve as reservoirs for transmission to unvaccinated cattle, as well as remaining susceptible to relapses and production losses associated with the carrier state.[69] In addition, because the vaccine requires an oil adjuvant and is composed of whole *Anaplasma* organisms admixed with erythrocyte stroma, immunized cattle develop antierythrocyte iso-antibodies which can cause isoerythrolysis in calves ingesting these colostral anti-bodies.[93-95] The overall incidence of isoerythrolysis is low; however, morbidity and mortality can be significant within individual, especially purebred herds.[96] Efforts to reduce neonatal isoerythrolysis have included alteration of the immunization schedule, production of the isolate in cattle with less common blood groups or in ovine erythrocytes, and conjugation of the immunogen to lipids to diminish the isoantibody response.[97-100] While this vaccine induces strong antibody titers to whole organism antigens and *Anaplasma*-specific lympho-cyte/macrophage responses, the significance of these responses in protection and which epitopes induce protection have not been described.[69]

Immunization of cattle using related whole organism preparations (also containing eryth-rocyte stroma) from other *A. marginale* isolates provided partial protection against challenge similar to that seen with the Anaplaz® vaccine.[101,102] The severity of morbidity upon challenge is reduced in most, but not all, vaccinates, and significant parasitemia and anemia developed in all vaccinated cattle upon challenge.[101] The results again indicate that more basic research in epitope identification and presentation is needed to develop a more effective vaccine.

V. DEVELOPMENT OF ANTIGENICALLY DEFINED VACCINES

The lack of progress in developing effective, widely used *Anaplasma* vaccines results from the complexity of the organism's invertebrate vector and vertebrate stages, antigenically variant isolates, and their ability to persist in the host. Yet, the protection-inducing ability of current methods of immunoprophylaxis, despite their other drawbacks, clearly indicate that effective immunization is feasible. Advances in the molecular biology of infectious agents have provided the opportunity to unravel the complexity of the host-parasite rela-tionship to identify conserved *Anaplasma* epitopes capable of inducing cross-protection. Once identified, protective epitopes can be efficiently produced and presented to the host immune system using genetically engineered viral vectors carrying key *Anaplasma* epi-topes.[103] This approach focuses the immune response on these key epitopes, does not in-corporate erythrocyte antigens, and, by using a live viral vector, overcomes the traditional poor efficacy of killed vaccines.[20,56,69,103] Erythrocyte culture of *Anaplasma* and purification of initial bodies are important tools in identifying protective antigens, but are unlikely to be competitive as a source of vaccine due to erythrocyte contamination, poor efficacy of killed vaccines, and high cost of production.[56,59,104-106] The current emphasis in anaplasmosis, as well as in other hemoparasitic diseases, is to clone and express genes coding for protective epitopes.

Our strategy is to identify and isolate initial body-surface proteins using a combination of biochemical and immunological techniques. Cattle immunized with individual surface proteins are challenged with homologous and heterologous isolates of *A. marginale*. Proteins that induce partial or complete protection are selected as candidates for incorporation in a genetically engineered vaccine. Next, full-length genes coding for the proteins or gene sequences coding only for the isolate-common, protective epitopes are cloned, and either the recombinant protein is retested as an immunogen or the gene is inserted directly into vaccinia for testing as a recombinant viral vector.

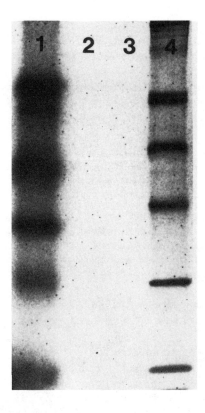

FIGURE 4. Identification of *A. marginale* initial body surface proteins recognized by neutralizing antibody.[104] Initial bodies purified from Florida isolate *A. marginale*-parasitized erythrocytes were surface radiolabeled with ^{125}I, detergent disrupted, and immunoprecipitated with either preimmune antibody (Lane 3) or antibody capable of neutralizing the infectivity of 10^{10} initial bodies (Lane 1).[104] Immunoprecipitated proteins were separated by SDS polyacrylamide gel electrophoresis and detected by autoradiography. Surface proteins of apparent molecular weights 105, 86, 61, 36, and 31 kDa were specifically recognized by neutralizing, but not preimmune sera (Lanes 1 and 3).[104]Uninfected erythrocyte membranes radiolabeled with ^{125}I under identical conditions were not recognized by the neutralizing antibody (Lane 2).[104] [14]C molecular weight standards are 200, 92.5, 69, 46, 30, and 14.3 kDa (Lane 4).

We selected the Florida isolate of *A. marginale* as the principal isolate for our studies, due to its high virulence and ability to induce widely cross-protective immunity.[34,55,69,84,88] The latter indicated that the Florida isolate may express more cross-protective epitopes than other isolates and increased our chances of detecting these epitopes. Surface radioiodination of purified initial bodies, followed by immunoprecipitation with a neutralizing antibody, identifies five major surface proteins with apparent molecular sizes of 105, 86, 61, 36, and 31 kDa (designated with the prefix AmF to indicate organism and isolate) (Figure 4).[104] The antibody was produced against purified, infective, and intact initial bodies and was shown to completely neutralize the infectivity of 10^{10} Florida isolate initial bodies for splenectomized calves (Table 5).[104] Our results clearly established that initial body epitopes capable of inducing protective antibody were present and that these five surface proteins most likely bear the epitopes responsible for the induction of the antibody. The key questions raised by this initial experiment: which protein or proteins bear the critical epitopes, were the epitopes shared by other isolates and *A. centrale,* and could cattle immunized with individual proteins mount a protective response? We have isolated, characterized, and tested both AmF 105

Table 5
ANTIBODY-MEDIATED NEUTRALIZATION OF
VIRULENT *A. MARGINALE* INITIAL BODIES[a]

Ab reactivity[b]	No. infected/challenged	No. dead/challenged
Preimmune	4/4	4/4
Anti-initial body	0/6	0/6

[a] Initial bodies were purified from Florida isolate-parasitized erythrocytes and incubated for 45 min at 20°C with antisera diluted 1:1 in RPMI 1640.[104] The mixture was inoculated intramuscularly into splenectomized calves.[104] Blood samples were collected daily for 75 d postinoculation in order to determine packed cell volume and presence of parasitized erythrocytes.

[b] Immune sera was produced by repeated immunization with 5×10^8 purified initial bodies in either complete or incomplete Freund's adjuvant.[104] This sera had an indirect immunofluorescent antibody titer of 1:16,000 to initial bodies. Preimmune sera was unreactive.[104]

and AmF 36 as protective immunogens. AmF 86, AmF 61, and AmF 31 are currently being isolated in our laboratories and have not been tested as immunogens to date.

A. AmF 105 as a Protective Immunogen

We have identified a neutralization-sensitive epitope on AmF 105 which was subsequently shown to be highly conserved among *A. centrale* and *A. marginale* isolates from Israel, Kenya, and the U.S. (Tables 2 to 4, reviewed in Section III).[40,45,47] The presence of this conserved epitope on 100% of initial bodies in all stages of acute infection provided strong rationale for testing AmF 105 as a cross-protective immunogen.[40,45] In addition, AmF 105 shares cross-reactive epitopes with tick midgut stages and is recognized by high-titer antibodies from effectively premunized cattle.[46,107] To test whether this subunit could induce protection in immunized cattle, we isolated AmF 105 by immunoaffinity chromatography on a MAb 15D2-Sepharose® 4B column (Figure 5).[47] Calves immunized with the purified protein developed high titers of antibody to AmF 105 and were significantly protected from challenge with either 10^8 purified Florida isolate initial bodies (Table 6) or 10^{10} Florida isolate-parasitized erythrocytes compared to identically challenged, ovalbumin-immunized calves.[47,108] Two of the AmF 105 immunized calves in each group did not show any parasitized erythrocytes in Wright's stained blood smears, while the other six developed only a transient parasitemia following a significantly prolonged prepatent period.[47,108] To test the ability of AmF 105 to induce cross-protective immunity, we challenged a third group of immunized calves with 10^{10} Washington-O-parasitized erythrocytes.[108] We chose the Washington-O isolate because it is a recent field isolate with clearly demonstrated antigenic, morphologic (Washington-O bears an appendage, Florida does not), and protein structural differences.[38,40] None of the five AmF 105-immunized calves developed parasitemia upon challenge, while all ovalbumin-immunized control calves were infected.[108] The more complete protection seen with heterologous challenge vs. homologous challenge with the Florida isolate is most likely due to the Washington-O isolate being less virulent, even in control calves, and therefore less likely to "breakthrough" in AmF 105-immunized calves.[108] These experiments clearly demonstrated that AmF 105 had strong potential as a protective immunogen and that production of AmF 105 via recombinant DNA expression was a necessary next step towards more complete characterization and testing.

Initial characterization of AmF 105 indicated that the affinity-purified immunogen occurred as a doublet on sodium dodecyl sulfate polyacrylamide gels (Figure 5).[47] Although true glycosylation would be highly unlikely in a prokaryote, the presence of strong noncovalent

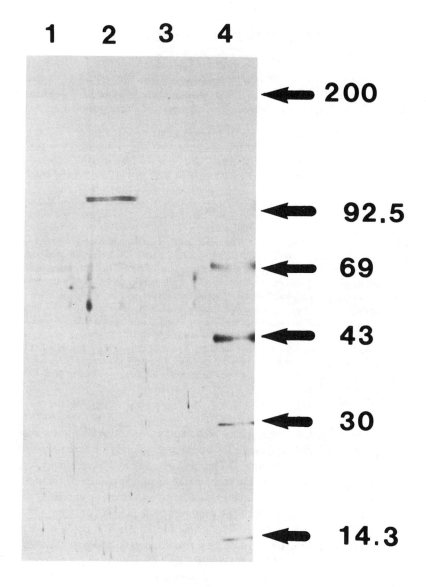

FIGURE 5. Purification of AmF 105 by immunoaffinity chromatography.[47] 10^{12} Florida isolate initial bodies were purified from parasitized erythrocytes, detergent disrupted, and bound to a MAb 15D2-Sepharose® 4B affinity column.[47] Following extensive washing, AmF 105 was eluted using 0.5% deoxycholate and 2 M KSCN.[47] Initial body proteins were separated by SDS polyacrylamide gel electrophoresis and detected by silver staining. Lane 2 contains initial body proteins prior to chromatography, Lane 3 contains purified AmF 105 following elution and dialysis.[47] The "doublet" composition of native AmF 105 comprising AmF 105U and AmF 105L can be noted in Lane 3.[110] Purified AmF 105 was subsequently tested as a protective immunogen in cattle.[47] Lane 1 was not loaded in order to control for silver staining artifacts, Lane 4 contains molecular weight standards (indicated in kilodaltons in the right margin).

interactions between surface membrane proteins and polysaccharides raised the possibility that binding of carbohydrate to the protein surface could account for the molecular weight heterogeneity and that neutralization may be directed against a carbohydrate moiety.[48,109] Examination of AmF 105 for bound carbohydrate residues has been uniformly negative using several different methods: periodic acid Schiff staining of purified AmF 105, metabolic

Table 6
**AmF 105 IMMUNIZATION INDUCES PROTECTION AGAINST
ANAPLASMOSIS[a]**

Group	No. infected/challenged	Mean days to 1% parasitemia (range)	Mean peak parasitemia	Mean low PCV
Nonimmunized	4/4	29(26—31)	4.2	23
Ovalbumin	5/5	33(31—35)	5.4	24.5
AmF 105	3/5[b]	b	<0.01[c]	31[c]
p[d]	ND[e]	≤0.01	≤0.01	≤0.01

[a] Calves were immunized with 100 μg of AmF 105 or 100 μg of ovalbumin emulsified in complete Freund's adjuvant and boosted three fold at 2-week intervals with 100 μg of antigen in incomplete Freund's adjuvant.[47] The AmF 105 vaccinates developed a titer of 10^4 to 10^5 as determined by radioimmunoassay with ^{125}I-AmF 105.[47] The calves and four nonimmunized control calves were challenged by intramuscular injection with 10^8 purified Florida isolate *A. marginale* initial bodies.[47] Blood samples were collected daily for 100 d postchallenge in order to determine packed cell volume and presence of parasitized erythrocytes. The mean number of days postchallenge to 1% parasitemia was calculated for all infected calves in each group.

[b] Two of the five AmF 105-immunized calves did not become infected and remained hematologically normal.[47] The three AmF 105 vaccinates that became infected developed a transient parasitemia of <0.01% on day 39 and did not develop clinical disease.[47]

[c] These hematologic values are based upon the three AmF 105 vaccinates with mild infection to demonstrate protection against disease afforded by AmF 105 immunization.[47]

[d] p values are calculated with the pooled t test to compare the responses of the AmF 105 and ovalbumin-immunized cattle.

[e] ND: not determined.

incorporation of radiolabeled carbohydrate precursors into AmF 105 during short-term *in vitro* cultivation of parasitized erythrocytes, carbohydrate-specific surface radiolabeling, and binding of AmF 105 to various lectins.[48] The lack of carbohydrate residues on AmF 105 suggested that the neutralizing MAbs are recognizing a peptide epitope.[48] Further indication that the AmF 105 neutralization-sensitive epitope is composed of amino acids was its sensitivity to complete protease digestion using pronase, proteinase-K, or trypsin.[48] Given that carbohydrate differences were not likely responsible for the Am 105 doublet, we considered two possibilities: (1) both proteins were derived from one gene and bear the neutralization-sensitive epitope reactive with MAbs 15D2/22B1 and were derived by differences in processing, or (2) the proteins were separate gene products with only one bearing the neutralization-sensitive epitope and the second protein being noncovalently bound to the first and, therefore, copurifying.[110] Complete separation of the two proteins — designated AmF 105U (upper) and AmF 105L (lower) — on 7.5% polyacrylamide gels containing 4 *M* urea followed by elution and reimmunoprecipitation with MAbs 15D2 or 22B1 indicated that AmF 105U bore the relevant epitope, while AmF 105L did not (Figure 6).[110] Peptide mapping using partial proteolysis of AmF 105U and AmF 105L demonstrated that the two proteins were structurally dissimilar and were most likely separate gene products (Figure 7).[110] Although only AmF 105U bears the demonstrated neutralization-sensitive epitope, we have shown that both AmF 105U and AmF 105L have surface-exposed epitopes, and therefore either or both could be responsible for protection in affinity-purified AmF 105-immunized cattle.[110]

Our approach to dissecting the antigenic basis of AmF 105 induced protection is to clone and express full-length genes coding for each protein in *Escherichia coli* and to test the recombinant-produced protein as a protective immunogen. The testing of full-length recombinant AmF 105U (rAmF 105U) and rAmF 105L homologous to native AmF 105U and L will allow determination of which components are required for a vaccine: rAmF 105U,

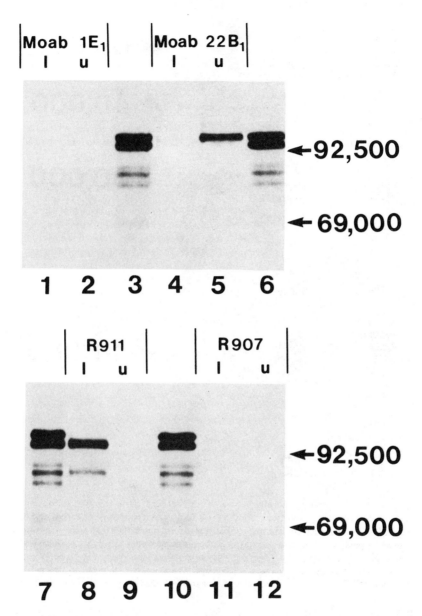

FIGURE 6. Antigenic dissimilarity of AmF 105U and AmF 105L. *A. marginale* initial
bodies were radiolabeled with ³⁵S-methionine during short-term *invitro* erythrocyte culture,
immunoprecipitated with MAb 22B1, and the precipitated AmF 105U + L detected by SDS
polyacrylamide gel electrophoresis with fluorography (Lanes 3, 6, 7, 10).[38,110] Individual
AmF 105U or AmF 105L bands were then excised, electroeluted, and reimmunoprecipitated
with different antibodies. The reimmunoprecipitates were separated by electrophoresis on
7.5% polyacrylamide SDS gels containing 4 M urea and detected by fluorography.[110] AmF
105L was reimmunoprecipitated with MAb 1E1 (an unrelated MAb against *Trypanosoma
brucei,* Lane 1); MAb 22B1 (neutralizing MAb to *A. marginale,* Lane 4); R911 (rabbit
antibody made to recombinant *E. coli* expressing AmF 105L); and R907 (preimmune rabbit
antibody).[110] AmF 105U was reimmunoprecipitated with MAb 1E1 (Lane 2), MAb 22B1
(Lane 5), R911 (Lane 9),and R907 (Lane 12).[110] These results demonstrate that AmF 105U
and AmF 105L are antigenically dissimilar and that only AmF 105U bears the neutralization-
sensitive epitope recognized by MAb 22B1.[110]

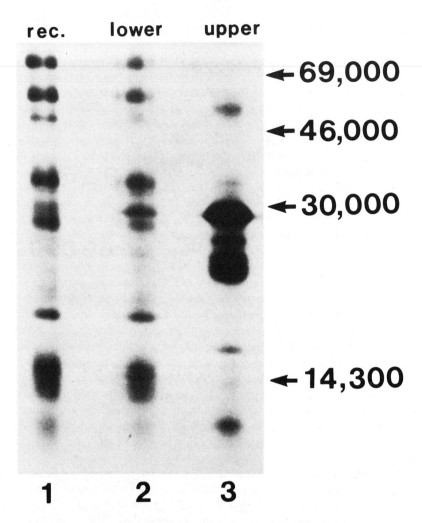

FIGURE 7. Structural dissimilarity between AmF 105U and native or recombinant AmF 105L.[110] Individual ^{35}S-radiolabeled AmF 105U, AmF 105L, or recombinant AmF 105L bands were excised, electroeluted, and subjected to digestion with 0.025 μg *Staphylococcus aureus* V8 protease.[110] The resulting fragments were separated on a 15% polyacrylamide gel and visualized by fluorography. Digestion patterns of recombinant AmF 105L (Lane 1) and native AmF 105L (Lane 2) demonstrate substantial homology, but are markedly dissimilar from AmF 105U (Lane 3).[110] Molecular weight markers are indicated by arrows (right margin).

rAmF 105L, or rAmF 105U + L. In the latter instance it may be critical to restore the noncovalent interaction between the integral membrane proteins.

We have cloned and expressed the AmF 105L gene in *E. coli* using pBR 322.[110] Full-length rAmF 105L is structurally and antigenically homologous with native AmF 105L and dissimilar to AmF 105U (Figure 7).[110] Recombinant AmF 105L is stably produced as approximately 1% of the *E. coli* proteins and appears to be expressed under its own promoter, as inserts in either orientation are equally expressed.[110] Production of antibody against rAmF 105L has enabled us to antigenically characterize AmF 105L independent of AmF 105U. The agglutination of initial bodies incubated with rabbit antibody to rAmF 105L confirms that AmF 105L has surface-exposed epitopes and meets criteria for a potential immunogen.[110] In addition, this antibody recognizes epitopes on 100% of initial bodies in parasitized erythrocytes throughout acute infection with any of the eight *A. marginale* isolates yet tested.[110] This demonstrates that similar to the conservation of the AmF 105U epitope detected by neutralizing MAbs 15D2 and 22B1, AmF 105L bears widely conserved epitopes. Unlike AmF 105U, the ability of these conserved epitopes to induce neutralizing antibody is unknown. Sera from the protected calves immunized with native AmF 105U + L have a high titer of antibody to rAmF 105L, indicating that the surface epitopes were immunogenic, but not necessarily protective in cattle.[111] The role of AmF 105L in inducing protective immunity and its effectiveness as an immunogen have not yet been determined. A pilot experiment in which 10^8 Florida isolate initial bodies were incubated with rabbit anti-rAmF 105L polyclonal antibody did not result in significant neutralization.[111] We are currently isolating rAmF 105L using monoclonal antibody affinity chromatography in order to test its efficacy as a protective immunogen in cattle.

Evidence to date indicates that very likely AmF 105U is important in inducing protective immunity and may be largely responsible for the protection seen in native AmF 105-immunized cattle.[47,110] AmF 105U bears the isolate- and species-conserved, neutralization-sensitive epitope, and this epitope is specifically recognized by high-titer antibody from the protected calves.[47,49] This epitope is conserved despite marked molecular size variation in the parent molecule among different isolates.[112] We have recently obtained expression of the AmF 105U gene in *E. coli* using pKK-233-2.[113] The protein is structurally identical to native AmF 105U and includes the neutralization-sensitive epitope recognized by MAbs 15D2 and 22B1.[113] In addition, sera from the native AmF 105-immunized cattle reacts strongly with recombinant Amf 105U protein.[113] Our efforts are presently directed at determining the role of AmF 105U in protective immunity and examining isolate and species differences in this surface protein and its coding gene. Testing recombinant-derived AmF 105U alone and with rAmF 105L as a protective immunogen remains the highest priority. In addition, we are determining the primary structure of the neutralization-sensitive epitope by DNA sequencing and synthetic peptide constructions. This will allow presentation of the neutralization-sensitive epitope to the host immune system in order to focus the response upon this epitope. If protective, this strategy could be used in viral vectors to generate an effective vaccine.

Critical to further development of AmF 105 as a vaccine component, is to understand the mechanism of protection in the native AmF 105 immunized cattle. We have determined that antisera from these protected cattle mediate initial body phagocytosis *in vitro* using bone marrow-derived macrophages.[108] Generation of recombinant AmF 105U and AmF 105L and their testing as protective immunogens will enable us to determine which component induces opsonizing antibody and how well opsonization correlates with protection. Other possible mechanisms of neutralization have not been closely examined in AmF 105U-immunized cattle. Understanding the mechanism of protection in subunit immunized cattle may facilitate selection and incorporation into a vaccine of only those epitopes that induce protective antibody.

B. AmF 36 as a Protective Immunogen

AmF 36 fulfills several criteria of a potential protective immunogen: (1) the protein has surface-exposed epitopes as demonstrated by lactoperoxidase-mediated radioiodination (Figure 4), (2) epitopes are conserved among *A. marginale* isolates in the U.S., Israel, and Kenya and *A. centrale* (Tables 2 and 4), and (3) AmF 36 epitopes are present in tick midgut stages of the organism (Figures 2 and 3). In addition, high titers of antibody to AmF 36 are present in cattle effectively premunized with different *A. marginale* isolates.[107] AmF 36, purified by monoclonal antibody affinity chromatography, elicits high titers of specific antibody in immunized calves, and these calves are significantly protected against challenge with either 10^8 Florida isolate initial bodies or 10^{10} Washington-O-parasitized erythrocytes compared to ovalbumin-immunized controls.[114] Two of the AmF 36-immunized calves in each challenge group did not become infected with *A. marginale,* and those AmF 36-immunized calves that became infected had significantly prolonged prepatent periods compared to control calves.[114] These experiments demonstrate that AmF 36 can induce protective immunity in cattle and is a candidate for inclusion in an antigenically defined vaccine. We are now screening gene libraries for expression of AmF 36 in *E. coli* in order to further characterize the molecule and test the ability of recombinant AmF 36 to induce protection in immunized cattle.

C. Progress in Subunit Vaccine Development

The current effort in hemoparasitic diseases is primarily to identify antigens relevant to protective immunity and to develop effective methods to present these antigens to the host. It is important to remember that testing of individual antigens as protective immunogens is quite different from testing an end-product vaccine. Testing of individual components is a critical but early step in vaccine development — the eventual vaccine may incorporate epitopes from several different surface proteins in order to achieve maximal efficacy. In anaplasmosis, two proteins, AmF 105 and AmF 36, have been identified as bearing protection-inducing epitopes.[47,108,110,114] In addition, at least three other initial body surface proteins, AmF 86, AmF 61, and AmF 31, are potential vaccine candidates.[104,107] Our strategy is to clone and express full-length genes coding for these surface proteins and to test recombinant-derived proteins as protective immunogens in cattle. Concomitantly, it is essential to identify isolate- and species-conserved epitopes and to determine the role of these epitopes in the induced protection.[45,47] Increased understanding of antigenic variation in key epitopes, the differences between tick-stage and erythrocyte-stage challenge, and mechanisms of immunity will be instrumental in selecting epitopes for improved anaplasmosis vaccines.

Although the steps remaining to develop, test, and deploy an effective anaplasmosis vaccine are numerous, they are becoming more clearly defined due to advances in the molecular biology of complex hemoparasites including *Anaplasma*. Similarly, the successes and failures of current *Anaplasma* vaccines indicate the requirements any vaccine must meet to be widely effective: (1) safe (unable to induce morbidity in vaccinates including antierythrocyte antibody); (2) induce strong protection; (3) induce cross-protection, requiring conserved epitopes; (4) not reliant upon a prolonged cold-chain; and (5) economical.[103] In addition, because different hemoparasites frequently share tick vectors and are coenzootic, to be truly effective and replace acaracide use as a principal means to control hemoparasitic disease, a vaccine should incorporate protective antigens for all regionally important hemoparasites. The availability of replicating viral vectors with large capacities for foreign genes and a wide host range provide an ideal vehicle to develop a multidisease vaccine.[103] Live recombinant vaccinia viruses expressing foreign genes are highly immunogenic in cattle, economically produced in tissue culture or animal skin, stable in freeze-dried form (continuous refrigeration is unnecessary), and simply administered by skin scratch — fulfilling all criteria for an effective vaccine in the tropics.[103,115,116] These breakthroughs in vaccine technology combined with advances in identifying protective *Anaplasma* immunogens makes development of an economical, widely cross-protective vaccine for anaplasmosis a realistic goal for the first time. Similar advances in identifying candidate vaccine antigens in the coenzootic arthropod-borne hemoparasites provide a promising future for integrated control over the devastating hemoparasitic diseases of cattle in the tropics.

ACKNOWLEDGMENTS

The original work referenced by the author was conducted at Washington State University and the University of Florida in collaboration with Travis C. McGuire, Anthony F. Barbet, William C. Davis, and Terry F. McElwain. This research has been supported by U.S.

Department of Agriculture special research grants 83-CRSR-2-2194 and 85-CRSR-2-2619, competitive research grants 85-CRCR-1-1908 and 86-CRCR-1-2247, and Agricultural Research Service, Animal Disease Research Unit cooperative agreement 58-9-AHZ-2-663, and the United States-Israel Binational Agricultural Research and Development Fund grants US-344-80 and US-846-84, and the U.S. Agency for International Development grant DPE-5542-GSS-7008-00.

REFERENCES

1. National Research Council, *Priorities in Biotechnology Research for International Development — Proceedings of a Workshop,* Directed by The Board on Science and Technology for International Development, Office of International Affairs, National Academy Press, Washington, D.C., 1982, 67.
2. **Lawrence, J. A., Foggin, C. M., and Norval, R. A. I.,** The effects of war on the control of diseases of livestock in Rhodesia (Zimbabwe), *Vet. Rec.,* 107, 82, 1980.
3. **Norval, R. A. I.,** Arguments against intensive dipping, *Zimbabwe Vet. J.,* 14, 19, 1983.
4. **Theiler, A.,** *Anaplasma marginale.* The marginal points in the blood of cattle suffering from a specific disease, in *Report of the Government Veterinary Bacteriologist 1908—1909,* Theiler, A., Ed., Transvaal Department of Agriculture, Transvaal, South Africa, 1910, 7.
5. **Richey, E. J.,** Bovine anaplasmosis, in *Current Veterinary Therapy Food Animal Practice,* Howard, R. J., Ed., W.B. Saunders, Philadelphia, 1981, 767.
6. **Kocan, K. M., Teel, K. D., and Hair, J. A.,** Demonstration of *Anaplasma marginale* Theiler in ticks by tick transmission, animal inoculation, and fluorescent antibody studies, *Am. J. Vet. Res.,* 41, 183, 1980.
6a. **Eriks, I. S., Palmer, G. H., McGuire, T. C., Allred, D. R., and Barbet, A. F.,** Detection and quatitation of *Anaplasma marginale* in carrier cattle by using a nucleic acid probe, *J. Clin. Microbiol.,* 1989, in press.
7. **Alderink, F. J. and Dietrich, R.,** Anaplasmosis in Texas: epidemiologic and economic data from a questionnaire survey, in *Proc. 7th Natl. Anaplasmosis Conf.,* Hidalgo, R. J. and Jones, E. W., Eds., Mississippi State University Press, State College, 1982, 27.
8. **Swift, B. L. and Thomas, G. M.,** Bovine anaplasmosis: elimination of the carrier state with injectable long-acting oxytetracycline, *J. Am. Vet. Med. Assoc.,* 183, 63, 1983.
9. **Kuttler, K. L. and Todorovic, R. A.,** Techniques of premunization for the control of anaplasmosis, in *Proc. 6th Natl. Anaplasmosis Conf.,* Jones, E. W., Ed., Heritage Press, Stillwater, OK, 1973, 106.
10. **Dikmans, G.,** The transmission of anaplasmosis, *Am. J. Vet. Res.,* 11, 5, 1950.
11. **Yeruhan, I. and Braverman, Y.,** The transmission of *Anaplasma marginale* to cattle by blood-sucking arthropods, *Refu. Vet.,* 38, 37, 1981.
12. **Pino, J. A.,** The neglected diseases of livestock, in *Babesiosis,* Ristic, M. and Kreier, J. P., Eds., Academic Press, New York, 1981, 545.
13. **McCallon, B. R.,** Prevalence and economic aspects of anaplasmosis, in *Proc. 6th Natl. Anaplasmosis Conf.,* Jones, E. W., Ed., Heritage Press, Stillwater, OK, 1973, 1.
14. **Mullenax, C. H.,** Estimated production and dollar losses due to stable enzootic anaplasmosis and babesiosis in the Colombian llanos 1981—1984, presented at Annu. Conf. of Anaplasmosis Research Workers, Baton Route, LA, September 21 to 23, 1986.
15. **Corrier, D. E. and Guzman, S.,** The effect of natural exposure to *Anaplasma* and *Babesia* infections on native calves in an endemic area of Colombia, *Trop. Anim. Health Prod.,* 9, 47, 1977.
16. **Glantz, M. H.,** Drought in Africa, *Sci. Am.,* 256, 34, 1987.
17. **Maas, J., Lincoln, S. D., Coan, M. E., Kuttler, K. L., Zaugg, J. L., and Stiller, D.,** Epidemiologic aspects of bovine anaplasmosis in semiarid range conditions of south central Idaho, *Am. J. Vet. Res.,* 47, 528, 1986.
18. **Lincoln, S. D., Zaugg, J. L., and Maas, J.,** Bovine anaplasmosis: susceptibility of seronegative cows from an infected herd to experimental infection with *Anaplasma marginale, J. Am. Vet. Med. Assoc.,* 190, 171, 1987.
19. **Norton, G. A.,** A strategic research approach for tick control programs, in *Ticks and Tick-Borne Diseases — Proc. of an International Workshop,* Sutherst, R. W., Ed., Argyle Press, Mentone, Victoria, Australia, 1987, 126.
20. **Wright, I. G. and Riddles, P. W.,** Biotechnological control of tick-borne disease, in *Expert Consultation on Biotechnology for Livestock Production and Health,* FAO, Rome, 1986.
21. **Magonigle, R. A. and Newby, T. J.,** Response of cattle upon reexposure to *Anaplasma marginale* after elimination of chronic carrier infections, *Am. J. Vet. Res.,* 45, 695, 1984.

22. **Corrier, D. E., Wagner, G. G., and Adams, L. G.,** Recrudescence of *Anaplasma marginale* induced by immunosuppression with cyclophosphamide, *Am. J. Vet. Res.,* 42, 19, 1981.
23. **Kuttler, K. L. and Adams, L. G.,** Influence of dexamethasone on the recrudescence of *Anaplasma marginale* in splenectomized calves, *Am. J. Vet. Res.,* 38, 1327, 1977.
24. **Jones, E. W., Norman, B. B., Kliewer, I. O., and Brock, W. E.,** *Anaplasma marginale* infection in splenectomized calves, *Am. J. Vet. Res.,* 291, 523, 1968.
25. **Jones, E. W., Kliewer, I. O., Norman, B. B., and Brock, W. E.,** *Anaplasma marginale* infection in young and aged cattle, *Am. J. Vet. Res.,* 29, 535, 1968.
26. **Carson, C. A., Sells, D. M., and Ristic, M.,** Cell-mediated immune response to virulent and attenuated *Anaplasma marginale* administered to cattle in live and inactivated forms, *Am. J. Vet. Res.,* 38, 173, 1977.
27. **Carson, C. A., Sells, D. M., and Ristic, M.,** Cell-mediated immunity related to challenge exposure of cattle inoculated with virulent and attenuated strains of *Anaplasma marginale, Am. J. Vet. Res.,* 38, 1167, 1977.
28. **Buening, G. M.,** Cell-mediated immune responses in calves with anaplasmosis, *Am. J. Vet. Res.,* 34, 757, 1973.
29. **Theiler, A.,** Gallsickness of imported cattle and the protective inoculation against this disease, *Agric. J. Union S. Afr.,* 3, 1, 1912.
30. **Ristic, M.,** Anaplasmosis, in *Bovine Medicine and Surgery,* Amstutz, H. E., Ed., American Veterinary Publications, Santa Barbara, CA, 1980, 324.
31. **Potgeiter, F. T. and van Rensburg, L.,** Infectivity, virulence and immunogenicity of *Anaplasma centrale* live vaccine, *Onderstepoort J. Vet. Res.,* 50, 29, 1983.
32. **Wilson, A. J., Parker, R., and Trueman, K. F.,** Experimental immunization of calves against *Anaplasma marginale* infection: observations on the use of living *A. centrale* and *A. marginale, Vet. Parasitol.,* 7, 305, 1980.
33. **Kreier, J. P. and Ristic, M.,** Anaplasmosis. XI. Immunoserologic characteristics of the parasites present in the blood of calves infected with the Oregon strain of *Anaplasma marginale, Am. J. Vet. Res.,* 24, 688, 1963.
34. **Kuttler, K. L., Zaugg, J. L., and Johnson, L. W.,** Serologic and clinical responses of premunized, vaccinated and previously infected cattle to challenge exposure by two different *Anaplasma marginale, Am. J. Vet. Res.,* 45, 2223, 1984.
35. **Ristic, M.,** Immunologic systems and protection in infections caused by intracellular blood protista, *Vet. Parasitol.,* 2, 31, 1976.
36. **Wagner, G. G.,** Immunoglobulin responses of cattle associated with recrudescent *Anaplasma marginale* infections, in *Proc. 7th Natl. Anaplasmosis Conf.,* Hidalgo, R. J. and Jones, E. W., Eds., Mississippi State University Press, State College, 1982, 307.
37. **Kreier, J. P. and Ristic, M.,** Anaplasmosis. X. Morphologic characteristics of the parasites present in the blood of calves infected with the Oregon strain of *Anaplasma marginale, Am. J. Vet. Res.,* 24, 676, 1963.
38. **Barbet, A. F., Anderson, L. W., Palmer, G. H., and McGuire, T. C.,** Comparison of proteins synthesized by two different isolates of *Anaplasma marginale, Infect. Immun.,* 40, 1068, 1983.
39. **Adams, J. H., Smith, R. D., and Kuhlenschmidt, M. S.,** Identification of antigens of two isolates of *Anaplasma marginale* using a western blot technique, *Am. J. Vet. Res.,* 47, 501, 1986.
40. **McGuire, T. C., Palmer, G. H., Goff, W. L., Johnson, M. I., and Davis, W. C.,** Common and isolate-restricted antigens of *Anaplasma marginale* detected with monoclonal antibodies, *Infect. Immun.,* 45, 697, 1984.
41. **Goff, W. L. and Winward, L. D.,** Detection of geographic isolates of *Anaplasma marginale* using polyclonal bovine antisera and microfluorometry, *Am. J. Vet. Res.,* 46, 2399, 1985.
42. **Kuttler, K. L. and Winward, L. D.,** Serologic comparisons of 4 *Anaplasma* isolates as measured by the complement-fixation test, *Vet. Microbiol.,* 9, 181, 1984.
43. **Kreier, J. P. and Ristic, M.,** Definition and taxonomy of *Anaplasma* species with emphasis on morphologic and immunologic features, *Z. Tropenmed. Parasitol.,* 23, 88, 1972.
44. **Carson, C. A., Weisiger, R. M., Ristic, M., Thurmon, J. C., and Nelson, D. R.,** Appendage-related antigen production by *Paranaplasma caudatum* in deer erythrocytes, *Am. J. Vet. Res.,* 35, 1529, 1974.
45. **Palmer, G. H., Barbet, A. F., Musoke, A. J., Katende, J. M., Rurangirwa, F., Shkap, V., Pipano, E., Davis, W. C., and McGuire, T. C.,** Recognition of conserved surface protein epitopes on *Anaplasma centrale* and *Anaplasma marginale* isolates from Israel, Kenya and the United States, *Int. J. Parasitol.,* 18, 33, 1988.
46. **Palmer, G. H., Kocan, K. M., Barron, S. J., Hair, J. A., Barbet, A. F., Davis, W. C., and McGuire, T. C.,** Presence of common epitopes between cattle (intraerythrocytic) and tick stages of *Anaplasma marginale, Infect. Immun.,* 50, 881, 1985.
47. **Palmer, G. H., Barbet, A. F., Davis, W. C., and McGuire, T. C.,** Immunization with an isolate-common surface protein protects cattle against anaplasmosis, *Science,* 231, 1299, 1986.

48. **Palmer, G. H., Waghela, S. D., Barbet, A. F., Davis, W. C., and McGuire, T. C.,** Characterization of a neutralization-sensitive epitope on the Am 105 surface protein of *Anaplasma marginale, Int. J. Parasitol.,* 17, 1279, 1987.

49. **Palmer, G. H., McGuire, T. C., and McElwain, T. F.,** Unpublished data, 1987.

50. **Kuttler, K. L.,** A study of the immunological relationship of *Anaplasma marginale* and *Anaplasma centrale, Res. Vet. Sci.,* 8, 467, 1967.

51. **Theiler, A.,** Further investigations into anaplasmosis of South African cattle, in *First Report of the Director of Veterinary Research,* Theiler, A., Ed., South African Department of Agriculture, Transvaal, South Africa, 1911, 7.

52. **Legg, J.,** Anaplasmosis cross-immunity tests between *Anaplasma centrale* (South Africa) and *Anaplasma marginale* (Australia), *Aust. Vet. J.,* 12, 230, 1936.

53. **Kuttler, K. L.,** Serological relationship of *Anaplasma marginale* and *Anaplasma centrale* as measured by the complement-fixation and capillary-tube agglutination tests, *Res. Vet. Sci.,* 8, 207, 1967.

54. **Corrier, D. E., Vizcaino, O., Carson, C. A., Ristic, M., Kuttler, K. L., Trevino, G. S., and Lee, A. J.,** Comparison of three methods of immunization against bovine anaplasmosis: an examination of post-vaccinal effects, *Am. J. Vet. Res.,* 41, 1062, 1980.

55. **Vizcaino, O., Corrier, D. E., Terry, M. K., Carson, C. A., Lee, A. J., Kuttler, K. L., Ristic, M., and Trevino, G. S.,** Comparison of three methods of immunization against bovine anaplasmosis: evaluation of protection afforded against field challenge exposure, *Am. J. Vet. Res.,* 41, 1066, 1980.

56. **McHardy, N.,** Immunization against anaplasmosis — a review, *Prev. Vet. Med.,* 2, 135, 1984.

57. **Lohr, K. F.,** Immunisierung gegen babesiose und anaplasmose von 40 nach Kenya importierten Charollais-Rindern und Bericht uber Erscheinungen der Photosensibilitat bei diesen Tieren, *Zentralbl. Veterinaermed. Reihe B,* 16, 40, 1969.

58. **Allen, P. C. and Kuttler, K. L.,** Effect of *Anaplasma marginale* infection upon blood gases and electrolytes in splenectomized calves, *J. Parasitol.,* 67, 954, 1981.

59. **Kessler, R. H., Ristic, M., Sells, D. M., and Carson, C. A.,** In vitro cultivation of *Anaplasma marginale:* growth pattern and morphologic appearance, *Am. J. Vet. Res.,* 40, 1767, 1979.

60. **Kessler, R. H. and Ristic, M.,** In vitro cultivation of *Anaplasma marginale:* invasion of and development in noninfected erythrocytes, *Am. J. Vet. Res.,* 40, 1774, 1979.

61. **Lawrence, J. A.,** The mechanical transmission of *Anaplasma marginale* under Rhodesian conditions, *Rhod. Vet. J.,* 8, 74, 1977.

62. **Weisenhotter, E.,** Research into the relative importance of Tabanidae (Diptera) in mechanical transmission. III. The epidemiology of anaplasmosis in a Dar-es-Salaam dairy farm, *Trop. Anim. Health Prod.,* 7, 15, 1975.

63. **Ewing, S. A.,** Transmission of *Anaplasma marginale* by arthropods, in *Proc. 7th Natl. Anaplasmosis Conf.,* Hidalgo, R. J. and Jones, E. W., Eds., Mississippi State University Press, State College, 1982, 395.

64. **Christensen, J. F. and Howarth, J. A.,** Anaplasmosis transmission by *Dermacentor occidentalis* taken from cattle in Santa Barbara County, California, *Am. J. Vet. Res.,* 27, 1473, 1966.

65. **Peterson, K. J., Raleigh, R. J., and Stroud, R. K.,** Bovine anaplasmosis transmission studies conducted under controlled natural exposure in *Dermacentor andersoni = (venustus)* indigenous area of eastern Oregon, *Am. J. Vet. Res.,* 38, 351, 1977.

66. **Wickwire, K. B., Kocan, K. M., Barron, S. J., Ewing, S. A., Smith, R. D., and Hair, J. A.,** Infectivity of three *Anaplasma marginale* isolates for *Dermacentor andersoni, Am. J. Vet. Res.,* 48, 96, 1987.

67 **Kocan, K. M., Goff, W., Stiller, D., Barbet, E. F., Edwards, W., Ewing, S. A., Hair, J. A., Barron, S. J., and McGuire, T. C.,** The development of *Anaplasma marginale* in salivary glands of *Dermacentor andersoni,* Proc. 69th Annu. Meet. of the Conf. of Research Workers in Animal Disease, Chicago, 1988, 41.

68. **Kocan, K. M.,** Personal communication, 1987.

69. **Ristic, M. and Carson, C. A.,** Methods of immunoprophylaxis against bovine anaplasmosis with emphasis on the use of the attenuated *Anaplasma marginale* vaccine, in *Immunity to Blood Parasites of Animals and Man,* Miller, L. H., Pino, J. A., and McKelvey, J. J., Eds., Plenum Press, New York, 1977, 151.

70. **Kuttler, K. L.,** Current anaplasmosis control techniques in the United States, *J. S. Afr. Vet. Med. Assoc.,* 50, 314, 1979.

71. **Potgeiter, F. T.,** Epizootiology and control of anaplasmosis in South Africa, *J. S. Afr. Vet. Med. Assoc.,* 50, 367, 1979.

72. **Norman, B. B.,** Current status of vaccination for the control of bovine anaplasmosis, in *Proc. 6th Natl. Anaplasmosis Conf.,* Jones, E. W., Ed., Heritage Press, Stillwater, OK, 1973, 103.

73. **Thompson, K. C., Todorovic, R. A., Mateus, G., and Adams, L. G.,** Methods to improve the health of cattle in the tropics: immunization and chemoprophylaxis against hemoparasitic infections, *Trop. Anim. Health Prod.,* 10, 75, 1978.

74. **Todorovic, R. A., Gonzalez, E. F., and Lopex, G.,** Immunization against anaplasmosis and babesiosis. II. Evaluation of cryopreserved vaccines using different doses and routes of inoculation, *Tropenmed. Parasitol.,* 29, 210, 1978.

75. **Franklin, T. E. and Huff, J. W.,** A proposed method of premunizing cattle with minimum inocula of *Anaplasma marginale, Res. Vet. Sci.,* 8, 415, 1967.

76. **Kliewer, I. O., Richey, E. J., Jones, E. W., and Brock, W. E.,** The preservation and minimum infective dose of *A. marginale,* in *Proc. 6th Natl. Anaplasmosis Conf.,* Jones, E. W., Ed., Heritage Press, Stillwater, OK, 1973, 39.

77. **Roby, T. O., Gates, D. W., and Mott, L. O.,** The comparative susceptibility of calves and adult cattle to bovine anaplasmosis, *Am. J. Vet. Res.,* 22, 982, 1961.

78. **Wilson, A. J.,** Observations on the pathogenesis of anaplasmosis in cattle with particular reference to nutrition, breed and age, *J. S. Afr. Vet. Assoc.,* 50, 293, 1979.

79. **Wilson, A. J. and Trueman, K. F.,** Some effects of reduced energy intake on the development of anaplasmosis in *Bos indicus* cross steers, *Aust. Vet. J.,* 54, 121, 1978.

80. **Callow, L. L.,** Tick-borne livestock diseases and their vectors. III. Australian methods of vaccination against anaplasmosis and babesiosis, *World Anim. Rev.,* 18, 9, 1976.

81. **Pipano, E.,** Bovine anaplasmosis and its control, in *Proc. 11th Int. Congr. Diseases of Cattle,* Mayer, E., Ed., Bergman Press, Haifa, Israel, 1980, 198.

82. **Pipano, E., Mayer, E., and Frank, M.,** Comparative response of Friesian milking cows and calves to *Anaplasma centrale* vaccine, *Br. Vet. J.,* 141, 174, 1985.

83. **Bigalke, R. D.,** The control of the ticks and tick-borne diseases of cattle in South Africa, *Zimbabwe Vet. J.,* 11, 20, 1980.

84. **Ristic, M.,** Anaplasmosis, in *Infectious Blood Diseases of Man and Animals,* Weinman, D. and Ristic, M., Eds., Academic Press, New York, 1968, 478.

85. **Ristic, M., Sibinovic, S., and Welter, C. J.,** An attenuated *Anaplasma marginale* vaccine, in *Proc. U.S. Livestock Sanitary Assoc.,* Blanton, E., Ed., U.S. Animal Health Association, Richmond, 1969, 56.

86. **Henry, E. T., Norman, B. B., Fly, D. E., Wichmann, R. W., and York, S. M.,** Effects and use of a modified live *Anaplasma marginale* vaccine in beef heifers in California, *J. Am. Vet. Med. Assoc.,* 183, 66, 1983.

87. **Garcia, A. O.,** Immunization against bovine babesiosis and anaplasmosis in the Cauca Valley: evaluation of monovalent and bivalent vaccination systems against babesiosis in natural conditions, *Rev. Inst. Colomb. Agric.,* 13, 739, 1978.

88. **Ristic, M. and Carson, C. A.,** An attenuatted *Anaplasma marginale* vaccine with emphasis on the mechanism of protective immunity, in *Tick-borne Diseases and their Vectors,* Wilde, J. K. H., Ed., Lewis Reprints, Tonbridge, England, 1978, 541.

89. **Kuttler, K. L. and Zaraza, H.,** Premunization with an attenuated *Anaplasma marginale,* in *Proc. 73rd Annu. Meet. U.S. Animal Health Assoc.,* Blanton, E., Ed., U.S. Animal Health Association, Richmond, 1969, 104.

90. **Popovic, N. A.,** Pathogenesis of Two Strains of *Anaplasma marginale* in *Dermacentor andersoni* Tick, Ph.D. thesis, University of Illinois, Urbana, 1968.

91. **Carson, C. A.,** Measurement of Cell-Mediated Immune Response in Cattle Induced by Virulent and Attenuated Strains of *Anaplasma marginale* Introduced in both Live and Killed Form and Correlation with Protective Immunity, Ph.D. thesis, University of Illinois, Urbana, 1975.

92. **Kuttler, K. L.,** Serial passage of an attenuated *Anaplasma marginale* in splenectomized calves, in *Proc. 73rd Annu. Meet. U.S. Animal Health Assoc.,* Blanton, E., Ed., U.S. Animal Health Association, Richmond, 1969, 131.

93. **Dennis, R. A., O'Hara, P. J., Young, M. F., and Dorris, K. D.,** Neonatal immunohemolytic anemia and icterus of calves, *J. Am. Vet. Med. Assoc.,* 156, 1861, 1970.

94. **Wilson, J. S. and Trace, J. C.,** Neonatal isoerythryolysis of the bovine, in *Proc. 74th Annu. Meet. U.S. Animal Health Assoc.,* Blanton, E., Ed., U.S. Animal Health Association, Richmond, 1971, 115.

95. **Hines, H. C. and Bedell, D. M.,** Some effects of blood antigens in *Anaplasma marginale* and other vaccines, in *Proc. 6th Natl. Anaplasmosis Conf.,* Jones, E. W., Ed., Heritage Press, Stillwater, OK, 1973, 82.

96. **Dimmock, C. K. and Bell, K.,** Haemolytic disease of the newborn in calves, *Aust. Vet. J.,* 74, 1, 1974.

97. **Searl, R. C.,** Use of an *Anaplasma* vaccine as related to neonatal isoerythrolysis, *Vet. Med. Small Anim. Clin.,* 75, 101, 1980.

98. **Carson, C. A., Sells, D. M., and Ristic, M.,** Cutaneous hypersensitivity and isoantibody production in cattle injected with live or inactivated *Anaplasma marginale* in bovine and ovine erythrocytes, *Am. J. Vet. Res.,* 37, 1059, 1976.

99. **Francis, D. H., Buening, G. M., and Amerault, T. E.,** Characterization of immune responses of cattle to erythrocyte stroma, *Anaplasma* antigen, and dodecanoic acid-conjugated *Anaplasma* antigen: humoral immunity, *Am. J. Vet. Res.,* 41, 362, 1980.

100. **Francis, D. H., Buening, G. M., and Amerault, T. E.,** Characterization of immune responses of cattle to erythrocyte stroma, *Anaplasma* antigen, and dodecanoic acid-conjugated *Anaplasma* antigen: cell-mediated immunity, *Am. J. Vet. Res.,* 41, 368, 1980.

101. **McHardy, N. and Simpson, R. M.,** Attempts at immunizing cattle against anaplasmosis using a killed vaccine, *Trop. Anim. Health Prod.,* 5, 166, 1973.

102. **Bannerjee, D. P., Sarup, S., and Gautam, O. P.,** Attempts to immunize cattle with *Anaplasma marginale* killed vaccine, *Harayana Vet.,* 20, 102, 1981.

103. **Moss, B. and Flexner, V.,** Vaccinia virus expression vectors, *Annu. Rev. Immunol.,* 5, 305, 1987.

104. **Palmer, G. H. and McGuire, T. C.,** Immune serum against *Anaplasma marginale* initial bodies neutralizes infectivity for cattle, *J. Immunol.,* 133, 1010, 1984.

105. **James, M. A., Montenegro-James, S., and Ristic, M.,** Isolation of an in vitro produced soluble *Anaplasma*-albumin complex, *Am. J. Vet. Res.,* 43, 1863, 1982.

106. **McCorkle-Shirley, S., Hart, L. T., Larson, A. D., Todd, W. J., and Myhand, J. D.,** High-yield preparation of purified *Anaplasma marginale* from infected bovine red cells, *Am. J. Vet. Res.,* 46, 1745, 1985.

107. **Palmer, G. H., Barbet, A. F., Kuttler, K. L., and McGuire, T. C.,** Detection of an *Anaplasma marginale* common surface protein present in all stages of infection, *J. Clin. Microbiol.,* 23, 1078, 1986.

108. **Palmer, G. H., Cantor, G. A., Pantzer, C., Barbet, A. F., and McGuire, T. C.,** Manuscript submitted, 1988.

109. **Osborn, M. J. and Wu, H. C. P.,** Proteins of the outer membrane of gram-negative bacteria, in *Ann. Rev. Microbiol.,* Starr, J. J., Ingraham, J. L., and Balows, A., Eds., Annual Reviews, Palo Alto, CA, 1980, 369.

110. **Barbet, A. F., Palmer, G. H., Myler, P. J., and McGuire, T. C.,** Characterization of an immunoprotective protein complex of *Anaplasma marginale* by cloning and expression of the gene coding for the Amf 105L polypeptide, *Infect. Immun.,* 55, 2428, 1987.

111. **Barbet, A. F., Palmer, G. H., and McGuire, T. C.,** Unpublished data, 1987.

112. **Oberle, S. M., Palmer, G. H., Barbet, A. F., and McGuire, T. C.,** Molecular size variations in an immunoprotective protein complex among isolates of *Anaplasma marginale, Infect. Immun.,* 56, 1567, 1988.

113. **Allred, D. R., McGuire, T. C., Palmer, G. H., McElwain, T. F., and Barbet, A. F.,** in preparation.

114. **Palmer, G. H., Oberle, S. M., Barbet, A. F., Goff, W. L., Davis, W. C., and McGuire, T. C.,** Immunization of cattle with a 36-kilodalton surface protein induces protection against homologous and heterologous *Anaplasma marginale* challenge, *Infect. Immun.,* 56, 1526, 1988.

115. **Mackett, M., Yilma, T., Rose, J. K., and Moss, B.,** Vaccinia virus recombinants: expression of VSV genes and protective immunization of mice and cattle, *Science,* 227, 433, 1985.

116. **Gillespie, J. H., Geissing, C., Scott, F. W., Higgins, W. P., Holmes, D. F., Perkus, M., Mercer, S., and Paoletti, E.,** Response of dairy calves to vaccinia viruses that express foreign genes, *J. Clin. Microbiol.,* 23, 283, 1986.

Chapter 2

COWDRIA VACCINES

J. D. Bezuidenhout

TABLE OF CONTENTS

I. INTRODUCTION

Cowdrioses, commonly known as heartwater, is a tick-borne disease affecting mainly cattle, sheep, and goats, but it also affects certain wild animals and mice.[1-3] It is caused by a rickettsia, *Cowdria ruminantium,* and has been transmitted experimentally by 11 species of ticks all belonging to the genus *Amblyomma.*[4]

The disease occurs in large parts of Africa south of the Sahara, Madagascar, Réunion, Mauritius, São Tomè, and on certain islands in the Caribbean.[5] Heartwater is regarded as the most important tick-borne disease in South Africa,[6] and next to East Coast fever, the most important tick-borne disease in Africa.[5]

Despite the fact that heartwater transmission was elucidated as far back as the beginning of this century,[7] and the causative organism described in 1925 by Cowdry,[4,8] control of the disease remains a serious problem. In search of better ways of control, research has been greatly hampered by the inability, until recently,[10,11] to cultivate the organism *in vitro.* Antigen for vaccination, serology, and other research aspects, therefore, had to be obtained from alternative, less suitable sources.

In the early years, even before the mode of transmission or the causative organism were described, it was recognised that animals which recovered from the disease acquired an immunity.[12,13] Since then, numerous attempts at producing a vaccine have been made. However, most of these early attempts were unreliable, unsuccessful, or impractical and included vaccination with gall and other body fluids, hyperimmune sera, modified attenuated "virus", repeated inoculation of sublethal doses of blood and tick suspension, both intravenously and subcutaneously, and the inoculation of formalized animal and tick tissues subcutaneously.[14-20]

It was also established earlier, but not fully realized, that young animals were not as susceptible to intravenous inoculation with virulent blood as older animals.[13] The use of virulent blood to stimulate immunity was followed up by Hutcheon[21] and Spreull.[16] However, the fact that most animals died from vaccination reactions that could not be controlled led to this method being abandoned, and the control of the disease was thereafter aimed at controlling the vector.[17] Later, Neitz and Alexander[22,23] utilized the innate resistance of young animals to vaccinate them against heartwater. The discovery of chemotherapeutic drugs with which heartwater reactions could be treated, made it possible to extend vaccination to older animals.[24-27]

More recently, an alternative vaccine was developed by using *Amblyomma hebraeum* nymphs infected with *C. ruminantium.*[28]

Although heartwater-infective brain tissue was found to be infective in the majority of cases when administered subcutaneously,[29-31] only blood and nymph suspensions have ever been produced and used commercially.

Both these *Cowdria* vaccines are far from ideal, and both their production and use are full of problems which limit their use. However, they are the only methods available to the farmer to artificially stimulate the immunity of susceptible animals, and, if handled with circumspection, are a useful tool in the prevention of the disease.

II. VACCINE PRODUCTION PROCEDURES

Methods for the production of *Cowdria* vaccines, originally developed at the Veterinary Research Institute, Onderstepoort, Republic of South Africa, have, with some modifications, been described by other researchers.[28,32-35]

Although the nymph suspension vaccine is considerably cheaper to produce, the production procedures for the nymph and blood vaccines are essentially the same and involve the following steps: isolation and selection of an isolate of *C. ruminantium* suitable for vaccination, production of a stock antigen, production of vaccine itself, quality control of both stock and vaccine antigens, and bottling and storage of both stock antigens and final vaccines.

A. Isolation and Selection of *Cowdria* Isolates for the Purpose of Vaccination

A number of ways have been described for the isolation of *C. ruminantium* from the field. These attempts can be grouped under the headings which follow.

1. From Animals

Blood[12,16,17,20,36] or organs such as lymph nodes,[37] spleen,[12] brain tissue,[29,30,38] and kidneys[39] of animals that suffered or died from heartwater were found to be suitable sources of *C. ruminantium*.

The intravenous inoculation of pooled blood from animals in endemic areas into susceptible animals has also been used to isolate *C. ruminantium*.[33]

2. From Ticks

Amblyomma ticks have been collected from animals in heartwater endemic areas, homogenized, and the supernate inoculated intravenously into susceptible animals.[40-42] Ticks, especially in their immature stages, have been collected and, after moulting, fed on susceptible animals in order to allow them to transmit heartwater.[43]

B. Selection of *Cowdria* Isolates

Once *C. ruminantium* has been isolated, infective material is snap-frozen with or without the addition of 5 to 10% dimethyl sulfoxide (DMSO).[28,44-47] It is then maintained in dry ice ($-60°C$), liquid nitrogen ($-196°C$), or in an ultralow-temperature electric freezer ($-70°C$). Further experiments are essential to establish its suitability as a vaccine strain. In this regard, the aspects which follow are important.

1. Purity

The isolate(s), especially those obtained from the blood or organs of sick or dead animals, but also to a lesser extent, from ticks, should be free of other pathogens such as viruses, protozoa, or other rickettsiae. Contamination of *C. ruminantium* isolates with other microorganisms apparently has not been a real problem in the past, and no specific methods to purify such isolates have been described. However, the use of nonsusceptible or resistant animals to identify possible contaminants and the use of specific antiprotozoal drugs and antibiotics are possibilities that should be looked into. Due to their selective susceptibility for certain organisms *Amblyomma* ticks with a well-established laboratory history and free of other rickettsiae may prove to be a valuable tool in the isolation of *C. ruminantium* from possibly contaminated isolates.

2. Pathogenicity

Some differences with regard to the virulence of *Cowdria* isolates have been described.[6,35,49-51] Indications at present are that animals will not develop a durable immunity unless they are vaccinated with an isolate which is pathogenic for the target animal.[52] Apart from young animals with an innate age resistance to heartwater (see later), a definite reaction is a prerequisite for the development of a durable immunity.[13,20] Immunity, on the other hand, does not depend on the virulence of a strain.[20,31] Isolates that give an early temperature reaction before the onset of clinical signs are more suitable and definitely helpful during the monitoring of vaccine reactions.[34] All attempts to attenuate the organism by serial passage in sheep and laboratory animals such as mice and ferrets have been unsuccessful or impractical.[31,52,54-56] Despite the fact that no tetracycline-resistant strains of *Cowdria* have so far been identified, it is very important that the isolate selected for vaccine purposes is susceptible to specific chemotherapeutic treatment.

3. Immunogenicity

A great number of *C. ruminantium* isolates have been made and their immunogenicity studied.[29,43,49,52,57-59] All these studies indicated that despite their geographical separation, all *Cowdria* isolates, with the exception of the Kümm isolate, were cross-protective. However, recently an absence of cross-protection between certain isolates of *Cowdria* was demonstrated.[60,61] It was also established that the Ball 3 isolate[51] does not protect against the Welgevonden[42] and Mali isolates.[62] Some of these findings may only be of academic importance, while others have definite practical implications.

Cross-protection trials with regard to heartwater are usually done with sheep or goats. Animals should preferably be inoculated twice with a homologous isolate before being challenged with a heterologous one. This is to ensure that they are solidly immune to the homologous strain, which is an important factor in the evaluation of the subsequent results.[5] However, large groups of animals immunized only once with a specific strain can be used to demonstrate the absence or presence of cross-protection by challenging them with other strains.[62]

Taking all the above into consideration, it is important to select as mild a strain as possible, but one which gives a temperature warning before the onset of clinical signs and has a wide range of cross-protection, especially against local strains. Furthermore, it must also be susceptible to treatment with available anti-*Cowdria* drugs.

C. Production and Storage of Stock Antigen

In the case of heartwater, a stock antigen is prepared and used to infect animals to be bled for vaccine production or for infecting larval ticks, in the case of the nymph vaccine. If prepared, tested, and stored properly, stock antigens are a sure source of *C. ruminantium* which will usually produce constant and similar reactions in recipient animals. This is not only important, but of great help in the planning of a program for vaccine production.

Stock antigens are usually prepared from the blood of reacting sheep, but there is no reason why tick stabilates or infective mouse tissue can not fulfill the same purpose.

Sheep from a heartwater-free area are stabled under tick-free conditions. Precautions should be taken to ensure that the sheep do not harbor any protozoa or viruses that may interfere with the production and use of the vaccine.[33] Examination of blood smears and vaccination against certain viruses, e.g., bluetongue are some methods that can be followed to ensure that no unwanted disease will be spread by the vaccine.

The sheep, in groups of about 10, are then inoculated intravenously with the same volume of inoculum containing the selected strain. Once the sheep develop a febrile reaction of above 40°C for 2 to 3 d (usually 9 to 10 d postinoculation), they are bled from the jugular vein, using a trocar, directly into a cold (4°C), sterile solution of buffered peptone-lactose (BLP) with 10% dimethyl sulfoxide (DMSO) and certain antibiotics. Sodium citrate is added during the preparation of the BLP at a rate of 20 g/l.[34] Bleeding continues until a 1:1 blood to BLP ratio is achieved. During bleeding, the blood/BLP mixture is kept in motion, and the container in which the blood is collected is surrounded by crushed ice to ensure a temperature of between 4 to 8 °C.

The blood is then, under a sterile hood, bottled in 10-ml plastic vaccine containers and sealed with rubber-lined metal caps. Except for approximately ten containers which are held back to test for the presence of other microorganisms, all other containers with virulent blood are then snap-frozen in liquid nitrogen (-196°C) or in a mixture of dry ice and alcohol (-60°C) and stored on dry ice or in liquid nitrogen or in ultralow-temperature electric deep freezers (-70°C). After freezing, the contents of some containers are also tested for infectivity, and in both instances the methods described for the quality control of the final vaccine (see later) are employed.

D. Preparation of the Blood Vaccine

Although sheep are most commonly used, goats and cattle are also, under certain conditions, suitable animals for the production of a *Cowdria* blood vaccine.[32] Sheep similar to those mentioned before are used, and by following the same methods described in the previous section, a batch of blood vaccine is prepared (approximately 3000 to 4000 doses). It is kept in liquid nitrogen or on dry ice until the quality control results are available. Only after it has passed the quality control tests is it issued as a blood vaccine.

Animals that were bled are treated immediately afterwards with oxytetracycline intravenously at a dosage rate of 10 mg/kg. Long-acting tetracyclines injected intramuscularly may also be used, but due to muscular damage caused by these products, animals are less suitable for slaughtering afterwards.[63]

E. Preparation of the Nymph Suspension Vaccine

Methods for the preparation of a tick-derived vaccine against heartwater were first described by Bezuidenhout.[38] A few years later, this vaccine temporarily replaced the blood vaccine normally issued by the Veterinary Research Institute, Onderstepoort.[64] However, during this period in which it was widely used under field conditions, the occurrence of serious shock in lambs and especially goat kids after intravenous injection of the vaccine[65] led to its withdrawal. In sheep and cattle the nymph suspension was found to be equal to the blood vaccine as regards immunogenicity.[28,66] Current research is aimed at eliminating the shock factor from the nymph vaccine. The shock factor is located in the supernatant after centrifugation at 30,000 × g for 20 min, while the *Cowdria* organisms are in the sediment. However, the viability and preservability of these organisms may not be as good as in uncentrifuged suspensions.[67] The following steps are involved in the preparation of a nymph suspension vaccine.

1. Establishment of a Colony of Amblyomma Ticks for the Purpose of Vaccine Production

Methods for the selection, rearing, and maintenance of *Amblyomma hebraeum* are available.[28,68] These methods will most probably also suit other *Amblyomma* species.

2. Infection of Larvae with C. ruminantium

Larvae are infected with heartwater by feeding them on reacting sheep that have been inoculated with the stock antigen described above. This is done in such a way that most larvae feed during the height of the febrile reaction of the sheep.[28,68]

It has been demonstrated that it is only during this relatively short period that blood of reacting animals is infective for ticks.[4,69] Only those larvae that drop during the height of the temperature reaction are used further in the production of a nymph vaccine. They are kept at 27°C until they have moulted and are ready to feed again; 2 to 20 weeks after moulting, batches of nymphae are fed on susceptible sheep in order to amplify the organisms in the ticks.[4]

3. Surface Sterilization of Engorged Nymphae

Nymphae that have engorged and detached from the host are collected, and all dead or damaged ticks are removed. The remainder of the ticks are counted and sterilized externally using methods previously described.[28,34] This entire washing procedure is performed in a closed system to prevent external contamination. A horizontal rocker is used to keep the nymphs in constant motion. Disinfectants and water are drawn into the tick container and, after certain periods, sucked out by means of a vacuum pump.

4. Preparation of an Infective Nymph Filtrate

Using a Polytron®* tissue homogenizer, the ticks are homogenized in the same container in which they were cleaned. BLP[70] containing sodium benzyl penicillin (240 mg/l), streptomycin sulfate (200 mg/l), Collistin (10,200 IU/l), and a silicone antifoam agent SAG-471®** (0,005 ml/l) are added at a rate of 15 ml for every 100 nymphae. The duration of homogenization is usually 1 min at full speed, but it may be necessary to increase the time if more than 1000 nymphae are homogenized at the same time. The whole procedure is carried out on crushed ice to ensure a temperature of approximately 4°C. To remove macerated tick integument, the tick suspension is then filtered through a series of 3 gauze or nylon filters (35 to 625 pores per square centimeter). After filtration, and while the filtrate is stirred on a magentic stirrer, more BLP is added to give a final concentration of five nymphae per milliliter. Before snap-freezing in liquid nitrogen or a mixture of dry ice and alcohol (96%), DMSO is added drop by drop (5 ml/100 ml filtrate). The DMSO may also be added into the BLP mixture before homogenization.

5. Dilution of Concentrated Filtrate

One container with 10 ml nymph filtrate is removed from the liquid nitrogen cylinder or cold storage facility and quickly thawed under running tap water. Twofold dilutions ranging from $1/_{80}$ to $1/_{320}$ of a nymph are made, and 2 ml of each dilution injected intravenously into a sheep. The undiluted concentrate is also cultured to establish the presence of any bacterial or fungal contamination.

If the infectivity of the nymph filtrate is satisfactory, i.e., still infective at at least $1/_{80}$ dilution and free of other contaminants, the concentration of nymphae needed for the preparation of a final vaccine is then determined. The highest dilution which gave a reaction in sheep is then multiplied or concentrated 20 times in order to ensure a final vaccine that may lose some infectivity without becoming completely ineffective.

For example, if the sheep that received the $1/_{80}$ and $1/_{160}$ dilution reacted, but the one that received the $1/_{320}$ dilution did not, then the following calculation is made to determine the concentration of nymphae needed in the final vaccine. Concentration of nymphae/dose of final vaccine $= 20 \times$ highest positive dilution of nymph filtrate $= 20 \times 1/_{160} = 1/_8$.

The number of nymphae per milliliter of concentrate (NC) and the concentration of nymphae per dose of final vaccine (ND) is then used to determine the volume concentrate (VC) required to prepare a specific number of doses of final vaccine (DFV).

$$VC = \frac{DFV \times ND}{NC}$$

6. Bottling of Final Nymph Vaccine

The required volume of frozen nymph suspension is thawed and diluted in BLP containing all the ingredients described above except for the antifoam agent, which is left out. The mixture, which is held on crushed ice, is continually mixed by means of a peristaltic pump and, at the same time, dispensed in plastic vaccine containers as either 2 doses of 2 ml or 5 doses of 2 ml per bottle. Immediately after the bottles have been capped they are dropped into liquid nitrogen or a mixture of dry ice and alcohol. The vials with the final vaccine are then stored on dry ice or in liquid nitrogen.

F. Quality Control of *Cowdria* Vaccines

Each batch of nymphal or blood vaccine has to pass a number of quality control checks before it is issued.[28,34] The purpose of this is to ensure that the vaccine is free of bacterial

---

* Polytron: Kinematica GmbH, CH 6005 Luzern, Steinhofhalde, Switzerland.

** SAG-471: Union Carbide Corporation, General Office, 270 Park Avenue, New York, 10017.

and fungal contaminants, pathogenic viruses, and toxins. At the same time, the infectivity of the vaccine is also tested at 1:10 and 1:20 dilution in two sheep each. Both sheep injected with a 1:20 dilution should show typical heartwater reactions. However, if they fail to react at this dilution, but the sheep which received the 1:10 dilution show a strong reaction, the vaccine is also suitable for issueing.

III. VACCINATION PROCEDURES

The vaccine is transported from the vaccine factory to the farm on dry ice or in liquid nitrogen. In instances where the blood vaccine is stored in bulk at co-operatives or similar outlets, it can also be transported on crushed ice in vacuum flasks in which it can retain its infectivity for at least 8 hours.[71]

Immediately prior to use it is thawed quickly under running tapwater or in a container with water at approximately 21 to 26°C.

All injections must be given intravenously. As shock is known to occur with both vaccines, but especially in young lambs and kids when injected with the nymph vaccine, care should be taken to inject the vaccine slowly. The following methods of vaccinations have been developed taking the species of animal, age, and number of animals into consideration:

A. Vaccination of Young Animals

Calves under the age of 1 month and lambs and kids under the age of 7 d usually possess a nonspecific resistance to heartwater.[22,33,72] There have been some indications that the natural resistance of goat kids may not be adequate to protect them from fatal heartwater vaccination reactions.[46] However, in a recent experiment it was found that the majority of Boer goat kids, vaccinated at an age of 2 to 12 d, were resistant to heartwater. Only 4 out of 74 showed a reaction and were treated; 3 months later they were challenged with the homologous strain and all were found to be immune.[73] This natural resistance to heartwater, also referred to as an age resistance, is not well understood at present, but is independent of the immune status of the mother animal.[22] A small percentage of calves may show a strong reaction and may even die if not treated in time. Animals should be carefully watched in order to identify such cases in time. This method of vaccination is usually performed in heartwater endemic areas to protect the calf crop against outbreaks of the disease. It is also the method with the lowest risk of losses due to vaccination reactions.

B. Infection and Treatment Method of Vaccination

This method of vaccination is usually practiced in older animals that have lost their innate age resistance to heartwater. Such animals usually react, and proper treatment is essential to avoid great losses. This procedure is also advocated in the case of valuable animals or when animals are brought into an endemic area from an area free of heartwater. There is a definite risk involved in using this procedure, and dedicated, close supervision is a prerequisite for success.

Animals are usually vaccinated in small groups (<20), and the rectal temperature of every animal is recorded daily from about 7 to 28 d after vaccination. It is strongly recommended that such temperatures should be plotted on graph paper in order to assist the observer in detecting any deviation from the normal. Some workers prefer to take temperatures twice a day, in the early morning and late in the afternoon,[74] while others advocate early morning temperatures only.[33] Temperatures should be taken every day at the same time, preferably before they are fed, and care should be taken to avoid chasing the animals around beforehand as this will elevate the body temperatures, making the evaluation thereof very difficult.

If during the incubation period the early morning body temperature rises above 39.5°C in cattle and goats and 40°C in sheep, a heartwater vaccination reaction should be expected.[65]

Treatment with short-acting oxytetracycline at 5 to 10 mg/kg or long-acting formulations at 20 mg/kg is then given intramuscularly on the first or second day during which the temperature is elevated. Daily treatment should continue until the temperature drops dramatically or returns to normal. In the case of short-acting formulations, the treatment is usually given for at least 2 d, even if the temperature has dropped. Temperatures that persist beyond 48 h after the last treatment are regarded as persistent temperature reactions,[74] and a further dose is given, preferably intravenously. Sulfadimidine was found to be effective in some instances where reactions persisted in spite of repeated oxytetracycline treatments.[74]

Babesioses or even anaplasmosis may sometimes complicate matters,[22,75] and the examination of blood smears is indicated when animals do not respond satisfactorily to treatment.

Relapses after treatment may occur, and it is advisable, especially in valuable animals, to monitor the temperature reaction for an additional period of 2 weeks after the last treatment.[33]

Treatment with tetracyclines does not interfere with the establishment of immunity,[27] unless given too early.[53]

C. The Block Method of Vaccination

The block method of vaccination,[75-78] also known as systematic treatment,[5,33] is widely practiced in South Africa, especially when immunizing large numbers of small kids and lambs.

This method involves the intravenous inoculation of heartwater infective material, usually blood, into a large number of animals and then treating them indiscriminately on a specific day or days, irrespective of whether they show a temperature reaction or not. Although this procedure makes immunization possible on a large scale, it has some serious drawbacks. It is generally accepted that, in animals which have lost their age resistance, a durable immunity to heartwater follows only after the animal has developed a definite febrile reaction.[20,53] During the block method, temperatures are usually not monitored, which results in a great uncertainty about the immunity of animals afterwards. If treatment is performed too early after inoculation, animals either do not develop an immunity[53] or, in some instances, they have to be retreated.[76-78] On the other hand, if treatment is given too late during the course of the reaction, it may result in heavy losses.[79]

As the incubation period of the disease is affected by the infectivity of the dose,[20,28,29,52] and probably other factors such as age and breed, it is very difficult to lay down a specific day for treatment which will prove to be effective in all cases. However, it is advised that exogenous cattle should be treated on day 14, indigenous cattle on day 16, sheep and Angora goats on day 11, and Boer goats on day 13 after vaccination.[53]

In order to determine the most suitable day for treatment of a large number of animals, a controlled block method is now being advocated in South Africa.[62] This procedure entails the monitoring of rectal temperatures of about 10% of the animals that have been vaccinated on a specific day; 1 d after one or more of this indicator sample have reacted, temperatures of all the animals are taken and all those above 40.5°C are marked and treated with tetracycline. In all cases treatment is repeated 2 d later. This procedure is repeated on the remaining animals for the following few days until all the animals have reacted. This method was tested in 80 Merino sheep. No losses occurred, and apart from two animals that did not react after vaccination, all the other animals were found to be immune to challenge with virulent blood 3 weeks after the initial vaccination.[62]

D. General Remarks Regarding the Vaccination Procedure

It is inadvisable to vaccinate pregnant animals, especially if their unborn are valuable. However, when temperatures are monitored and treatment given with the first rise in temperature, vaccination of pregnant ewes can be successful. A total of 43 Dorper ewes, 2 to 4 months pregnant, were vaccinated with virulent blood (Ball 3 isolate). They were all treated on the first day of reaction and only 1 animal aborted 3 weeks later, while the remaining 42 produced healthy offspring.[80]

A certain percentage of animals, especially cattle, fail to react after vaccination. This is usually low (2 to 5%),[33] but can sometimes be very high.[81] Apart from a possible persistence in the innate age resistance,[81] there appear to be other factors responsible for the phenomenon of nonreactors, but these factors are unknown at this stage. Nonreactors can be given a second vaccination which sometimes produces a reaction; however, the risk of shock increases with repeated inoculation of blood.[75,82]

Great differences in the duration of immunity in the different domestic animals, in the absence of natural challenge, have been reported.[5,83] The immunity in sheep may wane after 6 months,[57,84] but in other cases it may remain sufficient to protect animals against a fatal outcome for at least 4 years.[49]

The duration of immunity in goats is poorly documented and was only tested for a period of about 2 months after vaccination, during which time they proved to be sufficiently immune to protect them from fatal heartwater.[5,29] In Angora goats the degree of immunity seems to depend much on the time at which the animals were treated during the vaccination reaction. Early treatment resulted in a poor immunity, while animals treated on the second and third days of the reaction were immune to challenge 107 to 205 days later.[79] However, in the presence of natural tick challenge, the majority of Angora goats were found to be immune to challenge 1 years after vaccination.[85]

In cattle, the duration of immunity in the absence of natural challenge appears to be in the region of about 2 years.[86]

Although some animals may again be susceptible as early as 2 months after vaccination,[87] the possibility that these cases were due to other strains not protected by the vaccine strain cannot be excluded.

IV. DISCUSSION

The methods for production of heartwater vaccines and the immunization of animals against the disease are far from ideal. They are, however, the only available methods to protect animals against the disease, especially in endemic areas.

All the major problems regarding research on *Cowdria* are being addressed at the moment.[88] These include the search for a more suitable antigen,[11] an easier route of inoculation,[89] better stability and preservation,[90] effective serological and other techniques to determine exposure and immunity,[91-93] and the development of a *Cowdria* DNA recombinant vaccine.[94]

Hopefully, these studies will lead to the development of a *Cowdria* vaccine which will be safe and practical to use in areas where the disease prevails or where the danger of introducing it exists.

ACKNOWLEDGMENTS

I gratefully acknowledge the assistance of Drs. E. M. Nevill and J.L. du Plessis and Mrs. Susan Brett in the preparation of this article.

REFERENCES

1. **Van de Pypekamp, H. and Prozesky, L.,** Heartwater. An overview of the clinical signs, susceptibility and differential diagnosis of the disease in domestic ruminants, *Onderstepoort J. Vet. Res.,* 54, 263, 1987.
2. **Oberem, P. T. and Bezuidenhout, J. D.,** Heartwater in hosts other than domestic ruminants, *Onderstepoort J. Vet. Res.,* 54, 271, 1987.
3. **MacKenzie, P. K. I. and van Rooyen, R. E.,** The isolation and culture of *Cowdria ruminantium* in albino mice, in *Proc. Int. Congr. on Tick Biology and Control,* Whitehead, G. B. and Gibson, J. D., Eds., Tick Research Unit, Rhodes University, Grahamstown, South Africa, 1981, 33.
4. **Bezuidenhout, J. D.,** Natural transmission of heartwater, *Onderstepoort J. Vet. Res.,* 54, 349, 1987.
5. **Uilenberg, G.,** Heartwater (*Cowdria ruminantium* infection): current status, *Adv. Vet. Sci. Comp. Med.,* 27, 427, 1983.
6. **Neitz, W. O.,** Heartwater, *Bull. Off. Int. Epizool.,* 70, 329, 1968.
7. **Lounsbury, C. P.,** Tick heartwater experiments, *Agric. J. Cape G. H.,* 16, 682, 1900.
8. **Cowdry, E. V.,** Studies on the aetiology of heartwater. II. *Rickettsia ruminantium* (northern species) in the tissues of ticks transmitting the disease, *J. Exp. Med.,* 42, 253, 1925.
9. **Cowdry, E. V.,** Studies on the aetiology of heartwater. I. observation of a *Rickettsia ruminantium* (northern species) in the tissues of infected animals, 11th and 12th Rep. of the Director of Veterinary Education and Research, Union of South Africa, 1925, 180.
10. **Bezuidenhout, J. D., Paterson, C., and Barnard, B. J. H.,** *In vitro* cultivation of *Cowdria ruminantium,* *Onderstepoort J. Vet. Res.,* 52, 112, 1985.
11. **Bezuidenhout, J. D.,** The present state of *Cowdria ruminantium* cultivation in cell lines, *Onderstepoort J. Vet. Res.,* 54, 205, 1987.
12. **Dixon, R. W.,** Heartwater experiments, *Agric. J. Cape G. H.,* 14, 205, 1898.
13. **Dixon, R. W.,** Heartwater experiments, *Agric. J. Cape G. H.,* 15, 790, 1899.
14. **Dixon, R. W. and Spreull, J.,** Report on investigations into heartwater, *Natal Agric. J.,* 22, 857, 1899.
15. **Spreull, J.,** 1902; as cited in **Hutcheon, D.,** Heartwater in sheep and goats, *Agric. J. Cape G. H.,* 20, 633, 1902.
16. **Spreull, J.,** Heartwater inoculation experiments, *Agric. J. Cape G. H.,* 24, 433, 1904.
17. **Spreull, J.,** Heartwater, *J. Dep. Agric. Union S. Afr.,* 4, 236, 1922.
18. **Dixon, R. W.,** Heartwater sheep and goats, *Agric. J. Cape G. H.,* 36, 5, 554, 1910.
19. **Theiler, A. and DuToit, P. J.,** The transmission of tick-borne diseases by the intrajugular injection of the emulsified intermediary host itself, *Bull. Soc. Pathol. Exot.,* 19, 725, 1928.
20. **Alexander, R. A.,** Heartwater: the present state of our knowledge, *Rep. Dir. Vet. Serv. Anim. Ind. Onderstepoort,* 1, 89, 1931.
21. **Hutcheon, D.,** Heartwater in sheep and goats, *Agric. J. Cape G. H.,* 20, 633, 1902.
22. **Neitz, W. O. and Alexander, R. A.,** The immunization of calves against heartwater, *J. S. Afr. Vet. Med. Assoc.,* 2, 103, 1941.
23. **Neitz, W. O. and Alexander, R. A.,** Immunization of cattle against heartwater and control of tick-borne diseases, redwater, gallsickness and heartwater, *Onderstepoort J. Vet. Res.,* 20, 137, 1945.
24. **Neitz, W. O.,** Uleron in the treatment of heartwater, *J. S. Afr. Vet. Assoc.,* 11, 15, 1940.
25. **Haig, D. A., Alexander, R. A., and Weiss, K. E.,** Treatment of heartwater with Terramycin, *J. S. Afr. Vet. Med. Assoc.,* 25, 45, 1954.
26. **Weiss, K. E.,** Heartwater — its treatment and control, *Int. Vet. Bull.,* 1, 28, 1952.
27. **Simpson, R. M. and Wiley, A. J.,** The use of aureomycin hydrochloride in the treatment of heartwater, *Annu. Rep. Veterinary Department of Kenya for 1949,* 1951, 23.
28. **Bezuidenhout, J. D.,** The development of a new heartwater vaccine, using *Amblyomma hebraeum* nymphs infected with *Cowdria ruminantium,* in *Proc. Int. Congr. on Tick Biology and Control,* Whitehead, G. B. and Gibson, J. D., Eds., Tick Research Unit, Rhodes University, Grahamstown, South Africa, 1981, 41.
29. **Ilemobade, A. A.,** Study of Heartwater and the Causative Agent *Cowdria ruminantium* (Cowdry, 1925) in Nigeria, Ph.D. thesis, Ahmado Bello University, Zaria, 1976.
30. **Ilemobade, A. A. and Blotkamp, C.,** Heartwater in Nigeria. II. The isolation of *Cowdria ruminantium* from live and dead animals and importance of routes of inoculation, *Trop. Anim. Health Prod.,* 10, 39, 1978.
31. **Uilenberg, G.,** Heartwater disease, in *Diseases of Cattle in the Tropics,* Ristic, M. and McIntyre, I., Eds., Martinus Nijhoff, Netherlands, 1981, 345.
32. **Arnold, R. M. and Kanhai, K.,** Heartwater infection (cowdriosis of rickettsioses of ruminants), a manual for the preparation of infected blood for heartwater immunication, *F.A.O./MONAP, Maputo Vet. Res. Inst.,* 1979.
33. Anon., Methods of immunization against *Cowdria ruminantium,* in *Ticks and Tick-Borne Disease Control, A Practical Field Manual,* Food and Agriculture Organization, Rome, 2, 564, 1984.

34. **Oberem, P. T. and Bezuidenhout, J. D.,** The production of heartwater vaccine, *Onderstepoort J. Vet. Res.,* 54, 485, 1987.

35. **Camus, E. and Barrè, N.,** La Cowdriose (heartwater), *Rev. Elev. Med. Vet. Pays Trop.,* 1982.

36. **MacKenzie, P. K. I. and van Rooyen, R. E.,** The isolation and culture of *Cowdria ruminantium* in albino mice, in *Proc. Int. Congr. on Tick Biology and Control,* Whitehead, G. B. and Gibson, J. D., Eds., Tick Research Unit, Rhodes University, Grahamstown, South Africa, 1981, 33.

37. **du Plessis, J. L. and Kümm, N. A. L.,** Passage of *Cowdria ruminantium* in mice, *J. S. Afr. Vet. Assoc.,* 42, 217, 1971.

38. **Anon.,** Annual Report for 1950, Veterinary Department of Kenya, Government Printer, Nairobi, 1952, 16.

39. **Jongejan, F., van Winkelhoff, A. J., and Uilenberg, G.,** *Cowdria ruminantim* (Rickettsiales) in primary goat kidney cell cultures, *Res. Vet. Sci.,* 29, 392, 1980.

40. **Barrè, N., Camus, E., Birnie, E., Burridge, M. J., Uilenberg, G., and Provost, A.,** Setting up a method for surveying the distribution of Cowdriosis (heartwater) in the Caribbean, in *Proc. 13th World Congr. Diseases of Cattle,* South African Veterinary Association, Pretoria, 1984, 33.

41. **Burridge, M. J., Barrè, N., Birnie, E. F., Camus, E., and Uilenberg, G.,** Epidemiological studies on heartwater in the Caribbean with observations on tick-associated bovine Dermatophilosis, in *Proc. 13th World Congr. Diseases of Cattle,* South African Veterinary Association, Pretoria, 1984, 542.

42. **du Plessis, J. L.,** A method for determining the *Cowdria ruminantium* infection rate of *Amblyomma hebraeum:* effects in mice injected with tick homogenates, *Onderstepoort J. Vet. Res.,* 52, 55, 1985.

43. **MacKenzie, P. K. I. and McHardy, N.,** The culture of *Cowdria ruminantium* in mice: significance in respect of the epidemiology and control of heartwater, *Prev. Vet. Med.,* 2, 227, 1984.

44. **Weiss, K. E. Haig, D. A., and Alexander, R. A.,** Aureomycin in the treatment of heartwater, *Onderstepoort J. Vet. Res.,* 25, 41, 1952.

45. **Haig, D. A.,** Tick-borne rickettsioses in South Africa, *Adv. Vet. Sci.,* 2, 307, 1955.

46. **Thomas, A. D. and Mansveldt, P. R.,** Immunization of goats against heartwater, *J. S. Afr. Vet. Med. Assoc.,* 28, 163, 1957.

47. **Ramisse, J. and Uilenberg, G.,** Conservation d'une souche de *Cowdria ruminantium* par congelation, *Rev. Elev. Med. Vet. Pays Trop.,* 23, 313, 1970.

48. **Ilemobade, A. A., Blotkamp, J., and Synge, B. A.,** Preservation of *Cowdria ruminantium* at low temperature, *Res. Vet. Sci.,* 19, 337, 1975.

49. **Neitz, W. O., Alexander, R. A., and Adelaar, J. F.,** Studies on immunity to heartwater, *Onderstepoort J. Vet. Sci. Anim. Ind.,* 21, 243, 1947.

50. **Neitz, W. O.,** Tick-borne diseases as a hazard in the rearing of calves in Africa, *Bull. Off. Int. Epizoot.,* 62, 607, 1964.

51. **Haig, D. A.,** Note on the use of the white mouse for the transport of strains of heartwater, *J. S. Afr. Vet. Med. Assoc.,* 23, 167, 1952.

52. **du Plessis, J. L.,** Mice Infected with a *Cowdria ruminantium* — like agent as a model in the study of heartwater, D.V.Sc thesis, University of Pretoria, South Africa, 1982.

53. **du Plessis, J. L. and Malan, L.,** The block method of vaccination against heartwater, *Onderstepoort J. Vet. Res.,* 54, 493, 1987.

54. **Du Toit, P. J. and Alexander, R. A.,** An attempt to attenuate the virus of heartwater by passage, *Rep. Dir. Vet. Serv. Anim. Ind. Onderstepoort,* 1, 151, 1931.

55. **Adelaar, J. F.,** Personal communication in Henning, 1956.

56. **Ramisse, J.;** as cited in **Uilenberg, G.,** in *Diseases of Cattle in the Tropics,* Ristic, M. and McIntyre, I., Eds., Martinus Nijhoff, Netherlands, 1981, 345.

57. **Neitz, W. O.,** The immunity in heartwater, *Onderstepoort J. Vet. Sci. Anim. Ind.,* 13, 245, 1939.

58. **Van Winkelhoff, A. J. and Uilenberg, G.,** Heartwater. Cross-immunity studied with strains of *Cowdria ruminantium* isolated in West- and South Africa, *Trop. Anim. Health Prod.,* 13, 160, 1981.

59. **Uilenberg, G., Zikovic, D., Dwinger, R. H., Ter Huurne, A. A. H. M., and Pierie, N. M.,** Cross immunity between strains of *Cowdria ruminantium, Res. Vet. Sci.,* 35, 200, 1983.

60. **Logan, L. L., Birnie, E. F., and Mebus, C. A.,** Cross-immunity between isolates of *Cowdria ruminantium, Onderstepoort J. Vet. Res.,* 54 (Abstr.), 345, 1987.

61. **Jongejan, F., Uilenberg, G., Fransen, F. F. J., Gueye, A., and Niewenhuijs, J.,** Antigenic differences between stocks of *Cowdria ruminantium, Res. Vet. Sci.,* 1987, submitted.

62. **Olivier, J. A., Bezuidenhout, J. D., and du Plessis, J. L.,** Unpublished data, 1986.

63. **Petzer, I.-M., Giesecke, W. H., and van Staden, J. J.,** A comparative investigation on the tissue compatability in cattle of several oxytetracycline formulations for intramuscular administration, in *Proc. 13th World Conf. Diseases of Cattle,* South African Veterinary Association, Pretoria, 1984, 944.

64. **Bezuidenhout, J. D. and Oberem, P. T.,** New vaccine against heartwater, *S. Afr. Tydskr. Landbouwet.,* 81, 2, 1985.

65. **Van der Merwe, L.,** The infection and treatment method of vaccination against heartwater, *Onderstepoort J. Vet. Res.,* 54, 489, 1987.

66. **Bezuidenhout, J. D. and Spickett, A. M.,** An *in vivo* comparison of the efficacy of the heartwater blood and ground-up tick suspension vaccines in calves, *Onderstepoort J. Vet. Res.,* 52, 269, 1985.

67. **Bezuidenhout, J. D.,** Unpublished data, 1986.

68. **Heyne, I. H., Elliot, E. G. R., and Bezuidenhout, J. D.,** Rearing and infection techniques for *Amblyomma* species to be used in heartwater transmission experiments, *Onderstepoort J. Vet. Res.,* 54, 461, 1987.

69. **Barrè, N. and Camus, E.,** The reservoir status of goats recovered from heartwater, *Onderstepoort J. Vet. Res.,* 54, 435, 1987.

70. **Potgieter, F. T. and Bester, J. B.,** Freeze-drying of *Anaplasma marginale, Onderstepoort J. Vet. Res.,* 49, 179, 1981.

71. **Olivier, J. A.,** Unpublished data, 1986.

72. **du Plessis, J. L. and Malan, L.,** The non-specific resistance of cattle to heartwater, *Onderstepoort J. Vet. Res.,* 54, 333, 1987.

73. **du Plessis, J. L.,** Unpublished data, 1987.

74. **Van der Merwe, L.,** Field experience with heartwater, *(Cowdria ruminantium)* in cattle, *J. S. Afr. Vet. Med. Assoc.,* 50, 323, 1979.

75. **Fick, J. F. and Schuss, J.,** Heartwater immunization under field conditions in Swaziland, *J. S. Afr. Vet. Med. Assoc.,* 23, 9, 1952.

76. **Uilenberg, G.,** Etudes sur la cowdriose à Madagascar, *Rev. Elev. Med. Vet. Pays Trop.,* 33, 21, 1971.

77. **Poole, J. D. H.,** Flock immunization of sheep and goats against heartwater. II. Preliminary experiments on flock immunization of goats, *J. S. Afr. Vet. Med. Assoc.,* 33, 357, 1962.

78. **Poole, J. D. H.,** Flock immunization of sheep and goats against heartwater. I. Investigations regarding routine flock immunization of sheep, *J. S. Afr. Vet. Med. Assoc.,* 33, 35, 1962.

79. **du Plessis, J. L., Jansen, B. C., and Prozesky, L.,** Heartwater in Angora goats. I. Immunity subsequent to artificial infection and treatment, *Onderstepoort J. Vet. Res.,* 50, 137, 1983.

80. **Bezuidenhout, J. D. and Olivier, J. A.,** Unpublished data, 1986.

81. **du Plessis, J. L., Bezuidenhout, J. D., and Lüdemann, C. J. F.,** The immunization of calves against heartwater: subsequent immunity both in the absence and presence of natural tick challenge, *Onderstepoort J. Vet. Res.,* 51, 193, 1984.

82. **Barnard, W. G.,** Heartwater immunization under field conditions in Swaziland, *Bull. Epizoot. Dis. Afr.,* 1, 300, 1953.

83. **Stewart, C. G.,** Specific immunity in farm animals to heartwater, *Onderstepoort J. Vet. Res.,* 54, 34, 1987.

84. **du Plessis, J. L.,** The influence of dithiosemicarbazone on the immunity of sheep of heartwater, *Onderstepoort J. Vet. Res.,* 48, 175, 1981.

85. **Gruss, B.,** Problems encountered in the control of heartwater in Angora goats, *Onderstepoort J. Vet. Res.,* 54, 513, 1987.

86. **du Plessis, J. L. and Bezuidenhout, J. D.,** Investigations on the natural and acquired resistance of cattle to artificial infection with *Cowdria ruminantium, J. S. Afr. Vet. Med. Assoc.,* 50, 334, 1979.

87. **Arnold, R. M. and Asselbergs, M.,** Tick-borne diseases. Immunization of cattle imported into Mozambique, *World Anim. Rev.,* 40, 23, 1981.

88. **Bigalke, R. D.,** Future prospects and goal setting regarding research on heartwater, *Onderstepoort J. Vet. Res.,* 54, 543, 1987.

89. **Bezuidenhout, J. D., Olivier, J. A., Gruss, B., and Badenhorst, J. V.,** The efficacy of alternative routes for the infection and vaccination of animals against heartwater, *Onderstepoort J. Vet. Res.,* 54, 497, 1987.

90. **Logan, L. L.,** *Cowdria ruminantium:* stability and preservation of the organism, *Onderstepoort J. Vet. Res.,* 54, 187, 1987.

91. **du Plessis, J. L. and Malan, L.,** The application of the indirect fluorescent antibody test in research on heartwater, *Onderstepoort J. Vet. Res.,* 54, 319, 1987.

92. **du Plessis, J. L., Camus, E., and Oberem, P. T.,** Heartwater serology: some problems with the interpretation of results, *Onderstepoort J. Vet. Res.,* 54, 327, 1987.

93. **Stewart, C. G.,** Specific immunity in mice to heartwater, *Onderstepoort J. Vet. Res.,* 54, 343, 1987.

94. **Ambrosio, R. E., du Plessis, J. L., and Bezuidenhout, J. D.,** The construction of genomic libraries of *Cowdria ruminantium* in an expression vector, lambda gt 11, *Onderstepoort J. Vet. Res.,* 54, 255, 1987.

Chapter 3

BABESIA VACCINES ATTENUATED BY BLOOD PASSAGE AND IRRADIATION

Stuart M. Taylor

TABLE OF CONTENTS

I. ATTENUATED VACCINES

A. Introduction

The use of injections of infected whole blood to try to immunize cattle against contracting a severe tick-transmitted and unsupervised infection of babesiosis is almost as old as the recognition of the parasite itself and its relationship with the disease. Within 5 years of the description of the parasite in 1888 by Babes,[1] Smith and Kilborne had shown it to be the cause of serious illness of cattle in Texas,[2] and a few years later Hunt and Collins and Pound, in Australia, and Connaway and Francis in the U.S., had reported that the blood from infected cattle could be used to induce an artificial infection which rendered the animal immune to subsequent natural disease.[3-5] The process was rather haphazard, since the number of parasites used in the inoculum was unknown, and at one extreme could result in severe and fatal reactions to the vaccinal dose and at the opposite to a lack of induction of immunity. In order to reduce the possibility of the former, which must have been a frequent result when used in adult cattle, drugs were introduced in 1908 which could reduce the symptoms after immunization.[6] The process was subsequently used in many countries[7,8] for all species of bovine *Babesia*. It remained essentially the same until the late 1950s, although it had been modified and slightly improved in Australia from the 1930s onward by the use of blood from recovered but subclinically affected cattle, known locally as "bleeders", immunized against both *B. bovis* and *B. bigemina*.[9] This method reduced the number of severe reactions, but by 1964 it became obvious that the vaccine from such cattle was not always highly infective and that losses in vaccinated cattle were becoming economically important enough to stimulate research in vaccine improvements.[9-11] The majority of the experiments were carried out by workers studying *B. bovis* in Australia and involved attenuation of the parasite so that blood infected with it produced adequate protection with a minimum of side effects. The techniques used have since been applied in many countries in South America and in South Africa.

Attenuation of blood vaccines by passage has been attempted for the economically important species parasitizing cattle. The methods which have been examined are

1. Rapid passage in splenectomized normal hosts.
2. "Slow" passage in normal hosts.
3. Rapid passage in abnormal hosts.

B. Rapid Passage in Splenectomized Normal Hosts

1. Babesia bovis

Originally four strains of *B. bovis* isolated from cattle in different areas of Queensland were used.[12] Vaccine was prepared in splenectomized calves in such a way that a standard number of organisms could be injected into each animal requiring immunization. Titration of the infective dose required had taken place, and it had been found that while inocula containing as few as 10^3 *B. bovis* could be infective, even 10^5 parasites did not ensure adequate infection of every animal. A standard infective dose of 10^7 parasites had therefore been selected.

a. Collection Techniques

In order to facilitate regular adequate production of *B. bovis* for vaccine, calves which had been splenectomized between the ages of 2 and 8 weeks were infected intravenously each week with 10^9 *B. bovis*.[13,14] When a parasitaemia of 0.75 to 2% was observed, large volumes of blood were collected from the calf. In order to do this without causing excessive mortality of calves, a system of exchange transfusion was evolved.[13] The method involves surgical exposure of the carotid artery under local anesthesia. The artery is then clamped with two clamps and a longitudinal incision made between them. A cannula, which should be siliconized to reduce the possibility of clotting within it, is inserted towards the heart, the proximal clamp released, and parasitized blood collected from a tube attached to the cannula into a sodium citrate solution. The volume of sodium citrate used was 480 ml of a 4.5% solution in a 4-l flask. After collection of parasitized blood, 2 l of heparinized, uninfected blood is pumped into the carotid artery via a distally directed cannula, after taking care that any air in the cannula or tubing is displaced by blood by prior release of the distal clamp. The procedure may be repeated $\frac{1}{2}$ h and 24 h later. By that method, 8 l of heavily parasitized blood can be collected from one calf before the animal is treated with a babesicide and the wounds sutured. The collected blood is then examined to estimate the parasitemia before chilling to 4°C. The infectivity of the blood reduces daily in storage and, hence, diluent is added to it in quantities decreased by a factor of 1.5 before dispatch each day. The concentrated blood is not used for more than 1 week after collection.

The suitability of various diluents has been assessed. Initially some common solutions were tried, e.g., phosphate-buffered saline, buffered citrate, and Alsever's and Tyrode's solutions, but were found to be inferior to citrated bovine blood in that infectivity of diluted blood with the former group was reduced.[12] As a result, uninfected bovine blood was initially used. This was superseded by the use of a cell-free 25% plasma-based diluent, in which bovine plasma was diluted with a balanced salt solution containing electrolytes in the same ratio as plasma.[12] Later, after experiments by Farlow[15] which demonstrated that adequate glucose was necessary for maintaining parasite survival, a 10% serum mixture containing 0.1% glucose was used to replace plasma, and methods for the collection of large volumes of serum and its sterilization reported.[16]

b. The Relationship Between Number of Passages and Virulence

After several years of using vaccines prepared by rapid serial passage in splenectomized calves, it became apparent that changes in virulence, immunogenicity, and infectivity for *Boophilus microplus* were taking place, and these problems became the subject of detailed investigations.

Callow et al. examined three strains of *B. bovis,*[17] each of which at the start of the experiment had been subjected to frequent passages via intact cattle and were considered virulent. Each strain was then passaged rapidly through a series of splenectomized calves and the virulence tested in intact animals after different numbers of passages. The changes in virulence were quantified by the resulting cumulative parasitemia, the total increase in body temperature above normal, and the number of days during which parasites were observed. It was noted that between passages three and six, virulence tended to increase, but in one strain started to decrease after passage 9, but was not fully apparent until passage 30. With other strains, the reduction was more obvious after the 11th passage. The manifestations of reduced virulence were that parasites become more plentiful in venous blood, but their duration and the febrile response of the host were decreased.

Two other interesting observations were made. First, from 1968 onwards, vaccine failures began to be recorded in increasing numbers,[18] and after investigations in Australia and in South Africa, it was concluded that after many serial passages (C90) in splenectomized calves some strains lost their immunogenicity.[19] Second, it was found that virulence could be restored by a few passages in intact cattle. As a result of these findings, vaccine strains supplied in Australia are passaged from a minimum of 20 to a maximum of 30 times in splenectomized calves; after which, virulence is restored via intact cattle and a new series commenced.

c. Loss of Infectivity of Passaged Strains for Ticks

Suspicions were aroused during the middle of the 1960s that strains of *B. bovis* which had been transmitted many times to cattle by syringe with no natural cycling through ticks were gradually altered in some way which reduced their infectivity for the vectors. A comparison was made of a strain which had been syringe passed approximately 200 times with 2 recently isolated strains.[20] It was found that the passaged strain when given identical conditions to the field strains was not transmitted by *Boophilus microplus*.

In a further experiment, the infectivity of non-tick-infective strains of *B. bovis* during relapses of parasitemia to 11 months after primary infection was examined to ascertain whether during recrudescences any change or reversion to infectivity had taken place. It was found that the relapse parasites were still noninfective, and investigations were carried out to try to clarify which aspects of tick and parasite metabolism and structure were affected. Stewart examined the stages of *B. bovis* developing in the tick and noted that the loss of infectivity was associated with abnormal development of the parasites within the gut of the tick.[23] Further comparisons with infective and noninfective strains were carried out.[22] Unmodified strains were known to cause heavy mortality of engorged female ticks which also had hemolymph discolored by ingested bovine hemoglobin which had leaked from damaged epithelial cells in the gut of the tick. These effects were not found in ticks which had ingested modified avirulent strains, and it was concluded that the higher mortality rate was associated with epithelial damage and that virulent strains of *B. bovis* were pathogenic for both cattle and tick. It was also reported that virulence for ticks was lost between passages 33 to 40 and tick infectivity after approximately 60 passages.

Initially it was thought that the use of a vaccine which was not transmitted to ticks would be beneficial in that a gradual reduction of the number of infected ticks would take place and, hence, the incidence of infection reduced. Practical experience showed, however, that in an enzootic situation the presence of cattle which were not challenged after vaccination or which were not vaccinated led to a loss of enzootic stability with the consequential outbreak of serious disease when such cattle came into contact with infected ticks, either by their movement or the accidental transfer of the ticks to the areas in which challenge had declined. It has now become policy in Australia that only tick-infective vaccine strains are used.

d. Antigenic Changes in Passaged Strains

Curnow showed in 1973 by the use of an erythrocyte agglutination test that *B. bovis* parasites taken from blood during primary and relapse parasitemias were antigenically different and that when relapse parasites were transmitted via *Boophilus microplus*, they reverted to a common antigenic type thought to be specific to the strain involved.[24] Little further work took place to clarify the nature of antigenic differences between strains of *B. bovis* until the advent of modern protein electrophoretic and immunodiagnostic techniques. Using these, Kahl et al.[25] analyzed the antigens of three strains.

The strains investigated were virulent (after the 2nd syringe passage) and avirulent (after the 26th syringe passage) isolates of the same strain (K), and a separate virulent strain (L), which had been initially tick derived, but then subsequently syringe passaged via 5 splenectomized calves. Homologous sera from calves infected with these strains plus heterologous serum from a different avirulent strain were used for immunoprecipitation.

Summarized briefly, the methods used involved labeling *B. bovis*-infected erythrocytes in microaerophilous culture with ^{35}S-methionine for 10 h. Labeled cells were assayed for radioactivity, extracted with detergent (Triton® X), and centrifuged. The supernatants were retained and, after mixing aliquots with antisera, the resultant immune complexes, consisting of bovine immunoglobulins and babesial antigens, were isolated by passing through a column of rabbit anti-bovine IgG Sepharose® 4B. After elution from the columns, the radiolabeled proteins were subjected to two-dimensional electrophoresis using the method of O'Farrell.[26] Fluorographs of the gels were then made, and regions of dried gels which were of interest

Table 1
DIFFERENTIAL EXPRESSION
OF CANDIDATE VARIANT
ANTIGENS OF *BABESIA BOVIS*

	Antigen Expressed			
	K_A	K_V	O	L
K_A (Type A)	+	−	−	−
K_A (Type B)	+	−	−	−
K_A (Type D)	+	−	−	−
K_A (Type C)	−	−	+	−
K_V	−	+	+	−
K_V (post-ticks)	−	+	+	−
C_V	−	+	+	−
C_A	−	+	+	−
C_A (SX 27)	−	+	+	−
C_{NT} (SX 70)[a]	−	−	−	−
L_V	−	−	−	+
L_{AIR}[b]	−	−	−	+

[a] C_{NT}, avirulent C strain *B. bovis* after 70 Sx does not express K_{A1}, K_V, O, or L and is not tick transmissible.

[b] L_{AIR}, virulent L strain *B. bovis* (L_V, Kahl et al., 1982) made avirulent by irradiation (Wright et al., 1980).

From Kahl, L. P., Mitchell, G. F., Dalgliesh, R. J., Stewart, N. P., Rodwell, B. J., Mellors, L. T., Timms, P., and Callow, L. L., *Exp. Parasitol.*, 56, 222, 1983.

further analyzed by one-dimensional sodium dodecylsulfate electrophoresis with buffer containing dithiothrietol and staphylococcal V8 protease.

Differences were most apparent in the acidic regions of the gels. Both virulent and avirulent isolates of the K strain had dominant antigens of 65 and 34 kDa, plus a minor antigen of 56 kDa. The avirulent K strain was characterized by two antigens, designated K_{A1} and K_{A2} and of 47.5 and 43 kDa, respectively. The former was absent in both virulent strains K and L, but the 43-kDa antigen plus three low-molecular-weight (ca. 19 kDa) antigens were present. The virulent K strain, K_v had a characteristic antigen of 46.8 kDa, and the virulent L strain one of 40 kDa. Various minor antigens were common to all.

The authors postulated from these observations that the antigens present in the avirulent K strains which could induce protective immunity could be markers of host protection and avirulence, and specifically that the antigen of 43 kDa might be of special interest because of its dominance in avirulent strains and small presence in virulent isolates.

In order to examine antigenic differences between vaccinal and virulent strains which had been isolated after different forms of propagation, Kahl et al.[27] applied similar methods to avirulent and virulent K strains after passage through ticks and both intact and splenectomized calves and compared the antigens visualized with those of a separate strain (C) in virulent, avirulent, and non-tick-infective isolates, and another strain (L).

The results are summarized in Table 1. The same antigens were found in the avirulent K strain as in the first experiment, i.e., 65 kDa, 47.5 kDa (K_{A1}), and 43 kDa (K_{A2}), but after different forms of passage alterations took place. After tick transmission, K_{A2} decreased; after three syringe passages via intact calves, K_{A1} was lost, but a different antigen, type O of 46.8 kDa, appeared. After tick transmission of the latter, K_{A1} reappeared, K_{A2} decreased, and O was absent. The antigens of the virulent K strain were not altered by tick passage, i.e., K_{A1} remained absent and K_{A2} minor, and the strain remained characterized by antigen O and there was increased representation of a polypeptide of M_r 35 kDa. The C strain had similarities to the K strain, i.e., there was a marker protein of 65.8 kDa which was slightly less active than the K marker of 65 kDa. In the C strain, which was not transmissible to ticks, there was no K_{A1}, but K_{A2} was present.

These two experiments clarified some of the possibilities of antigen change which can take place with different methods of passage in the production of live vaccines for *Babesia* species. They also led to speculation and investigation into the specific processes which take place in both the tick and in the body of cattle which effect the changes.

e. Genetic Studies of Passage-Attenuated Avirulent Vaccinal and Virulent Strains of B. bovis

After the analyses of the polypeptide antigens had shown that some were specific to particular isolates, depending on their sources and treatments prior to isolation, some of the molecular events leading to the formation of the antigens were examined. Cowman et al.[28] constructed a cDNA library from polyadenylated RNA of the same avirulent *B. bovis,* designated K_A, and identified clones which hybridized selectively to cDNA from polyadenylated RNA of the other isolates of the K and C strains previously described,[25-27] plus two additional strains L and S. Although the hybridization patterns were similar with different probes, a few which hybridized strongly with K_A but not with K_V were purified, rescreened, and DNA examined by Southern blotting. Three clones, designated pK4, pK5, and pK6, were of particular interest.

The transcripts from pK5 were predominantly expressed in avirulent isolates, and it was considered that this clone might be a marker of avirulence. Of the other clones, pK4 revealed two major RNA species of 1.2 and 0.8 kilobases (kb) in isolates K_A, K_V, and C_V, in contrast to the L and S isolates which contained only a 1.0-kb species. The remaining clone pK6 showed transcripts of different sizes in each strain. From these results the authors hypothesized that most isolates arising from field infections were composed of a heterogenous mixture of subpopulations and that one of the uses of their observations might be the possibility of distinguishing between such infections at the DNA level.

Further studies of clone pK4 were carried out because it had shown cross-reactivity to cDNA from some virulent isolates.[29] As previously observed, it had hybridized to two species of polyadenylated RNA from the K_A isolate of *B. bovis*. The two longest clones, designated BabR1.2cDNA and BabR0.8cDNA, were analyzed to determine their relationship to the RNA species by probing Northern blots with fragments derived from the 5' and 3' ends of both clones. Clones from the 5' end of BabR1.2cDNA hybridized identically to clone pK4. In contrast, those from the 3' end hybridized only to the 1.2-kb RNA. A probe from the 3' end of the BabR0.8cDNA clone hybridized to the 0.8-kb RNA species, but not to that of the 1.2-kb RNA, and in addition identified a new RNA of 1.0 kb. It was concluded from these observations that the two clones of pK4 were homologous at the 5' end, but differed at the 3' end.

Segments of five genomic fragments of the BabR locus were isolated and restriction maps of EcoR1 digests carried out. It was observed that a fragment designated Babr9.7 contained three tandem repeats with identical restriction sites and that two other fragments, BabR7.7 and BabR5.7, contained two and one copies of this repeat, respectively. The tandem repeats were examined in more detail and it was found that the BabR1.2cDNA was identical to the corresponding length of BabR9.7. It was concluded that the sequences of the 1.2-kb RNA were contiguous and located within the tandem. Further, similar investigations produced the conclusion that the avirulent vaccinal strain K_A consisted of a major subpopulation containing the BabR9.7 fragment. However, in another similar experiment which examined *B. bovis* populations produced by infections of one parasite isolated by limited dilution, Gill et al.[30] found that the BabR pattern of these cloned parasites was considerably less complex than that previously observed with K_A. The most striking difference was the minor presence of the 9.7-kb fragment and the major presence of three other BabR bands (6.5, 3.25, and 0.75 kb). In addition, although most of the lines were avirulent, some were more virulent than the original K_A from which they had been cloned. As a result, the authors concluded that although the BabR locus is a suitable marker of parasite subpopulations, it has little role in parasite virulence. Their general conclusion that K_A contains a number of different *B. bovis* subpopulations, the balance of which can be altered by selection pressure during passage when the immune system of the host, be it splenectomized or intact cattle or a tick vector, reacts with the ability of the parasites to vary genetically was confirmed.

2. Babesia divergens

Blood taken from carrier cattle which had recovered from infection with *B. divergens* was used in Sweden from 1926 to 1928 in the same manner as the early Australian workers had used *B. bovis*.[8] From 1928 until the 1960s acute reaction blood from experimentally infected calves replaced carrier blood,[31] and similar practices were followed in other European countries.[32] Few controlled studies were carried out, but Ryley reported that rapid passages of infected blood through an unspecified number of splenectomized calves resulted in an increase in the virulence of the strain involved,[33] which had originally been isolated from a clinical case in Northern Ireland, and that very rapid passage with parasites collected early in parasitemia enhanced this effect. A few years later, Purnell et al.,[34] with a different Irish isolate of *B. divergens*, compared the effects of injecting primary reaction and carrier blood taken at intervals up to 196 d after infection into cows and intact and splenectomized calves, but found no differences in the resultant reactions to the infections.

The only documented experiment with a known number of rapid passages via splenectomized calves was reported by Taylor et al.[35] These authors compared the effects of infection with a strain of *B. divergens* after 6 and 15 rapid passages through splenectomized calves. Both juvenile (18 months of age) and aged adult cows were infected with the same number of parasites from both passage series. No differences in the clinical effects on cattle of the same age groups were noted, and it was concluded that neither number of rapid passages had any attenuating effect on the strain involved. This view was reinforced by the same authors (unpublished observations), who noted that a further series of 26 rapid passages via splenectomized calves had no discernible effect on the virulence of the parasite.

3. Babesia bigemina

Attenuation of the species by multiple rapid syringe passages through splenectomized calves in the same manner as applied with *B. bovis* was attempted by Callow and Tammemagi (reported by Dalgliesh et al.).[36] The strain used was passaged at 2- to 9-d intervals through a series of 35 splenectomized calves. While some reduction of virulence was noted after 18 such passages, the strain used was still considered virulent after the completed series, and the attention of these authors turned to other methods of attenuation of the organism. Despite these findings, *B. bigemina* collected after syringe passage through splenectomized calves is used as a vaccine, in some cases after cryopreservation, and given simultaneously with *B. bovis* and *A. marginale*,[42] the inoculation being intravenous. In other reports,[43-44] it has been given subcutaneously, but in many instances therapeutic or chemoprophylactic babesicides have been required to control severe reactions.[45]

4. Babesia ovis

The use of a passaged vaccine for this parasite of sheep was reported by Kyurlov.[37] After 4 syringe passages via splenectomized lambs which were simultaneously treated with tetracycline, infected blood was used to immunize 86 weaned lambs. Treated lambs suffered only a febrile reaction and were resistant to subsequent artificial challenge with three times the number of infected erythrocytes as had been used for vaccination. A vaccine dose of 3.5×10^7 infected erythrocytes was thought to be the optimum for immunization.

C. Attenuation by "Slow" Passage
1. Babesia bigemina

Initially bovine babesiosis in Australia had been attributed to *B. bigemina,* since parasitemias caused by it are generally higher than those produced by *B. bovis* infection, and in vaccines up to 1964 both species were included. When the new attenuated *B. bovis* vaccine was produced at that time, *B. bigemina* was not included. There continued to be an occasional need for it for vaccination of cattle moved from tick-free to *B. bigemina*-enzootic areas and for controlling rare outbreaks of disease caused by it.[38] As previously described, rapid syringe passage through splenectomized animals proved unsuccessful.[36] However, Tidswell, Sergent et al, and Kemron et al.[39-41] had all observed earlier that blood from cattle in the carrier state of infection with *B. bigemina* when injected into other cattle induced a much less severe reaction than blood from the same animal during a primary infection. Using that information, Dalgliesh et al.[36] investigated the effects of various forms of passage on *B. bigemina* and their suitability in the production of a live vaccine.

A strain of *B. bigemina* which had been passaged through both the tick *Boophilus microplus* and 27 cattle during a period of 6 years was used to infect a splenectomized calf. It was then subjected to rapid syringe passage through 35 splenectomized calves which failed to produce a significant attenuation of virulence, followed by passage via *B. microplus* to an intact calf which was splenectomized 2 weeks after the primary infection. Blood was collected from it during relapse parasitemias to infect a further series of seven intact calves which carried the infection for 3 to 16 weeks before they themselves were splenectomized. From the resultant relapse parasitemias, blood was collected for further passages and eventually for vaccine production. Aliquots of blood from the calves at various stages of the relapse parasitemias had been collected and stored frozen. The parasites in such isolates were tested for virulence after one passage in a splenectomized calf to produce a large number of infected cells by injecting 10^8 cells intravenously into susceptible cattle, and for induction of immunity by heterologous challenge. When produced in this manner, the vaccine for *B. bigemina* proved to be of low stable virulence.

D. Attenuation by Rapid Passage in Abnormal Hosts
1. Babesia divergens

This organism is unique amongst the *Babesia* species which infect cattle in that it can be passaged and maintained in an abnormal but unsplenectomized host, the Mongolian gerbil *(Meriones unguicalatus).* The relationship was first noted almost by accident when organisms from a splenectomized infected man were injected into what was one of the few laboratory rodents available in the hospital at the time, but it has since been fully described.[46-47] The possibility of attenuation of the virulence of *B. divergens* for cattle by multiple passage through gerbils was investigated by Murphy et al.[48] These authors subjected a strain of *B. divergens* to 61 rapid passages through intact gerbils which were infected with 10^7 parasitized erythrocytes. After the last passage, five splenectomized calves were infected with parasitized blood (5.6×10^7 infected erythrocytes) and a further three similiar calves were infected with blood taken from the first group. The parasite remained highly infective for cattle and was, if anything, more virulent. In an examination of the possibility of antigenic changes, the organism was apparently unaltered by the treatment, and the authors concluded that alternative means of attenuation of *B. divergens* would be required.

II. IRRADIATED BABESIAL VACCINES

A. Introduction

Studies of the effects of ionizing radiation on protozoa began shortly after the discovery of the X-ray itself. Halberstaedter observed that irradiation of *Trypanosoma gambiense* could produce a loss of infectivity, but not of motility of the organism, and that there was a wide variation in resistance to the same dose of radiation within the population exposed to it.[49-50] A later observation with the poultry parasite *Plasmodium gallinaceum* reported that high doses of X-rays could prevent multiplication and that injection of treated parasites produced some degree of immunity.[50] Thereafter, the immunogenic properties of various irradiated parasites was studied with the first report specific to *Babesia* species being that of Phillips.[52-54] Investigations into the chemical and biological events associated with the treatments were also carried out. In a review, Little presented evidence that the basic lesion caused by the heat or excitation of the ionizing radiation caused free hydroxyl radical formation which, depending on the dose involved, could result in a range of effects from temporary impairment of cellullar functions to cell death.[55] Not all cells were equally susceptible, and certain phases of cell development, notably those involved with DNA synthesis and mitosis, were most affected.

The initial report by Phillips used the model of *B. rodhaini* infection of rats and mice[54] and examined both the exposure to radiation required to make the parasite noninfective and the immunogenicity of infected blood irradiated at different levels. In the former, 2×10^9 *B. rodhaini*-infected erythrocytes were irradiated from a ^{60}Co-650 source and injected into groups of mice. The radiation doses were 50, 100, 200, 250, 400 and 800 G-rays (Gy). Parasites irradiated with the two highest doses were observed for 48 h after inoculation, but not thereafter. Those mice injected with blood exposed to 50 Gy suffered from severe disease and died on days 5 or 6 after inoculation. The intermediate doses of 100 to 250 Gy caused a proportional lengthening of the period before patent parasitemia was observed, and in the two higher-dose (200 and 250 Gy) groups some mice survived the infection. It was concluded that for *B. rodhaini* exposure a radiation dose of 400 Gy or more rendered the parasite noninfective.

In subsequent experiments, rats and mice were inoculated with one or two doses of irradiated infected erythrocytes. The irradiation levels ranged from 400 to 800 Gy. It was observed that one inoculum of inactivated parasites delayed the onset of parasitemia resulting from a superimposed challenge infection of 10^6 parasites 17 d later and that two inocula of 2×10^9 irradiated cells given 6 d apart protected intact rats from a challenge infection of 2×10^5 *B. rodhaini* given 19 d after the primary injection and delayed the onset of parasitemia in splenectomized rats.

B. rodhaini was also used by Irvin et al. who compared the effects of radiation on that species with those *B. major* and *B. divergens*.[55] The method of quantification used was by the incorporation of tritiated (^{3}H)hypoxanthine in cultures for 24 h after various doses of irradiation, since a previous study had shown that it correlated well with the metabolic activity of the parasites. All three parasites in infected blood were given doses of and radiation ranging from 2.5 to 1200 Gy. Infected blood was diluted to 1×10^8 infected cells in 5 ml of culture medium (Eagles minimum essential medium plus 100 IU benzylpenicillin ml^{-1}, 100 μg ml^{-1} streptomycin, 20 μg ml^{-1} mycostatin, 50 mM HEPES buffer, and 10% fetal calf serum) maintained at 37°C. After irradiation, tritiated hypoxanthine was added at the rate of 2 μCi ml^{-1}. After 24 h of incubation, the uptake of (^{3}H)hypoxanthine was measured by scintillation counting and comparison with those obtained from noninfected blood.

It was observed that there was a linear reduction in hypoxanthine uptake and lengthening of prepatent period proportional to increasing doses of radiation above 5 Gy and that there was a rapid fall in infectivity with doses greater than 400 Gy. There was also a difference in radiosensitivity between the parasites, with the cattle species *B. major* and *B. divergens* being more resistant than *B. rodhaini*. The authors postulated that the infectivity of irradiated blood vaccines might be assessed using this method and remove the necessity for calibration by cattle inoculation.

Further effects of irradiation on *B. bovis* were reported by Wright et al.[58] who measured the protease concentration in both virulent parasites and the same strain after irradiation with 350 Gy. They observed that irradiated parasites produced negligible amounts of protease in comparison to those fully virulent, but that parasite doubling times in both populations were the same. They concluded that the production of the enzyme was not necessary for parasite multiplication in the host, but that its presence was an indicator of the virulence of the strain.

The effects of irradiation on the transmittability of *B. bovis* infections to the tick *Boophilus microplus* was investigated by Wright et al.[59] Intact steers were infected with 10^8 and calves with 10^7 parasites in whole blood. Some of the cattle that were infected with cells had been irradiated with 350 Gy. Larval ticks were fed on the infected cattle during the period of patent parasitemia, and the engorged females which subsequently dropped off were collected, incubated at 28°C and 95% relative humidity, and the eggs produced were allowed to hatch. After hatching, batches of larvae were allowed to feed on previously uninfected cattle which were monitored for *B. bovis* infection. None of the cattle exposed to tick lines originally fed on groups infected with irradiated parasites developed babesiosis as did those emanating from ticks from nonirradiated infections. Recrudescences for 1 year after were similar, and the authors were of the opinion that the nonreversion to tick transmissibility was therefore permanent.

Two theories on the mode of action of irradiated babesial vaccines have been proposed. First, Purnell and Lewis were of the opinion that the protective effect was due to the combined inoculation of large numbers of parasites killed by irradiation simultaneously with a small number of viable organisms which had survived.[60] The antigens present in the dead parasites were thought to stimulate a sufficient immune response to restrict the subsequent multiplication of the live *Babesia* to nonlethal numbers. In order to reach that conclusion they carried out two experiments. In the first, splenectomized calves were inoculated with 10^{10} *B. divergens* killed by an irradiation dose of 500 Gy plus 10^8 irradiated at 250 Gy. The resultant reactions were less than those observed with infection with parasites irradiated at 250 Gy by themselves. In the second experiment, similar results were obtained with 10^{10} dead parasites in combination with 10^4 fully viable, nonirradiated organisms.

This view was refuted by Wright et al.[61] In a series of experiments, they indicated that irradiation at a dose of 350 Gy had produced a predominantly avirulent population which had a multiplication rate similar to nonirradiated parasites and in which pathogenic protease-producing parasites had been selectively inactivated. Their argument was substantiated by the lack of serological response in cattle injected with large numbers of killed parasites in the absence of a suitable adjuvant.[62,63] These authors also concluded that since the avirulence seemed permanent, irradiation of virulent *Babesia* might be a less expensive and laborious method of attenuation of *B. bovis* than multiple rapid passage through splenectomized calves.

Irradiation as a means of attenuation has been examined with the most important species affecting cattle and to a limited extent for *B. ovis* in sheep.

B. Specific Irradiated *Babesia* Vaccines

1. Babesia bovis

The first experiments were carried out by Mahoney et al.[62] in 1973; the aim was to calibrate the dose of irradiation which would provide the best combination of attenuation and immunogenicity. Splenectomized calves were used as they were thought to give a more sensitive assessment of the immunity induced.

Blood from a heavily infected calf was collected into 1% of the disodium salt of EDTA in physiological saline and divided into aliquots of 10 ml which were stored at 4°C before, during, and after gamma irradiation. Six groups of calves were infected intravenously with 10^8 *B. bovis* irradiated with doses of 20, 40, 80, 160, 240, and 320 Gy, and a seventh group with nonirradiated blood. A progressive lengthening of the incubation period proportional to the irradiation dose of the blood was observed, with the range from 2 to 3 d at 20 Gy to 11 d at 320 Gy. In each of the groups injected with blood irradiated with more than 80 Gy, some calves survived the infection.

In a second experiment of similar design, infected blood was irradiated with doses from 200 to 400 Gy and injected at doses of 10^8 into calves. Control groups for comparison of prepatent periods were injected with 10^1 to 10^8 parasites. A similar result on the relationship of irradiation level and prepatent period was observed, apart from an anomaly, with these parasites given doses of 280 Gy, for which the period was the same as 10^1 nonirradiated organisms.

The doubling time of parasite multiplication was significantly longer for irradiated than for nonirradiated, being 9.3 and 7.9 h, respectively. Two of the controls and ten given irradiated blood survived the infections and were subsequently challenged 1 month later with 10^6 virulent organisms; all previously infected calves survived, whereas naive control calves succumbed to the challenge infection.

In the final part of the experiment, groups of calves were injected with infected blood irradiated at 300, 500, and 700 Gy, 10^{10} freeze-dried parasites, and nonirradiated blood. Only calves given 300-Gy-irradiated blood developed a detectable parasitemia and serologically positive reactions. All were challenged with 10^6 *B. bovis* 7 weeks later. All the calves previously infected with blood irradiated at 300 Gy or nonirradiated survived; three out of four at 500 Gy were similar, but all the other calves were severely affected.

This group of experiments showed that most *B. bovis* were inhibited by a level of gamma irradiation of 320 Gy, but a small percentage of the population could withstand that exposure and required 500 Gy to bring about complete inactivation, and that immunity to reinfection was evident only in calves in which parasites had been able to multiply. As a result, the Australian workers decided that an irradiation dose of 350 Gy might be the most suitable for the *B. bovis* strains with which they were working, although in a further paper they used various doses in an investigation of the relationship between pathogenicity and irradiation level.[64] In it, two experiments were undertaken; in the first, groups of splenectomized calves were inoculated with infected blood (10^8 cells) irradiated at 200, 300, 400, and 500 Gy, respectively, as well as another with a nonirradiated aliquot of the same infection. Various pathophysiological parameters and parasite multiplication times were then monitored. Only those calves given blood irradiated at 500 Gy survived, although the disease took significantly longer to cause fatalities in groups receiving irradiated parasites. In contrast to the previous experiment cited, there were no significant differences in the parasite doubling time.

In the second experiment, intact, 2-year-old steers were used. They were injected intravenously with 10^8 *B. bovis* irradiated with 350 Gy; controls were given the same number of parasites without irradiation. The cattle given irradiated blood developed only mild transient babesiosis with significant reductions in the pathophysiological changes as compared to the controls which were severely affected.

In a continuation of these investigations, the same research group examined some aspects of the immunity induced by irradiated *B. bovis*.[61] Two year old intact steers were used. In one experiment they were divided into groups which were inoculated with various strains of *B. bovis* which had either been irradiated at 350 Gy, were nonirradiated, or which had been irradiated and then stored in liquid nitrogen vapor for 10 months. Some were challenged with 10^8 virulent organisms 3 weeks or 10 months after the initial immunizing infections. The majority of cattle given 10^8 *B. bovis* irradiated with 350 Gy suffered only mild infections; controls given nonirradiated organisms became severely affected. It was estimated that of the 10^8 infected cells injected, only 2.5×10^3 survived irradiation, and that there was a change in the virulence, because cattle given the latter number of infected virulent cells died. Cattle challenged after both intervals were, with one exception, protected against clinical babesiosis.

As a result of these experiments the authors concluded (1) that 350 Gy of gamma radiation was a suitable irradiation level to render *B. bovis* avirulent, but of sufficient immunogenicity to protect against subsequent severe clinical symptoms of babesiosis, (2) that irradiation was a method of reducing the virulence of a strain without affecting the multiplication rate, but (3) that the occasional severe reactions and failure to stimulate immunity warranted further investigation.

2. Babesia divergens and Babesia major

The first reported experiment with these two European cattle piroplasms was that of Brocklesby et al. with *B. major*.[65] Parasitized erythrocytes (5×10^8) were exposed to radiation doses of 0, 100, 300, and 500 Gy and injected intravenously into four pairs of splenectomized calves; simultaneously four other calves were inoculated with 5×10^4, 5×10^5, 5×10^6, and 5×10^7 nonirradiated infected erythrocytes from the same source. The subsequent reactions varied from severe for those which received 0 and 100 Gy, to mild for 300 Gy, and inapparent at 500 Gy. Only the two control calves given the highest infective doses were affected by babesiosis. These, plus four of the calves given cells irradiated at 300 and 500 Gy, were challenged with 4.1×10^8 erythrocytes infected with the same strain; unfortunately the results of the challenge infection were marred by the simultaneous appearance of *Eperythrozoon tuomi,* but the authors concluded from their observations that the initial infection had stimulated some resistance to the organism.

There followed a series of parallel experiments from the same workers involving both *B. major* and *B. divergens,* which examined the protection stimulated initially in splenectomized and subsequently in intact calves.[66-69] Primary infections were by intravenous injection of parasites irradiated at levels from 240 to 400 Gy, and challenges were by injection of nonirradiated organisms from both homologous and heterologous strains. As a result of this work, it became apparent that irradiation of both *Babesia* species with doses of 240 to 320 Gy produced mild reactions in calves infected with the treated organisms. The calves were immune to subsequent challenge infections. Conversely, calves given infected cells irradiated with 360 and 400 Gy produced inconsistent results with little or no primary reaction or stimulation of resistance.

The logical extension to these experiments was to examine the efficacy of immunization with irradiated *Babesia* against natural tick-transmitted infections. Taylor et al. carried out two such experiments, one with homologous and the other with heterologous challenge.[70-71] In the first, two groups of intact, 6-month-old calves were infected with 1.5×10^{10} erythrocytes irradiated with either 250 or 300 Gy. After 4 weeks these calves plus another control group were transferred to a pasture known to have a large population of *Ixodes ricinus* infected with *B. divergens,* from which an isolate was used to produce the blood irradiated for immunization. In the ensuing 2 months, all of the control group contracted babesiosis, six of the ten suffering severe clinical disease. None of the pretreated calves were clinically affected, although several had detectable but insignificant parasitemias.

That experiment was followed up with a similar experiment, except that although the strain irradiated was the same, the challenge infection was presumed to be heterologous, since the source of isolation was approximately 100 mi from the area of the vaccine strain. The cattle used were older — 18-month-old heifers and a 3-year-old bull. They were injected with 3.36×10^{10} infected cells irradiated with 300 Gy gamma irradiation. During the 3 weeks following inoculation most of the heifers were observed to have low parasitemias and no clinical reactions; the bull, on the other hand, had a noticeable parasitemia (1%) and a reduction of packed cell volume from 0.3 to 0.231. l^{-1}. At 3 weeks after inoculation, the vaccinates plus ten controls were moved to tick-infested grazing. As in the first experiment, the controls all subsequently became infected with *B. divergens,* two severely, whereas the pretreated cattle were clinically unaffected.

As a result of the observation of the more severe reaction after immunization of the adult bull, the same group decided to further examine reactions in adult animals. In an experiment which utilized strains of *B. divergens* which had been subjected both to multiple rapid passages through splenectomized calves and irradiation,[35] the effects were compared in juvenile and adult intact cattle. A group of adult cows were injected with 10^9 infected erythrocytes which had been collected after the 15th passage and then irradiated with 300 Gy. No parasites were observed in blood smears until the 18th day after inoculation, after which all the cattle suffered severe clinical babesiosis. The prepatent period before clinical disease was observed was the same as another group given 10^4 nonirradiated cells of the same isolate. It was concluded that although irradiation had reduced the number of viable parasites infected, it had no effect on the virulence of the *B. divergens* strain used and that it was unlikely that a safe and practical *B. divergens* vaccine could be produced by irradiation of the parasite.

3. Babesia bigemina

In the earliest reports of the effects of irradiation on *Babesia* species, Castro and Canabez examined the effects on a Uruguayan isolate of radiation doses ranging from 10 to 100 Gy.[72] They observed that the organism lost infectivity for a nonsplenectomized, 2-year-old bovine with irradiation of 250 Gy, but was infective in a younger splenectomized animal. This indicated that *B. bigemina* might be more radiation sensitive than the other bovine species of *Babesia*. This observation was further explored by Bishop and Adams, working with a Colombian isolate.[73] Four groups of intact calves were respectively injected intravenously with 10^{10} *B. bigemina* irradiated with 240, 360, 480, and 600 Gy and their reactions monitored and compared with control groups given nonirradiated parasites (10^4, 10^7, and 10^{10}). All the calves given organisms irradiated with 240 Gy developed patent parasitemias, albeit after prolonged prepatent periods; only one of the calves given blood treated with 360 Gy became infected, and none of those with blood irradiated more heavily.

All the calves were challenged with 10^{10} *B. bigemina* 4 weeks after the initial inoculation. Although some calves died from what apparently were anaphylactic reactions to the intravenous challenge, calves previously given irradiated parasites, including those with blood irradiated at 480 and 600 Gy, resisted the development of severe clinical infections. The authors concluded that after this means of inactivation, the parasites might be more immunogenic than those killed by other methods.

A later report with an Indian strain of *B. bigemina* irradiated with doses of 500 and 600 Gy and injected (2×10^{10}) into calves twice with a 15-d interval between inoculations showed that the two injections of what were apparently inactivated parasites induced protection against the worst effects of subsequent challenge infections.[74]

More detailed studies of the relationship between irradiation treatment and *B. bigemina* have been reported by Wright et al.[75,76] An Australian strain of *B. bigemina* was subjected to irradiation with 250, 300, and 350 Gy, and 1×10^8 parasites from each treatment plus some which were nonirradiated injected into different groups of splenectomized calves. Amongst those given treatment parasites, only one animal in the group injected with blood irradiated at 250 Gy developed a patent parasitemia, indicating that the radiation sensitivity of the *B. bigemina* used was such that a dose of 250 Gy was close to the maximum dose which would not inactivate all the parasites. In a follow-up experiment, the same strain was irradiated with doses of 200, 240, and 280 Gy, and the same number of parasites injected. In this case, no significant differences in parasitemia or prepatent period were observed. A second strain was then treated in a similar manner. In this instance there was no difference between nonirradiated parasites and those dosed with 250 Gy, and a slight parasitemia developed in one animal injected with organisms dosed with 300 Gy. The cattle used were challenge infected with the first strain, and all became diseased although those at 250 Gy were least affected. The authors concluded that heterologous cross-protection was more difficult to induce than with *B. bovis,* that *B. bigemina* is less amenable to attenuation than the former parasite, and that strain differences in radiation sensitivity make it unlikely that a satisfactory solution will be found.

4. Babesia ovis

There has been one report of the use of irradiated *B. ovis* to try to stimulate protective immunity in sheep.[77] Infected erythrocytes were irradiated with doses of radiation of 200, 250, 320, 350, 400, and 500 Gy and injected into sheep. The effects were monitored and the sheep challenged with untreated parasites 1 month later. The authors observed that although treatment at 200 and 250 Gy lowered the virulence of the parasite, the effect was not sufficient to prevent serious disease. On the opposite extreme, doses of greater than 320 Gy had little effect, but did not stimulate immunity to challenge. It was concluded that irradiation treatment with 300 Gy was optimum for the isolate used.

III. PROBLEMS ASSOCIATED WITH ATTENUATED AND IRRADIATED *BABESIA* VACCINES

The major advantage of attenuated vaccines for *B. bovis* and *B. bigemina* is that they are available; they are also fairly easy and cheap to produce. Unfortunately there are problems associated with whole blood vaccines. Amongst them are a relatively short shelf life, (7 d), the possibilities of accidental transmission of other pathogens, e.g., bovine leukosis,[78] of occasional severe reactions to vaccination, loss of immunogenicity, unless the strains used are carefully monitored, and the induction of antibodies to blood constituents with resultant neonatal isoerythrolysis.[79-81] The danger from the latter has apparently been lessened with the use of cell-free diluents with passage-attenuated vaccines,[15-16] but would be a more likely sequelae with the use of irradiated vaccines which as used experimentally have involved the injection of much larger volumes of blood. As a result of these drawbacks, most research on babesial vaccines is now concentrated on the isolation of parasitic antigens for development with recombinant technology.

REFERENCES

1. **Babes, V.,** Sur l'hemoglobinurie bacterienne du boeuf, *C.R. Acad. Sci.,* 107, 692, 1888.
2. **Smith, T. and Kilborne, F. L.,** Investigations into the nature, causation and prevention of Texas or Southern cattle fever, *U.S. Dep. Agric. Bur. Anim. Ind. Bull.,* 1, 1, 1893.
3. **Hunt, J. S. and Collins, W.,** Tick fever in Queensland, *J. Comp. Pathol. Ther.,* 10, 88, 1897.
4. **Pound, C. J.,** Tick fever. Notes on the inoculation of bulls as a preventative against tick fever at Rathdowney and Rosedale, *Queensl. Agric. J.,* 1, 473, 1897.
5. **Connaway, J. W. and Francis, M.,** Texas fever. Experiments made by the Missouri experiment station and the Missouri state board of agriculture in co-operation with the Texas experiment station in immunising northern breeding cattle against Texas fever for the southern trade, *Mo. Agric. Exp. Stn. Bull.,* 48, 1, 1899.
6. **Nuttall, G. H. F. and Graham-Smith, G. S.,** Notes on the drug treatment of canine piroplasmoses, *Parasitology,* 1, 220, 1908.
7. **Theiler, A.,** Gall sickness of imported cattle and the protective inoculation against this disease, *Agric. J.,* 3, 1, 1912.
8. **Klarin, E.,** Protective inoculation against piroplasmoses in 1928, *Skand. Veterinartidskr.,* 19, 41, 1929.

9. **Seddon, H. R.,** Diseases of Domestic Animals in Australia, Part 4, Commonwealth of Australia Service Publ. No. 8, Department of Health, Canberra, 1952.
10. **Callow, L. L. and Tammemagi, L.,** Infectivity and virulence of blood from animals either recovered or reacting to *Babesia argentina, Aust. Vet. J.,* 43, 248, 1967.
11. **Tsur, I.,** Immunisation trials against bovine babesiasis. I. Vaccination with blood from latent carriers, *Refu. Vet.,* 18, 103, 1961.
12. **Callow, L. L.,** The control of babesiasis with a highly infective attenuated vaccine, in *Proc. 19th World Veterinary Congr.,* Vol. 1, Litoarte, Mexico, 1971, 357.
13. **Callow, L. I. and Mellors, L. J.,** A new vaccine for *Babesia argentina* infection prepared in splenectomised calves, *Aust. Vet. J.,* 42, 464, 1966.
14. **Callow, L. L.,** Vaccination against bovine babesiosis, in *Immunity to Blood Parasites in Animals and Man,* Miller, L. H., Pino, J. A., and McKelvey, J. J., Eds., Plenum Press, New York, 1977, 121.
15. **Farlow, G. E.,** Differences in the infectivity of *Babesia* incubated in plasma and serum and the role of glucose in prolonging viability, *Int. J. Parasitol.,* 6, 513, 1976.
16. **Rodwell, B. J., Timms, P., and Parker, R. J.,** Collection and sterilisation of large volumes of bovine serum and its use in vaccines against bovine babesiosis and anaplasmosis, *Aust. J. Exp. Biol. Med. Sci.,* 58, 143, 1980.
17. **Callow, L. L., Mellors, L. J., and McGregor, W.,** Reduction in virulence of *Babesia bovis* due to rapid passage in splenectomised cattle, *Int. J. Parasitol.,* 9, 333, 1979.
18. **Callow, L. L. and Dalgliesh, R. J.,** The development of effective safe vaccination against babesiasis and anaplasmosis in Australia, in *Ticks and Tick-Borne Diseases,* Proc. of Symp. 56th Australian Veterinary Assoc., Johnston, L. A. Y. and Cooper, M. G., Eds., Australian Veterinary Association, Townsville, Queensland, May 1979, 4.
19. **De Vos, A. J.,** Immunogenicity and pathogenicity of three South African strains of *Babesia bovis* in *Bos indicus* cattle, *Onderstepoort J. Vet. Res.,* 45, 119, 1978.
20. **O'Sullivan, P. J. and Callow, L. L.,** Loss of infectivity of a vaccine strain of *Babesia argentina* for *Boophilus microplus, Aust. Vet. J.,* 42, 252, 1966.
21. **Dalgliesh, R. J. and Stewart, N. P.,** Failure of vaccine strains of *Babesia bovis* to regain infectivity for ticks during long standing infections in cattle, *Aust. Vet. J.,* 53, 429, 1977.
22. **Dalgliesh, R. J., Stewart, N. P., and Duncalfe, F.,** Reduction in pathogenicity of *Babesia bovis* for its tick vector, *Boophilus microplus,* after rapid blood passage in splenectomised calves, *Zentralbl. Parasitenkd.,* 64, 347, 1981.
23. **Stewart, N. P.,** Differences in the life cycle between a vaccine strain and an unmodified strain of *Babesia bovis* (Babes 1889) in the tick *Boophilus microplus* (Canestrini), *J. Protozool.,* 25, 497, 1978.
24. **Curnow, J. A.,** Studies on antigenic changes and strain differences in *Babesia argentina* infections, *Aust. Vet. J.,* 49, 279, 1973.
25. **Kahl, L. P., Anders, R. F., Rodwell, B. J., Timms, P., and Mitchell, G. F.,** Variable and common antigens of *Babesia bovis* parasites differing in strain and virulence, *J. Immunol.,* 129, 1700, 1982.
26. **O'Farrell, P. H.,** High resolution two dimensional electrophoresis of proteins, *J. Biol. Chem.,* 250, 4007, 1975.
27. **Kahl, L. P., Mitchell, G. F., Dalgliesh, R. J., Stewart, N. P., Rodwell, B. J., Mellors, L. T., Timms, P., and Callow, L. L.,** *Babesia bovis:* proteins of virulent and avirulent parasites passaged through ticks and splenectomised or intact calves, *Exp. Parasitol.,* 56, 222, 1983.
28. **Cowman, A. F., Timms, P., and Kemp, D. J.,** DNA polymorphisms and subpopulations in *Babesia bovis, Mol. Biochem. Parasitol.,* 11, 91, 1984.
29. **Cowman, A. F., Bernard, O., Stewart, N., and Kemp, D. J.,** Genes of the protozoan parasite *Babesia bovis* that rearrange to produce RNA species with different sequences, *Cell,* 37, 653, 1984.
30. **Gill, A. C., Conman, A. F., Stewart, N. P., Kemp, D. J., and Timms, P.,** *Babesia Bovis:* molecular and biological characteristics of cloned parasite lines, *Exp. Parasitol.,* 63, 180, 1987.
31. **Bodin, I. and Hlidar, G.,** Vaccinations in Sweden against piroplasmosis, Proc. 9th Nordic Veterinary Congr., *Nord. Veterinaermed.,* 14 (Suppl. 2), 328, 1962.
32. **Hinaidy, H. K.,** Bovine babesiasis in Austria. IV. Studies with killed vaccines, *Berl. Muench. Tieraertzl. Wochenschr.,* 94, 121, 1981.
33. **Ryley, J. F.,** The chemotheropy of babesia infections, *Ann. Trop. Med. Parasitol.,* 51, 38, 1957.
34. **Purnell, R. E., Brocklesby, D. W., Stark, A. J., and Young, E. R.,** Reactions of splenectomised calves to the inoculation of blood containing *Babesia divergens* from an infected animal during its reaction and carrier phases, *J. Comp. Pathol.,* 88, 419, 1978.
35. **Taylor, S. M., Kenny, J., and Mallon, T.,** The effect of multiple rapid passage on strains of *Babesia divergens:* a comparison of the clinical effects on juvenile and adult cattle of passaged and irradicated parasites, *J. Comp. Pathol.,* 93, 391, 1983.
36. **Dalgliesh, R. J., Callow, L. L., Mellors, L. J., and McGregor, W.,** Development of a highly infective *Babesia bigemina* of reduced virulence, *Aust. Vet. J.,* 57, 8, 1981.

37. **Kyurlov, N.,** Testing a live vaccine against babesiasis in sheep, *Vet. Med. Nauki,* 14, 25, 1977.

38. **Dalgliesh, R. J.,** Australian tick fevers and vaccination, *Malay. Minist. Agric. Lands Bull.,* 146, 32, 1977.

39. **Tidswell, F.,** Second Report on Protective Inoculation Agaist Tick Fever, New South Wales Department of Public Health, Government Printer, Australia, 1900.

40. **Sergent, E., Donatien, A., Parrot, L., and Lestoquard, F.,** Etudes sur les Piroplasmoses Bovines, Institut Pasteur d'Algerie, Algeria, 1945.

41. **Kemron, A., Hadani, A., Egyed, M., Pipano, E., and Newman, M.,** Studies on bovine piroplasmasis caused by *Babesia bigemina.* The relationship between the number of parasites in the inoculum and the severity of the response, *Ref. Vet.,* 21, 112, 1964.

42. **Todorovic, R. A., Gonzalez, E. F., and Garcia, O.,** Immunisation against anaplasmosis and babesiasis. III. Evaluation of immunisation under field conditions in the Cauca river valley, *Tropenmed. Parasitol.,* 30, 43, 1979.

43. **Camoens, J. K.,** The preparation of a vaccine against *Babesia bigemina* and *Babesia argentina, Malay. Vet. J.,* 6, 196, 1978.

44. **Karbe, E. and Kimotho, G.,** Efficacy of immunisation with live *Babesia bigemina* in cattle, *Kenya Vet.,* 2, 19, 1978.

45. **Thompson, K. C., Todorovic, R. A., Mateus, G., and Adams, L. G.,** Methods to improve the health of cattle in the tropics: immunisation and chemoprophylaxis against haemoparasitic infectious, *Trop. Anim. Health Prod.,* 10, 75, 1978.

46. **Lewis, D. and Williams, H.,** Infection of the Mongolian gerbil with the cattle piroplasm *Babesia divergens, Nature (London),* 27, 170, 1979.

47. **Lewis, D., Young, E. R., Baggott, D. G., and Osborne, G. D.,** *Babesia divergens* infection of the mongolian gerbil: titration of infective dose and preliminary observations on the disease produced, *J. Comp. Pathol.,* 91, 565, 1981.

48. **Murphy, T. M., Gray, J. S., and Langley, R. J.,** Effects of rapid passage in the gerbil *(Meriones unguiculatus)* on the course of infection of the bovine piroplasm *Babesia divergens* in splenectomised calves, *Res. Vet. Sci.,* 40, 285, 1986.

49. **Halberstaedter, L.,** Experimentelle Untersuchungen An Trypanosomen Uber Die Biologische Strahlenwirkung, *Berl. Klin. Wochenschr.,* 51, 252, 1914.

50. **Halberstaedter, L.,** The effects of X-rays on trypanosomes, *Br. J. Radiol.,* 11, 267, 1938.

51. **Ceithaml, J. and Evans, E. A.,** The biochemistry of the malaria parasite. IV. The in vitro effects of X-rays upon *Plasmodium gallinaceum, J. Infect. Dis.,* 78, 190, 1946.

52. **Corradetti, A., Verolini, F., and Bucci, A.,** Resistenza a *Plasmodium berghei* di ralti albuii precedemente immunizzatio con *P. berghei* irradiato, *Parassitologia,* 8, 133, 1966.

53. **Duxbury, R. E. and Sadun, E. H.,** Resistance produced in mice and rats by inoculation with irradiated *Trypanasoma rhodesiense, J. Parasitol.,* 55, 859, 1969.

54. **Phillips, R. S.,** Immunity of rats and mice following injection with ⁶⁰Co irradiated *Babesia rodhaini* infected red cells, *Parasitology,* 62, 221, 1971.

55. **Little, J. B.,** Cellular effects of ionising radiation, *N. Engl. J. Med.,* 278, 308, 1968.

56. **Irvin, A. D., Young, E. R., and Adams, P. J. V.,** The effects of irradiation on *Babesia* maintained in vitro, *Res. Vet. Sci.,* 27, 200, 1979.

57. **Irvin, A. D. and Young, E. R.,** Further studies on the uptake of tritiated nucleic acid precursors by *Babesia* spp. of cattle and mice, *Int. J. Parasitol.,* 9, 109, 1979.

58. **Wright, I. G., Goodger, B. V., and Mahoney, D. F.,** Virulent and avirulent strains of *Babesia bovis:* the relationship between parasite protease content and pathophysiological effect of strain, *J. Protozool.,* 28, 118, 1981.

59. **Wright, I. G., Mirre, G. B., Mahoney, D. F., and Goodger, B. V.,** Failure of *Boophilus microplus* to transmit irradiated *Babesia bovis, Res. Vet. Sci.,* 34, 124, 1983.

60. **Purnell, R. E. and Lewis, D.,** *Babesia divergens:* combination of dead and live parasites in an irradiated vaccine, *Res. Vet. Sci.,* 30, 18, 1981.

61. **Wright, I. G., Mahoney, D. F., Mirrie, G. B., Goodger, B. V., and Kerr, J. D.,** The irradiation of *Babesia bovis.* II. The immunogenicity of irradiated blood parasites for intact cattle and splenectomised calves, *Vet. Immunol. Immunopathol.,* 3, 591, 1982.

62. **Mahoney, D. F., Wright, I. G., and Ketterer, P. J.,** *Babesia argentina:* the infectivity and immunogenicity of irradiated blood parasites for splenectomised calves, *Int. J. Parasitol.,* 3, 209, 1973.

63. **Mahoney, D. F. and Wright, I. G.,** *Babesia argentina:* immunisation of cattle with a killed antigen against infection with a heterologous strain, *Vet. Parasitol.,* 2, 273, 1976.

64. **Wright, I. G., Goodger, B. V., and Mahoney, D. F.,** The irradiation of *B. bovis.* I. The difference in pathogenicity between irradiated and non-irradiated populations, *Z. Parasitenkd.,* 63, 47, 1980.

65. **Brocklesby, D. W., Purnell, R. E., and Sellwood, S. A.,** The effect of ionising radiation on intraerythrocytic stages of *Babesia major, Br. Vet. J.,* 128, iii, 1972.

66. **Purnell, R. E., Brocklesby, D. W., and Stark, A. J.,** Protection of cattle against *Babesia major* by the inoculation of irradiated piraplasms, *Res. Vet. Sci.,* 25, 388, 1978.

67. **Purnell, R. E., Lewis, D., and Brocklesby, D. W.,** *Babesia major:* protection of intact calves against homologous challenge by the injection of irradiated piroplasms, *Int. J. Parasitol.,* 9, 69, 1979.

68. **Lewis, D., Purnell, R. E., and Brocklesby, D. W.,** *Babesia divergens:* the immunisation of splenectomised calves using irradiated piroplasms, *Res. Vet. Sci.,* 26, 220, 1979.

69. **Lewis, D., Purnell, R. E., and Brocklesby, D. W.,** *Babesia divergens:* protection of intact calves against heterologous challenge by the injection of irradiated piroplasms, *Vet. Parasitol.,* 6, 297, 1980.

70. **Taylor, S. M., Kenny, J., Purnell, R. E., and Lewis, D.,** Exposure of cattle immunised against redwater to tick challenge in the field: challenge by a homologous strain of *B. divergens, Vet. Rec.,* 106, 167, 1980.

71. **Taylor, S. M., Kenny, J., Purnell, R. E., and Lewis, D.,** Exposure of cattle immunised against redwater to tick-induced challenge in the field: challenge by a heterologous strain of *Babesia divergens, Vet. Rec.,* 106, 385, 1980.

72. **Castro, E. R. and Canabez, F.,** Biological properties and characteristics of *Babesia bigemina:* effects of ionising radiation on infectivity of infected whole blood, *Bol. Chil. Parasitol.,* 23, 30, 1968.

73. **Bishop, J. P. and Adams, L. G.,** *Babesia bigemina:* immune response of cattle inoculated with irradiated parasites, *Exp. Parasitol.,* 35, 35, 1974.

74. **Prasad, K. D. and Banerjee, D. P.,** Trial of irradiated infected blood as vaccine against babesiosis in cattle, *Indian J. Anim. Sci.,* 53, 773, 1983.

75. **Wright, I. G., Schuntner, C. A., and Goodger, B. V.,** Application of nuclear and related techniques to the diagnosis and control of tick-fome parasitic diseases of livestock, in *Nuclear and Related Techniques in Animal Production and Health,* Proc. IAEA Symp., IAEA, Vienna, 1986, 341.

76. **Wright, I. G., Goodger, B. V., Schuntner, C. A., Waltisbuhl, D. J., and Duzgun, A.,** Use of nuclear techniques in the study of some tick-borne haemoparasitic diseases, 1987, in press.

77. **Khalacheva, M. and Kararizova, L.,** Effect of ionising radiation on the virulence and the immunogenic properties of *Babesia ovis, Vet. Med. Nauki,* 14, 31, 1977.

78. **Hugoson, G., Vennstrom, R., and Henriksson, K.,** The occurrence of bovine leukosis following the introduction of babesiosis vaccination, in *Proc. 3rd Int. Symp. on Comparative Leukaemia Research,* S. Karger, Basel, 1968, 157.

79. **Dimmock, C. K. and Bell, K.,** Haemolytic disease of the newborn in calves, *Aust. Vet. J.,* 46, 44, 1970.

80. **Langford, G., Knott, S. G., Dimmock, C. K., and Derrington, P.,** Haemolytic disease of newborn calves in a dairy herd in Queensland, *Aust. Vet. J.,* 47, 1, 1971.

81. **Dimmock, C. K.,** Blood group antibody production in cattle by a vaccine against *Babesia argentina, Res. Vet. Sci.,* 15, 305, 1973.

Chapter 4

CULTURE-DERIVED *BABESIA* EXOANTIGENS AS IMMUNOGENS

Sonia Montenegro-James, I. Kakoma, and M. Ristic

TABLE OF CONTENTS

I. INTRODUCTION

A. Historical Perspective

It took nearly a century from the discovery of the etiology[1] and the vector[2] of *Babesia* species to the time that the continuous microaerophilous stationary phase (MASP) method for the cultivation of *Babesia bovis* was successfully developed.[3] This method afforded numerous opportunities for *in vitro* studies.[4-6] More importantly, a series of studies utilizing cell culture-derived antigens demonstrated their immunoprophylactic effectiveness against the disease.[7-14]

Animals vaccinated with the culture-derived soluble immunogens of *B. bovis*,[7-12] *B. bigemina*,[12] and *B. canis*[13,14] were clinically protected against tick and needle challenge. Consequently, a vaccine against babesiosis based on organism-free antigens, hereafter referred to as exoantigens, has been developed.[15] The latter term describes a group of proteinaceous substances released into the plasma of infected animals or into the supernatant medium of *in vitro* cultures of *Babesia* organisms. An alternative procedure, still widely practiced, is the use of a live attenuated *B. bovis* vaccine.[16]

Prior to the development of cell culture systems, exoantigens were derived from blood plasma or serum of animals acutely infected with various *Babesia* species.[17-21] These studies attempted to ascertain the immunogenic potential of these proteins and showed that only transient and partial protection could be achieved, presumably due to complex *in vivo* interaction. With the development of the *in vitro* system for exoantigen production,[3] this constraint has been circumvented.

The availability of culture-derived exoantigens has facilitated various *in vitro*[4-6] and clinical[7-14] studies and the recent development of a commercial vaccine against canine babesiosis.[14] In this review we will discuss the experience gained over the last decade of research on development of safe and efficacious vaccines against babesiosis utilizing culture-derived exoantigens. Table 1 summarizes the historical milestones in the struggle against babesiosis including the use of exoantigens.

B. Impact of Babesiosis on Animal Health

Babesiosis remains one of the major obstacles in the development of the livestock industry in tropical and semitropical countries.[22] Prior to the eradication campaign initiated in the U.S. in 1906, losses due to the disease were estimated in millions of dollars in North America alone.[22] Although the American endeavor was commendable, it is unfortunately not realistic for less-developed countries (LDCs), particularly due to economic constraints. Thus, the problem of hemotropic diseases persists in LDCs and 70% of reporting countries worldwide have bovine babesiosis.[23] According to Lombardo,[24] hemotropic diseases cause a loss of at least $875 million/year in Latin America alone. A significant component of this is attributable to babesiosis. Losses due to babesiosis occur mainly through weight loss, reduced milk production, and mortality. If one extrapolates the overall impact of babesiosis worldwide, approximately 1 billion cattle are exposed to the disease.[22] Table 2 summarizes some of the major losses due to bovine babesiosis.

Table 1
SOME HISTORICAL MILESTONES IN BABESIOSIS RESEARCH, INCLUDING APPLICATIONS OF EXOANTIGENS TO THE CONTROL OF BABESIOSIS

Year	Parameter	Key refs.
1888	Etiology	Babes[1]
1893	Vector	Smith and Kilbourne[2]
1966	Attenuated *B. bovis* vaccine	Callow[16]
1965—1971	Exoantigen concept	Ristic et al.[17-19]
1980	Continuous *in vitro* cultivation	Levy and Ristic[3]
1981—1983	Immunization of cattle with exoantigens	Smith et al.[7,8] Kuttler et al.[9,10] Montenegro-James et al.[11,12]
1981—1987	Characterization and quantitation of exoantigens	Smith et al.[8] James et al.[5] Montenegro-James et al.[11] Montealegre et al.[68]
1984	Commercialization of *B. canis* culture-derived exoantigen vaccine	Moreau et al.[14]
1987	Large-scale field and overall efficacy evaluation	Montenegro-James et al.[12]

Table 2
IMPACT OF BABESIOSIS ON ANIMAL HEALTH[a]

Category	Type of loss	Remarks
Mortality	50%, particularly pure bred cattle imported into endemic areas	Asia, Africa, Latin America, e.g., Mexico: approx. 280 million pesos[b]; Australia: 4¢/head/day—total: Queensland, Australia, $644,000
Production	Meat[a] Milk[a] Hides[a] Abortion[a] Infertility	Total loss in U.K.: $8000,000/year; U.S.: $500,000/year; other countries hard to estimate
Quarantine and surveillance	Patrol Dipping Preimmunization & veterinary services Chemotherapy	New South Wales, Australia, approximately $4.5 million/year

[a] Data adapted from McCosker.[22]
[b] If no tick control practiced.

C. Current Status of *Babesia* Antigens with Potential for Vaccinal Use

According to their origin and/or location in the host cell, *Babesia* antigens can be categorized into two groups: (1) soluble exoantigens, also termed metabolic antigens,[25] released from living parasites and parasitized cells. They are found in acute plasma of infected animals or in the supernatant fluids of *Babesia in vitro* cultures; and (2) somatic antigens extracted from merozoites or parasitized cells. These surface or internal antigens can be prepared as crude or purified parasite extracts.

Due to the enormous difficulties in separating the parasite from erythrocytic material, the use of certain preparations of *Babesia* antigens in immunodiagnosis and immunoprophylaxis has been greatly impeded.[26,27] Apparently the association of the parasite with the host erythrocyte is so complex that complete separation is virtually impossible.

Several studies on the use of crude *B. bovis* extracts as immunogens have been reviewed.[28] Generally, crude or partially purified infected erythrocyte lysates have been used in a regime of three immunizations given at a 2-week interval with Freund's adjuvant. Some degree of protection in intact or splenectomized cattle has been achieved following homologous and heterologous challenge.[29-36] In one study, precipitation of crude infected erythrocyte lysate with protamine sulfate resulted in soluble and fibrinogen-associated antigens (FAA) both of which were capable of inducing protective immunity to challenge exposure.[31] The FAA coprecipitate with host fibrinogen or parasite-modified fibrinogen which makes them immunogenic.[34] Wright et al.[35] produced three monoclonal antibodies against a partially purified lysate antigen. Antigens affinity-purified by the respective monoclonal antibodies had molecular weights of 44, 100, and 1300 kDa. Only the 44-kDa polypeptide, a minor component of the crude *B. bovis* lysate, induced protective immunity against homologous challenge. Further purification of this antigen by gradient gel electrophoresis resulted in a 29-kDa protein that conferred strong homologous strain immunity.[33] In another study,[36] Western blot analysis of an esterase isolated by affinity chromatography revealed a major band at 20 kDa. Bovine antibody against this esterase showed distinct cytoplasmic and membrane fluorescence of *B. bovis*-infected erythrocytes. This antigen was found to induce protective immunity against homologous challenge.[36]

A high degree of erythrocytic contaminants has been demonstrated in most parasitic extracts.[25,29-34] Some antigens are probably bound to carrier host proteins such as fibrinogen. This protein contributes to the pathogenesis of the disease by facilitating the adherence of infected erythrocytes to the capillary endothelium.[37]

Recently, Goodger et al.[38] isolated a heparin-binding fraction from *B. bovis*-infected erythrocytes. By Western blot analysis, a distinct band at 180 kDa and a doublet at 140 kDa were observed. Animals vaccinated with this fraction were protected against homologous challenge. Heparin-binding babesial proteins are located on or near the external membrane of host cells, as revealed by the immunofluorescence pattern of infected erythrocytes.[38] Since surface epitopes are considered important for induction of protective immunity, heparin-binding proteins may constitute relevant target antigens.

Clearly, tremendous research efforts have been devoted to the identification of merozoite-derived antigens. However, the feasibility of their practical application as immunogens has been limited by technicalities such as (1) the presence of contaminant host components; (2) the apparent need for complete Freund's adjuvant; (3) the common usage of at least three immunization doses, and (4) the use of immunocompromised animals (splenectomized) for testing their effectiveness.

Until recently the only vaccines against babesiosis consisted of living organisms which conferred protection by infection. However, since many disadvantages of this procedure have been recognized,[29] alternative methods for immunoprophylaxis are needed. The latter include selection of antigens that elicit an efficient protective immunity and the development of improved adjuvant formulations and/or novel delivery systems. Adjuvants should be composed of materials that can be safely used in food-producing animals.

Culture-derived *Babesia* exoantigen-containing immunogens devoid of erythrocytic components have been proposed as a practical and realistic approach for the control of babesiosis.[15] Culture-derived exoantigens, being essentially free from erythrocytic stromal antigens, do not induce formation of isoantibodies associated with such complications as neonatal isoerythrolysis.[39] Furthermore, the abundant supply of parasite antigens from supernatant fluids of *Babesia*-infected cell cultures makes the system a practical source of immunogens. The immunogenicity of culture-derived exoantigens, when supplemented with a potent and acceptable adjuvant, has been observed in independent trials conducted in Mexico,[7,8,40] the U.S.,[9,10] Australia,[39] France,[14] and Venezuela.[11,12,41] Vaccination parameters and corresponding data of some of these trials are summarized in Table 3.

In search of vaccines that are safe, efficacious, and cost-effective, immunogens comprising soluble exoantigens are prime candidates to satisfy these important criteria.

Studies on exoantigen characterization are in progress in order to identify and purify the relevant antigens necessary for induction of protective immunity. The availability of modern molecular techniques should facilitate the attainment of this goal.

II. *IN VITRO* CULTIVATION OF *BABESIA* SPECIES

A. Current Status of the Propagation of *Babesia* Parasites in Culture

The basis for the *in vitro* propagation of *Babesia* species originated from the candle jar method developed by Trager and Jensen for *Plasmodium falciparum*.[42] Short-term cultivation by that system was achieved for three *Babesia* species,[43] and in 1978 Erp et al.[44] reported the first successful attempt to grow *Babesia* parasites in suspension cultures in which *in vivo* conditions were simulated. Minor modification of this method resulted in the first continuous *in vitro* propagation of *B. bovis* by the spinner flask system.[45] The limitations of the latter technique, such as a low percentage of parasitized erythrocytes (PPE) and the relatively large volumes required, were overcome by the MASP system.[3] In its original description, the technique, based on a *B. bovis* Mexican isolate,[7] consisted essentially of a layer of infected erythrocytes at approximately 10% packed cell volume (PCV) in a medium containing 40% normal bovine serum (NBS) and 60% medium 199 supplemented with 25 mM N-2-hydroxyethylpiperazine-N-2-ethane sulfonic acid (HEPES) and antibiotics titrated to pH 7.0. The culture was maintained at a low oxygen tension provided by a ratio of 0.62 ml of suspension per square centimeter at 37°C under an atmosphere of 5% CO_2 in humidified air. The authors recommended daily medium replacement and subculturing every 2 to 3 d. The MASP system has provided a significant contribution in the development of improved *in vitro* propagation systems for *Babesia* parasites. However, this technique is labor intensive, utilizes a high serum concentration, and is generally suited for microanalytical studies.[4,6] Thus, it will require substantial modifications for use in a large-scale culture system.

In order to optimize the MASP method, Palmer et al.[46] recommended the use of zwitterionic buffer TES (2,2-hydroxy-11-bis-hydroxymethylaminoethane sulfonic acid) at a 40-mM concentration, storage of normal bovine erythrocytes in glucose-supplemented Puck's Saline G, and careful selection of suitable donor serum. The critical features of the MASP system: pH of 7.0, temperature of 37°C, and an atmosphere of 5% CO_2 in humidified air remained the same. Under these conditions a PPE level of up to 30% can be achieved. Later, Rodriguez et al.[47] were able to synchronize and clone *B. bovis* in such cultures, provided the oxygen tension was maintained at 2%. To date, culture techniques are available for short-term or continuous *in vitro* cultivation of various *Babesia* spp. such as *B. bigemina*,[49] *B. canis*,[13,14] *B. equi*,[50,51] *B. divergens*,[52] *B. microti*,[53] and *B. rodhaini*.[54]

Substantial progress has been made in continuous propagation of *Babesia* parasites. However, for mass production of immunogens, there is room for improvements, e.g., (1) establishment of precise requirements for *Babesia* spp. *in vitro*, (2) development of alternate

Table 3
VACCINATION PARAMETERS AND CORRESPONDING DATA USING
CULTURE-DERIVED *BABESIA BOVIS* IMMUNOGENS

	Mexico (Smith et al., 1984)	U.S. (Kuttler et al., 1982, 1983)	Australia (Timms et al., 1983)	Venezuela (Montenegro-James et al., 1985, 1987)
Vaccination				
Dose (ml-eq)	28	10	4	10
Saponin (mg)	1	1	2	1.5—3
Interval (2 doses; weeks)	2	3	2	3—4
Cattle	15 months, Holstein	1 year, Hereford	2 years, *Bos taurus*	Yearly, Holstein crossbred
Challenge				
Time (weeks)	12	7.25	10	12, 52
Dose (PE)	1×10^3	1×10^8	4×10^8	1×10^9
Type	Tick, homologous	Needle, homologous	Needle, heterologous	Needle, homologous & heterologous
Max % PCV	47[a] (53)[b]	39[a], 52[a] (63)[b]	40[a] (38)[b]	26[a], 36[a] (55)[b]
Death (D) or treatment (T)	0/4[a] (2/4)D[b]	0/21[a] (6/12)[b]D	1/5[a] (5/5)[b]T	0/18[a] (2/12)[b]D

[a]Vaccinated.
[b]Controls.

host cell systems,[48] and (3) large-scale cell culture, such as those developed for malaria parasites.[55] In this section we will review those *Babesia* culture systems that have been used primarily for exoantigen production.

B. *Babesia bovis* and *Babesia bigemina*

1. *Babesia bovis*

At the Instituto de Investigaciones Veterinarias (IIV) in Venezuela, a modified MASP technique has been implemented to facilitate the propagation of local and other *Babesia* isolates. Cultures are initiated using defibrinated blood of infected donor cattle at parasitemias between 1 to 2%. Infected erythrocytes are suspended to a final packed cell volume of 10 to 15% in a medium consisting of 70% medium 199 and 30% normal adult bovine serum, supplemented with 25 mM HEPES, 2 g glucose, 50 mg hypoxanthine, 30 mg reduced glutathione, 10 mg adenosine, 10^5 U penicillin G, and 10 mg streptomycin per liter. Adult *Bos taurus* cattle are used as a source of serum and fresh blood. Culture suspensions are titrated to pH 7.2, aliquoted at a ratio of 0.62 ml/cm² and incubated at 37°C under an atmosphere of 3% CO_2 in humidified air. The medium is changed daily and subcultures carried out at 48- to 72-h intervals. Prior to use, normal erythrocytes are washed once in incomplete medium and maintained for 1 week in the same media. These conditions have facilitated the propagation of three *B. bovis* isolates from Venezuela and single isolates from Mexico, Colombia, Argentina, Ecuador, and Australia.[11,12,56] The adaptability of *B. bovis* isolates to *in vitro* growth is quite variable. *Babesia* organisms are highly susceptible to *in vitro* changes, particularly with regard to fluctuations in CO_2, pH, and the presence of leukocytes. Optimal culture conditions should provide an environment of reduced oxygen tension, prevent the toxic effects of CO_2, and maintain pH with a daily change of medium. The latter fastidious requirement can be circumvented by placing the culture at 4°C for a 2-d period. Parasites remain viable and resume growth normally.

It has been observed that the kinetics of *B. bovis* growth during primary cultures vary from that of subsequent subcultures which lack the characteristic exponential growth until a period of adaptation and/or selection occurs.[56] Once the isolate is adapted to *in vitro* conditions, the rates of parasite multiplication are greater than those observed *in vivo*.[57] One of the most consistent observations on *in vitro* cultures of *B. bovis* parasites is a shift in morphology from ring to pyriform. In the host animal, most parasites are mainly of the ring form, whereas in culture the pyriform type predominates (Figure 1, Table 4). Ristic[58] observed that after 20 consecutive blood passages in splenectomized animals, a similar switch in parasite morphology occurred and was correlated with the selection of less-virulent parasites. Recently, Yunker et al.[59] reported that an *in vitro*-maintained, equine serum-adapted *B. bovis* strain lost virulence for splenectomized calves. Similarly, a *B. bovis* isolate was observed to lose infectivity for ticks after 6 months in culture.[60]

The development of a continuous cultivation system for *B. bovis* has facilitated research on parasite morphological characteristics,[58] the role of serum components,[61] the chemo-therapeutic efficacy of drugs,[15] and offered unprecedented potential for immunoprophylactic control of bovine babesiosis.[62]

2. *Babesia bigemina*

The *in vitro* cultivation of *B. bigemina*[49] has allowed relatively short-term growth at low PPEs. Culture requirements include washing of normal and infected bovine erythrocytes in sterile Vega and Martinez phosphate-buffered saline solution, and the use of a 10 to 15% (v/v) erythrocyte suspension in medium 199 with 20% normal bovine serum. Since *B. bigemina*-infected erythrocytes are extremely fragile, efforts to restore depleted erythrocyte ATP levels and maintain the elasticity of the erythrocyte membrane have been helpful in preventing lysis.[63] Thus, prior to culture initiation, infected erythrocytes are pretreated with

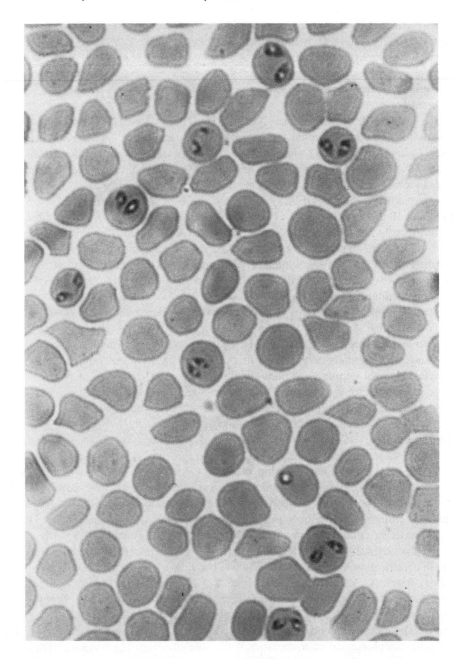

FIGURE 1.   Giemsa-stained thin blood film of *B. bovis*-infected from MASP *in vitro* cultures, showing typical "pyriform" parasite morphology.

PIGPA buffer[64] containing 50 mM sodium pyruvate, 50 mM inosine, 100 mM glucose, 500 mM Na$_2$H PO$_4$, and 5 mM adenosine in 0.9% sodium chloride (w/v), pH 7.2. One volume of buffer is added to ten volumes of infected blood and incubated for 1 h at 37° C. Erythrocytes are then washed twice with incomplete medium. The recommended culture medium and conditions are identical to those described for *B. bovis*. A substantial number of extraerythrocytic *Babesia* are commonly observed, and consequently the actual rate of parasite multiplication is greater than recorded. Whether the presence of extracellular merozoites is indicative of suboptimal culture conditions is not known. In general, the *in vitro* cultivation

Table 4
VIRULENCE AND MORPHOLOGIC CHARACTERISTICS OF VARIOUS
GEOGRAPHICAL ISOLATES OF *BABESIA BOVIS*

B. bovis isolate	Origin	Virulence	In vivo		In vitro	
			Max PPE	Morphology	Max PPE	Morphology
Venezuela (Yaracuy)	Adult carrier	Moderate	8.3%	90% R, 10% P	8.6%	20% R, 80% P
Venezuela (Guarico)	Acute case	High	4.0%	84% R, 16% P	5.0%	10% R, 90% P
Venezuela (Zulia)	Young carrier	High	4.3%	84% R, 16% P	3.5%	20% R, 80% P
Argentina	Tick derived	High	16.0%	70% R, 30% P	5.0%	20% R, 80% P
Colombia	High passage	Low	8.3%	85% R, 15% P	7.0%	20% R, 80% P
Ecuador	Tick derived	High	3.1%	70% R, 30% P	3.5%	20% R, 80% P
Australia	High passage	Low	6.0%	97% R, 3% P	8.1%	90% R, 10% P
Mexico	Tick derived	High	1.6%	80% R, 20% P	12.2%	20% R, 80% P

of *B. bigemina* has not been as successful as that of *B. bovis,* particularly with regard to the maximum PPE attainable. However, these culture systems still provide an abundant source of soluble exoantigens that can be used as effective immunogens.

C. *Babesia canis* and *Babesia equi*

1. *Babesia canis*

Molinar et al.[13] utilized the spinner flask system[45] for *in vitro* propagation of *B. canis* parasites. Briefly, media was RPMI-1640 supplemented with 25 m*M* HEPES and 20% normal dog serum adjusted to pH 7.4 before use. Infected (3 to 4 PPE), defibrinated blood was washed three times in incomplete media and kept at 4°C until use. Cultures are set at 20% PCV at 38°C in a humidified atmosphere of 5% CO_2 in air. The cells are maintained in suspension by gentle magnetic stirring. Every 48 h the cultures are centrifuged at 400 × *g* for 10 min at 4°C and fresh medium added. An average of 3 PPE was achieved by this method. Later, Moreau and Laurent[14] adapted the MASP system for *B. canis* cultures. Essentially, lower serum concentration, use of heparin as an anticoagulant instead of defibrinated blood, and alternate maintenance of cultures at 4 and 37°C are recommended.

2. *Babesia equi*

In the vertebrate host, *B. equi* organisms exist in intraerythrocytic and exoerythrocytic stages.[65] Macro- and microschizonts as in *Theileria* spp. have been observed in lymph node biopsies taken from *B. equi*-infected horses.[65] Extracellular merozoites from lymphocyte cultures are able to infect horse erythrocytes *in vivo* and *in vitro*.[50] Tick-derived sporozoites appear to actively invade lymphocytes, but do not invade equine erythrocytes.[65]

Recently, Montenegro-James et al.[51] successfully propagated *B. equi in vitro* utilizing the modified MASP system. Reactivation of a *B. equi* strain isolated in Venezuela was established by inoculation of an infected blood stabilate into a splenectomized donkey. At 1 week postinoculation the parasitemia was 1.8%, when infected blood was aseptically collected to initiate cultivation. Cultures were initiated at a 10% hematocrit in a medium consisting of 70% medium 199 and 30% normal donkey serum. Erythrocyte cultures were incubated at 37°C in an atmosphere of 3% CO_2 in humidified air. A maximal PPE of 3.5% was achieved at 48 h. Subcultures were conducted every 48 h by diluting the infected culture with normal donkey erythrocytes from a susceptible donor. Cultures were terminated after three subcultures. No intracellular microschizonts were found in the buffy coat of centrifuged, infected blood samples.

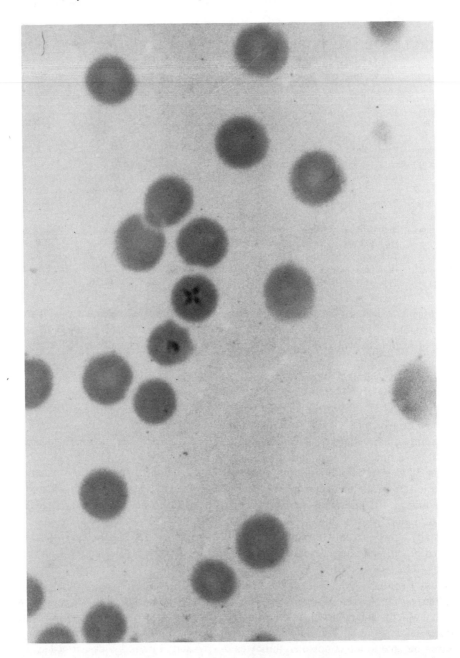

FIGURE 2.   Giemsa-stained *B. equi*-infected erythrocytes from MASP *in vitro* cultures, depicting "maltese cross" parasite forms.

Intraerythrocytic division and multiplication of *B. equi in vivo* have been described.[66] The parasite appears characteristically pleomorphic; ring-, comma-shaped and amoeboid forms are common. A unique morphologic feature of *B. equi* is the "maltese cross", consisting of four pear-shaped merozoites. It has been reported[67] that *in vivo* most parasites are pleomorphic with less than 2% of infected cells exhibiting the maltese cross morphology. In contrast, *in vitro* conditions appear to favor the maltese cross formation[51] (Figure 2).

A *B. equi* exoantigen has been demonstrated by gel precipitation tests in the serum of horses with acute babesiosis.[17] The characterization of culture-derived *B. equi* exoantigens will be discussed in a subsequent section of this chapter.

III. QUANTITATION AND CHARACTERIZATION OF CULTURE-DERIVED *BABESIA* EXOANTIGENS

A. The Two-Site Immunoassay for Exoantigen Quantitation

A two-site immunoassay (EIA) has been developed for detection and quantitation of culture-derived *B. bovis*[68] and *B. bigemina*[12] exoantigens. The basic procedure is an adaptation of a previously described two-site EIA.[69] Principally, the assay consists of four major steps: (1) purified specific antibody is bound to microtiter wells, (2) culture supernatants containing exoantigens are added, (3) enzyme-labeled specific antibody is applied, and (4) appropriate substrate is added and the resulting amount of color is measured. The bovine IgG used in this assay was isolated from hyperimmune sera of cattle that had been previously immunized with either culture-derived *B. bovis* or *B. bigemina* exoantigens and later needle challenged with virulent *B. bovis*- or *B. bigemina*-infected blood (IFA titers = 1:163,840 for *B. bovis,* and 1:40,960 for *B. bigemina).* Immune IgG was fractionated by caprylic acid[70] or ammonium sulfate precipitation[71] and further purified by anion-exchange chromatography. Anti-*Babesia* IgG was then conjugated to horseradish peroxidase (HRPO)[72] or alkaline phosphatase (AP)[73] for *B. bovis* and *B. bigemina,* respectively. Results of the assay can be expressed in EIA units:

$$\text{EIA unit or P/N ratio} = \frac{\text{OD}_{414} \text{ infected culture supernatant fluid}}{\text{OD}_{414} \text{ uninfected culture supernatant fluid}}$$

or as absolute OD_{414} readings. The typical reactivity pattern is depicted in Figure 3 with the respective titration of 3-d culture supernatant pools containing *B. bovis* or *B. bigemina* exoantigens. The two-site EIA has proved particularly useful for detecting antigenic activity during purification procedures.[74,75] Furthermore, with this assay the kinetics of antigen release into the supernatant fluids of *in vitro* cultures can be monitored. For instance, 24-h cultures showed the greatest antigenic activity, indicating the exoantigen synthesis and release occur predominantly at the early growth phase of the organism (Table 5 and Figure 4 for *B. bovis* and *B. bigemina,* respectively). Taylor et al.[76] have observed a similar pattern of exoantigen release *in vivo* using a two-site EIA to detect circulating antigens during *Plasmodium yoelii* infections of mice.

The EIA has also been employed as an analytical tool in the characterization of exoantigens following treatment with a variety of enzymes (Figure 5). Of particular importance is the capability of the two-site EIA for detecting variations in the antigenic composition among *B. bovis* isolates. Figure 6 depicts the differences in reactivity between homologous and heterologous *B. bovis* strains using anti-*B. bovis* IgG to a Venezuelan strain as a capture antibody and conjugate. The homologous reaction was consistently higher than those observed in the heterologous system. These findings are in agreement with cross-protection studies, since heterologous antibody levels were consistently lower for all strains analyzed (derived from different Latin American geographical regions).[11] In addition, the two-site EIA is applicable to monitoring the stability, i.e., shelf life, of babesial immunogens under a variety of storage conditions. For example, treatment of antigen with 0.1% formalin significantly reduces antigenicity and immunogenicity. Exoantigens remain stable when stored at $-70°C$ and after lyophilization. Another potential application of the test is the detection of parasite-specific antigens in the vector,[77] a factor of major epidemiological significance.

B. Characterization of *Babesia bovis* and *Babesia bigemina* Exoantigens

Intraerythrocytic parasites were first observed to release soluble antigens into the plasma of infected hosts when Eaton[78] reported soluble malarial antigen in the serum of monkeys

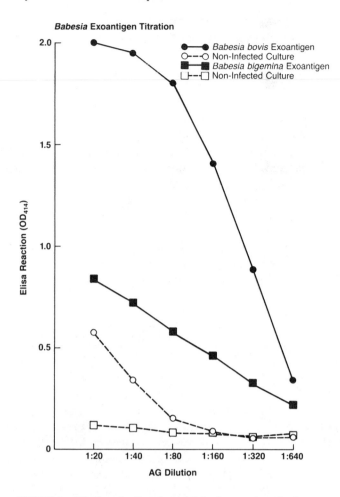

FIGURE 3.   Titration of culture-derived *Babesia* exoantigens by the two-site EIA using bovine anti-*B. bovis* (Mexican isolate) IgG.

infected with *Plasmodium knowlesi*. Since then, well-documented evidence has indicated that parasite-specific or parasite-modified host antigens are released *in vivo*[16-21] and *in vitro*[12,13,26,40,58,68] as soluble proteins. A series of our earlier studies analyzed the immunogenic properties of soluble antigens isolated from the serum of horses, dogs, and rats with acute babesiosis.[16-19] Similar studies with soluble antigens in acute plasma of cattle infected with *B. bigemina* and *B. bovis* investigated their immunogenic potential.[20,21] *In vivo*, however, these antigens appear to be altered by the formation of immune complexes, enzymatic degradation processes, and other physiologic elements, thus compromising their usefulness as immunogens.

The entry of *Babesia* merozoites into host erythrocytes is probably similar to that described for *Plasmodium* spp.[79,80] Penetration into erythrocytes involves 5 steps: (1) parasite contact with and attachment to an erythrocyte, (2) proper orientation of the apical end of the parasite, (3) fusion of parasitic and erythrocytic membranes, (4) release of rhoptry contents by the parasite, and (5) invagination of the erythrocyte, followed by shedding of the merozoite surface coat and parasite entry.

Miller et al.[81] reported the presence of a fibrillar surface coat on *P. knowlesi* merozoites. The surface coat was shed as merozoites penetrated the erythrocytes. Similarly, *B. microti* and *B. bigemina* merozoites have been shown to possess a thick surface coat.[82,83] The role

Table 5
REPRODUCIBILITY OF THE EIA AND ITS USE IN ASSESSING THE KINETICS OF EXOANTIGEN RELEASE INTO *BABESIA BOVIS IN VITRO* CULTURES

Conditions of supernatant fluid tested[a]	EIA units (n = 6)			Coefficient of variation (%)
	Mean	SE	Range	
Pooled	4.53	0.025	4.43—4.61	1.24
Lyophilized	2.92	0.0092	2.90—2.98	0.70
Freshly collected	4.18	0.022	4.10—4.25	1.17
Cultured				
24 h	4.69	0.023	4.61—4.79	1.12
48 h	3.73	0.028	3.64—3.81	1.70
72 h	3.76	0.026	3.68—3.84	1.38

[a] Protein concentration of all samples tested was 100 µg/ml.

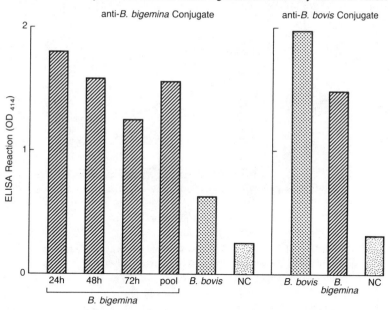

FIGURE 4. Cross-reactivity between *B. bovis* and *B. bigemina* exoantigens as measured by a two-site EIA. Antigenic reactivity measured in absorbance (optical density at 414 nm) units. Exoantigens of left and right panels reacted with anti-*B. bigemina* and anti-*B. bovis* capture antibodies and enzyme conjugates, respectively. (From Montenegro-James, S., Toro, M., Leon, E., Lopez, R., and Ristic, M., *Parasitol. Res.*, 74, 142, 1987. With permission.)

of the surface coat in protecting the merozoite was demonstrated by Brooks and Kreier[84] who postulated that the coat is associated with parasite-erythrocyte binding sites prior to invasion of susceptible red blood cells. Antibodies may interfere with this mechanism.[8,84]

Culture-derived soluble *Babesia* exoantigens are parasite-free, proteinaceous moieties that are derived from the merozoite surface coat.[8] The merozoite surface coat is the target of humoral immune effector mechanisms and elicits the production of antibodies that react directly with the parasite, causing aggregation and a complement-mediated lysis of mero-

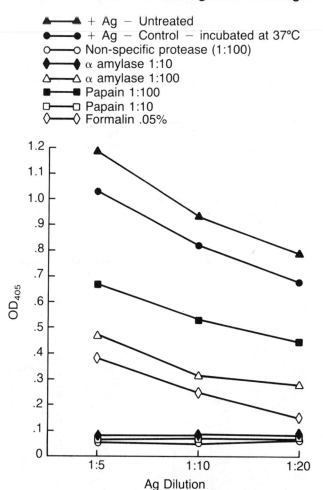

**Effect of Enzymes and Formalin on Soluble
Culture derived *Babesia bigemina* Exoantigen**

▲——▲ + Ag – Untreated
●——● + Ag – Control – incubated at 37°C
○——○ Non-specific protease (1:100)
◆——◆ α amylase 1:10
△——△ α amylase 1:100
■——■ Papain 1:100
□——□ Papain 1:10
◇——◇ Formalin .05%

FIGURE 5.   Titration of *B. bigemina* antigenic reactivity by the two-site
EIA following treatment of culture-derived exoantigen with various enzymes and formalin.

zoites.[8] These data support the hypothesis that a major constituent of culture-derived exoantigens is the merozoite surface coat.

Culture-derived *B. bovis* exoantigens were first characterized using rabbit polyspecific
anti-*B. bovis* antibodies.[5] Three major antigens were found to have apparent molecular
weights between 37 to 40 kDa, fast electrophoretic mobility in the albumin and alpha regions,
and were proteinaceous as determined by sensitivity to proteolytic enzymes. Immunopre-
cipitates derived from crossed immunoelectrophoresis of *B. bovis* exoantigens were subsequently used to produce rabbit monospecific antibodies to the individual *B. bovis* antigens.
One of the antigens was shown to be directly associated with the parasite, while two antigens
were localized within the membrane and cytoplasm of the infected erythrocyte.[85]

Isolation of babesial exoantigens from a complex protein mixture (40% serum supplement
for *B. bovis* and *B. equi*, 20% serum concentration for *B. bigemina* and *B. canis)* represents
a major obstacle in the identification and characterization of immunogens. Preparative purification of culture-derived *Babesia* exoantigens has been conducted by selective precipi-

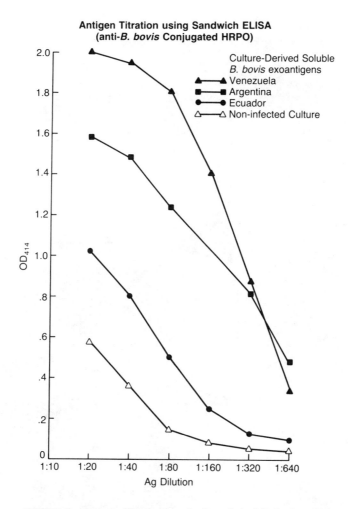

Antigen Titration using Sandwich ELISA
(anti-*B. bovis* Conjugated HRPO)

Culture-Derived Soluble
B. bovis exoantigens
▲——▲ Venezuela
■——■ Argentina
●——● Ecuador
△——△ Non-infected Culture

OD$_{414}$

Ag Dilution

FIGURE 6. Two-site EIA titration of culture-derived *B. bovis* exoanti-gens from strains isolated in Venezuela, Argentina, and Ecuador. Capture antibody and conjugate were prepared from an IgG fraction of hyperim-mune anti-*B. bovis* (Venezuelan isolate) serum.

tation with ammonium sulfate and anion-exchange chromatography.[74,75,86] Salt precipitation reduces the concentration of serum contaminants (albumin, immunoglobulins, etc.) and has been used as an initial purification step prior to anion-exchange chromatography. Using a sequential combination of pH gradient anion-exchange and size exclusion high-performance liquid chromatography (HPLC), Montealegre[74] isolated a relatively pure *B. bovis* (Mexican isolate) antigenic fraction. Western immunoblotting of the partially purified pH 4.0 fraction revealed three proteins with apparent molecular weights between 40 and 50 kDa when reacted with hyperimmune anti-*B. bovis* IgG.

Exoantigens from the Mexican isolate of *B. bigemina* have been recently purified by cation-exchange chromatography.[75] Following further isolation by Sephadex® G-200 mo-lecular sieve chromatography, most *B. bigemina* antigenic activity was eluted at a relative molecular weight of 60 kDa. Western blot analysis under native nondenaturing conditions demonstrated three antigens using sera from vaccinated or experimentally infected animals. The molecular weights of these antigens were estimated to be 64, 60, and 53 kDa.[75] More recently, Montenegro-James et al.[87] conducted further analyses of culture-derived *Babesia* antigens. *B. bovis* and *B. bigemina* antigens (exoantigens and merozoite extracts) of Mexican

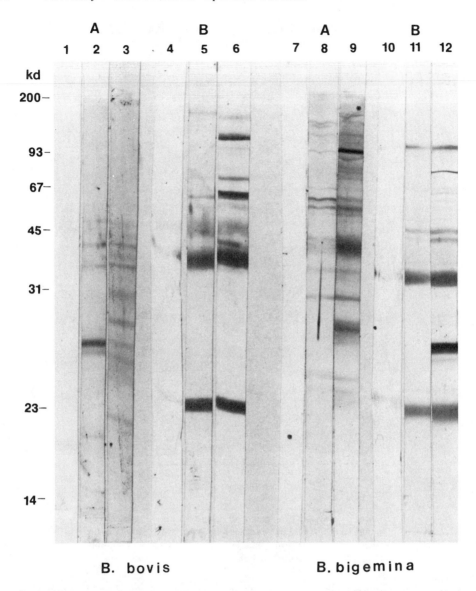

FIGURE 7.   Western blot analysis of *B. bovis* and *B. bigemina* proteins following electrophoresis on a reducing 10% SDS-PAGE gel, nitrocellulose transfer, and development with homologous bovine antisera (postvaccination, lanes 2, 5, 8, and 11; postchallenge, lanes 3, 6, 9, and 12) to *B. bovis* and *B. bigemina* exoantigens. Proteins were reacted with normal bovine serum in lanes 1, 4, 7, and 10. Antigen preparations are represented as follows: A, merozoite extracts; B, exoantigen-containing supernatant fluids from infected cultures.

isolates were analyzed by SDS-PAGE followed by Western immunoblotting. Electrophoresis was performed on 10% acrylamide gels under reducing conditions according to the method of Laemmli.[88] Proteins were transblotted onto nitrocellulose and reacted with postvaccination and challenge sera as recommended.[89] Postvaccination serum revealed 3 and 6 parasite proteins in the *B. bovis* exoantigen (exo) and merozoite (mz) antigen preparations, respectively, whereas 4 and 13 bands were found in the respective *B. bigemina* exo and mz antigen preparations. The molecular weight (M_r) ranges were: 164 to 55 kDa, *B. bovis* exo; 40 to 20 kDa, *B. bovis* mz; 92 to 37 kDa, *B. bigemina* exo; and 143 to 24 kDa, *B. bigemina* mz (Figure 7). Bovine serum obtained after vaccination and challenge revealed a greater spectrum of parasite proteins: *B. bovis* (9 exo and mz) ranging from 164 to 36 kDa (exo) and 49 to 20 kDa (mz). Regarding *B. bigemina,* 14 and 18 protein bands were revealed for exo and mz preparations, respectively. The M_r ranged from 92 to 26 kDa (exo) and 164 to 24 kDa (mz) (Figure 7).

Interestingly, the number and M_r range of *B. bovis* proteins of the Venezuelan isolate differ from those found with the Mexican isolate as reported here. Montenegro-James et al.[12] found at least six protein bands with an apparent M_r range of 82 to 18 kDa. Interstrain antigenic differences exist within the *B. bovis* species and have been well documented.[90-93] In a previous study designed to assess the degree of heterologous strain immunity, cattle vaccinated with a Venezuelan *B. bovis* immunogen failed to show significant protection against challenge with the Mexican strain.[11] The difference in the antigenic spectrum, i.e., presence of strain-specific proteins between these two isolates, may explain the lack of heterologous cross-protection. Nevertheless, the M_r range of proteins from both Venezuelan and Mexican *B. bigemina* isolates is similar (92 to 37 kDa Mexican isolate; 86 to 44 kDa Venezuelan isolate). *B. bovis* parasites show a high degree of antigenic heterogeneity. On the other hand, *B. bigemina* seems to be more antigenically stable.[75,94]

Serological cross-reactivity between *B. bovis* and *B. bigemina* has been observed.[95-100] It appears that one or more antigens may be common to *B. bovis* and *B. bigemina*. Wright et al.[100] have demonstrated polypeptides of similar M_r for both species. In Western blot analysis of culture-derived exoantigens, Montenegro-James et al.[87] obtained similar results using homologous and heterologous *Babesia* spp. systems (Figure 8). Of 12 *B. bigemina* proteins, 11 were identified with the heterologous anti-*B. bovis* serum; a polypeptide of 89 kDa was species specific. Conversely, anti-*B. bigemina* serum reacted with five of the eight *B. bovis* proteins detected in the homologous system. Proteins of apparent M_r 125, 109, and 75 kDa appeared to be unique to *B. bovis* (Figure 8).

C. Characterization of Culture-Derived *Babesia canis* and *Babesia equi* Exoantigens
1. Babesia canis
Molinar et al.[13] reported that *B. canis* antigens contained in the culture supernatant fluids were highly heterogeneous, with a major component of approximately 900 kDa presumably composed of multimeric protein aggregates. Recently, Bissuel et al.[101] observed a major 44-kDa polypeptide in *B. canis* culture medium precipitable with immune serum. However, Western blot analysis of sera from dogs of varying immunological status failed to show any correlation between particular antigenic reactivity and protective efficacy.

2. Babesia equi
Earlier reports[17,18] characterized soluble antigens found in the plasma of horses with acute infection. Those antigens were found to be proteinaceous, closely associated with serum globulins, and serologically reactive with sera of rats and dogs recovered from *B. rodhaini* and *B canis* infections, respectively. Subsequently, the protective potential of plasma-derived *B. equi* exoantigens was investigated.[102] More recently, Montenegro-James et al.[51] utilized culture-derived *B. equi* exoantigens to immunize susceptible donkeys. Two 10-ml-equivalent (eq) doses supplemented with 3 mg saponin were inoculated subcutaneously on days 0 and 30. Serum was collected weekly and analyzed by Western immunoblotting and the indirect fluorescent antibody (IFA) test. Hyperimmune serum from a naturally infected animal was used in the identification of antigens contained in a merozoite antigen preparation. Western immunoblots revealed 16 protein bands ranging from 108 to 20 kDa, with 7 major bands found at 82, 61, 53, 42, 40, 33, and 26 kDa (Figure 9). Postvaccinal sera collected at 21, 49, 57, and 63 d served as a source of antibodies which identified four protein bands at 61, 53, 39, and 30 kDa. Antibodies reactive with a protein band of apparent M_r 39 kDa persisted for almost 6 months postvaccination, when IFA titers reached baseline values. Studies are

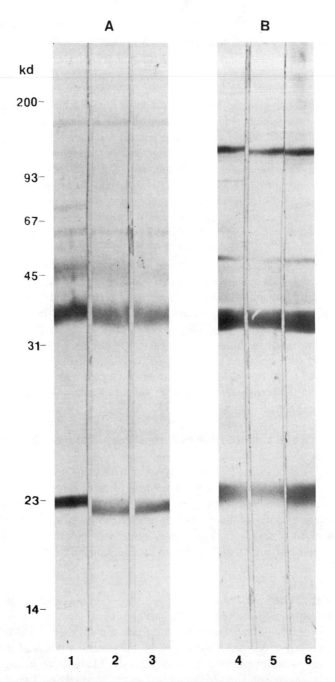

FIGURE 8.   Western blot analysis of *B. bovis* (A) and *B. bigemina* (B) exoantigens following development with bovine antisera to homologous (lanes 1 and 4) and heterologous (lanes 2, 3, 5, and 6) *Babesia* spp.

in progress to analyze the protective capabilities of culture-derived *B. equi* immunogen following challenge.

IV. USE OF *BABESIA* EXOANTIGENS AS IMMUNOGENS

A. Formulation of *Babesia* Exoantigen-Containing Immunogen and Vaccination Regime

The MASP[3] erythrocyte culture system is initiated using either *B. bovis* or *B. bigemina*

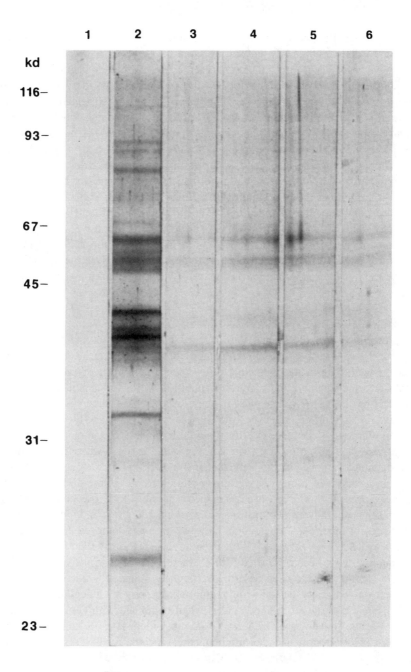

FIGURE 9.   Western blot analysis of *B. equi* merozoite proteins following development
with donkey antisera to *B. equi*. Antigens were reacted with normal donkey serum (lane 1),
serum from naturally infected animal (lane 2), and antisera collected sequentially after vaccination with *B. equi* exoantigens (lanes 3, 4, 5 and 6: days 14, 21, 49, and 63 post-second-
vaccination, respectively).

organisms as seed material. All parasite strains are maintained by transmission in *Boophilus
microplus* ticks under laboratory conditions and/or by no more than four syringe passages
in splenectomized calves. Initiation of *B. bovis, B. bigemina,* and *B. canis* cultures requires
fresh, infected blood because suboptimal growth results from use of cryopreserved *Babesia*.[46]
Reactivation of each strain is established separately by inoculation of infected blood stabilates

into splenectomized animals. Infections are monitored by examining thin peripheral blood smears and by determining body temperatures and PCVs. When a preestablished parasitemia level (1 to 3%) is reached, infected blood is aseptically collected to begin *in vitro* cultivation. Healthy, susceptible calves (or dogs for *B. canis*) are maintained in tick-free isolation units to serve as blood and/or serum donors. Cultures are set up at a 10% hematocrit in a medium consisting of 70 or 80% medium 199 and 30 or 20% adult bovine serum (*B. bovis* and *B. bigemina,* respectively). Media composition is detailed in Section II.B. Erythrocyte cultures are incubated at 37°C in an atmosphere of 3% CO_2 in humidified air. An average PPE over a 72-h culture cycle is 8% for *B. bovis* and 3% for *B. bigemina.* Briefly, the steps followed in the preparation of *Babesia* exoantigens for use as immunogens for bovine babesiosis are as follows: (1) initiate MASP cultures for a 3-d period, (2) collect supernatants daily, replenishing cultures with fresh medium, (3) centrifuge, pool, filter (0.45 μm), and lyophilize, and (4) add 3.0 mg saponin (Quil A) per dose and store at 4°C.

Culture-derived vaccines are advantageous in that storage and shipment may be easily facilitated by preparation in lyophilized form. The optimal dosage has been established at 10 ml-eq, and the vaccination regime consists of two subcutaneous inoculations at a 4-week interval.

B. Efficacy of *Babesia* Exoantigens as Immunogens

The protective capacity of such immunogens has to be rigorously tested before mass application. Factors such as the immunogenicity of different strains, duration of immunity, minimal protective dose, the protective efficacy of a combined *B. bovis-B. bigemina* immunogen, vaccine stability, and immunopotentiation with various adjuvants must be investigated. During the last 6 years, most of these aspects have been studied as a result of a cooperative research program between the University of Illinois and the Veterinary Research Institute in Maracay, Venezuela. A critical appraisal of field vaccination trials began 3 years ago on private- and government-owned farms selected in areas where large-scale dairy and beef production is crucial for the Venezuelan livestock industry. Much progress has been made in the analysis of protective immunity conferred by culture-derived *Babesia* exoantigens. Advances in this area have been made possible by close collaboration between university scientists, government institutes in tropical countries, and industrial laboratories.

At the present time, exoantigen-containing vaccines are being used against *B. bovis, B. bigemina,* and *B. canis* parasites. With each vaccine, extensive laboratory and field immunization trials have been conducted. The characteristics of an effective inactivated vaccine for bovine babesiosis have been described elsewhere,[27] and it is encouraging to note that *Babesia* exoantigens meet the following requirements: prevent clinical disease, are effective against different parasite strains, two doses when combined with Quil A saponin adjuvant induce protective immunity for at least 13 months, there is no resultant isoimmunization against host blood groups, and the immunogens are stable for 3 years when stored at 4°C and can be prepared in relatively large quantities. Continued optimization of currently available culture-derived immunogens will best guarantee the successful health management of food-producing animals in the tropics and other areas where babesiosis is endemic.

C. Comparative Evaluation of *Babesia bovis* and *Babesia bigemina* Exoantigens as Immunogens

In the tropics and subtropics, *B. bovis* and *B. bigemina* coexist, resulting in mixed infections.[22] Host susceptibility, management practices, and the virulence of local strains influence the relative importance of each species within a given endemic area.[16] Thus, cattle in Australia are usually vaccinated only against *B. bovis,* which is responsible for more than 90% of babesiosis outbreaks.[16] However, in Israel,[103] India,[104] and South Africa,[105] cattle are also vaccinated against *B. bigemina* because of its pathogenicity.

Table 6
ESSENTIAL INDICATORS[a] OF PROTECTION INDUCED IN VACCINATED CATTLE[b] *(B. BOVIS* AND *B. BIGEMINA* EXOANTIGENS IN 10, 5, AND 1 ml-EQ / DOSE, RESPECTIVELY) TO CHALLENGE EXPOSURE TO HETEROLOGOUS *B. BOVIS* AND *B. BIGEMINA* ORGANISMS

Immunogen	Group	Duration of parasitemia (days)	Max % PCV reduction	Weight gain 3 weeks post challenge (kg)
Babesia bovis	10 ml-eq	1.7 ± 0.5	29.0 ± 2.0	9.5 ± 2.0
	5 ml-eq	3.5 ± 0.5	39.0 ± 6.3[c]	4.25 ± 1.6
	1 ml-eq	3.0 ± 0.4	38.0 ± 4.1[c]	4.0 ± 1.4
	Controls	7.0 ± 0.5	43.4 ± 5.1	Loss
Babesia bigemina	10 ml-eq	2.5 ± 0.5	25.0 ± 2.8	7.0 ± 1.4
	5 ml-eq	2.25 ± 0.8	29.0 ± 4.3	6.25 ± 2.4
	1 ml-eq	2.75 ± 1.1	22.0 ± 6.1	6.5 ± 2.3
	Controls	11.0 ± 1.0	46.0 ± 4.3	Loss

[a] Mean ± standard error.
[b] Crossbred animals.
[c] Only values not significantly different than those recorded in control groups, $p < 0.05$.

Studies in Bolivia,[106] Colombia,[107] Mexico,[96] and Venezuela[108] have shown that *B. bigemina* infections are highly prevalent. Since dairy herds commonly graze in tick-infested pastures throughout these endemic areas, vaccination against *B. bigemina* may be an important consideration for the livestock industry. In two subsequent immunization experiments, Montenegro-James et al.[12] determined the comparative efficacy of culture-derived *B. bovis* and *B. bigemina* exoantigens to heterologous challenge, established the antigenic threshold for induction of protective immunity, and assessed the efficacy of a combined *B. bovis-B. bigemina* immunogen with a dose-response analysis in highly susceptible, purebred cattle.

Studies on the efficacy of *B. bigemina* and *B. bovis* exoantigens when administered as individual immunogens were conducted on six vaccinated groups of four animals each (yearling crossbred cattle). The animals received either 10, 5, or 1 ml-eq of culture-derived *B. bigemina* or *B. bovis* immunogens, respectively. The immunizations were administered subcutaneously on days 0 and 21. Eight nonvaccinated cattle were included as negative controls. At 3 months after the second dose, all animals were challenged with 1×10^9 virulent *B. bovis* or *B. bigemina* organisms of heterologous strains. The immune response was assessed by comparing parasitemia levels, PCV reductions, and weight gains in vaccinated and control groups (Table 6). Significant protection against heterologous challenge was observed in animals vaccinated with 10 ml-eq of either *B. bovis* or *B. bigemina* exoantigens. Most characteristic was the rapid clearance of parasites by vaccinated animals (duration of parasitemia: *B. bovis* vaccinates = 1.7 d, controls = 7.0 d; *B. bigemina* vaccinates = 2.5 d, controls = 11.0 d). However, at 5- and 1-ml-eq doses, *B. bovis* exoantigen was less efficient as determined by the maximum percent PCV reduction (10 ml-eq = 29%, 5 ml-eq 39%, 1 ml-eq = 38%). Nevertheless, the return to normal PCV levels was rapid, suggesting vascular congestion rather than erythrocyte destruction. Vaccinated cattle also eliminated parasites at a faster rate and gained weight after challenge as compared with control animals (Table 6). It is interesting to note that *B. bigemina* exoantigens

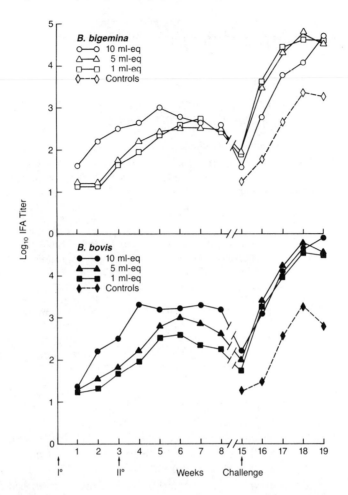

FIGURE 10. Mean log anti-*B. bovis* and anti-*B. bigemina* IFA titers following vaccination with 10, 5, and 1 ml-eq of *B. bovis* and *B. bigemina* immunogens, respectively. (From Montenegro-James, S., Toro, M., Leon, E., Lopez, R., and Ristic, M., *Parasitol. Res.*, 74, 142, 1987. With permission.)

induced considerable protection at all dosage levels, as indicated by the clinical responses. These data showed that *B. bigemina*-vaccinated cattle were more resistant to infection with heterologous strains.

With *B. bovis*, persistence of infection and parasitemic relapses attributed to antigenic variation have been demonstrated. It has been observed that the emerging variants revert to the basic antigenic types after transmission by the vector tick.[94] Relapses have been observed with *B. bovis* over at least 4 years, whereas relapses to *B. bigemina* cease within 2 months.[95] As observed in the Western blot antigenic analysis (see Figure 8, Section III.B), *B. bovis* parasites have a wider antigenic spectrum as compared to *B. bigemina*. Accordingly, immunization studies have indicated that with *B. bovis* the protective response is more strain specific than it is with *B. bigemina* parasites.[12]

The kinetics of IFA responses observed during the vaccination and challenge periods are shown in Figure 10. After vaccination, *B. bigemina* exoantigens elicited relatively lower antibody titers than did *B. bovis* antigens. Maximal antibody levels were achieved with the 10-ml-eq dosage in a dose-dependent pattern (*B. bovis:* 10 ml-eq = 1:2560, 5 ml-eq = 1:1122, 1 ml-eq = 1:501; *B. bigemina:* 10 ml-eq = 1:1071, 5 ml-eq = 1:501, 1 ml-eq

= 1:562). *B. bovis* antibody levels induced by the 5-ml-eq dose were intermediate between those obtained in response to the 10- and 1-ml-eq doses. *B. bigemina* exoantigens administered in 5- and 1-ml-eq dosages elicited similar antibody responses throughout the vaccination period. In summary, these experiments indicate that vaccination with *B. bovis* and *B. bigemina* exoantigens can effectively protect animals against virulent challenge. The important characteristics of host protective immunity include rapid elimination of parasites and the ability to withstand clinical manifestations of infection, as demonstrated by gains in body weight and clinicopathological parameters. The 10-ml-eq dosage appears to be optimal for ensuring complete immunity to heterologous *B. bovis* challenge. Finally, identification of the best cross-protective isolates would be advantageous in the selection of strains for vaccine production.

D. Efficacy of a Combined *Babesia bovis-Babesia bigemina* Immunogen

Simultaneous vaccination against *B. bovis* and *B. bigemina* with a combined immunogen is recommended in the tropics, where susceptible cattle are concurrently affected by these parasites. In order to optimize future babesiosis vaccination regimen utilizing culture-derived exoantigens, an evaluation of simultaneous vaccination with a combined *B. bovis-B. bigemina* immunogen was undertaken.[12] Animals used in this study were highly susceptible, purebred Holstein cattle imported from the U.S. to Venezuela. Highly productive breeds are commonly imported into Venezuela to improve the quality of native livestock. Susceptible, imported cattle are the principal targets for babesiosis and anaplasmosis, as they experience severe morbidity and mortality upon introduction into endemic areas.[109] A total of 23 Holstein cattle, approximately 15 months of age, were used in this study. Cattle were divided into three groups of five animals each and given the combined *B. bovis-B. bogemina* immunogen containing either 10, 5, or 1 ml-eq for each *Babesia* species. The vaccine was administered subcutaneously on days 0 and 21. At 3 months after the second dose, all vaccinated animals were challenged intravenously with 1×10^9 virulent parasites of each *Babesia* species; 8 nonvaccinated animals received the same number of either *B. bovis* (n = 4) or *B. bigemina* (n = 4) organisms. Body temperature, PCV levels, platelet counts, parasitemia, IFA titers, and selected erythrophilic proteins were monitored throughout the experiment. Body weights were determined monthly prior to challenge and biweekly post-inoculation. A comparison of several essential indicators of protection afforded by three dosages of the combined *B. bovis-B. bigemina* immunogen is shown in Table 7. Immunization with the *B. bovis-B. bigemina* exoantigens protected cattle against simultaneous, homologous *B. bovis* and *B. bigemina* challenge exposure, as indicated by the significant differences between the values recorded for vaccinated and each of the control groups of animals (p <0.05). Optimal responses were observed in cattle vaccinated with the 10-ml-eq dosage. Only transient pathophysiological reactions were observed in this group after challenge. Likewise, parasitemias were rapidly cleared. When compared with the efficacy observed with the 5- and 1-ml-eq dosages, significant differences were observed in the duration of *B. bigemina* parasitemia (10 ml-eq = 2.5 d, 5 ml-eq = 5.3 d, 1 ml-eq = 4.0 d), maximum percent PCV reduction (10 ml-eq = 27.7%, 5 and 1 ml-eq = 36.0%), and weight gain 3 weeks postchallenge (10 ml-eq = 4.5 kg, 5 ml-eq = 1.7 kg, 1 ml-eq = 1.4 kg). No marked differences were observed between the efficacy of the 5- and 1-ml-eq dosages. Nonvaccinated control animals showed acute clinical disease following challenge infection with either *B. bovis* or *B. bigemina* organisms. The clinical manifestations were characterized by ataxia, labored respiration, profuse salivation, anorexia, anemia, and hemoglobinuria *(B. bigemina)*. Antibody levels following vaccination with the combined *B. bovis-B. bigemina* immunogen at 10, 5, and 1 ml-eq/dose are depicted in Figure 11. Anti-*B. bovis* IFA titers after vaccination were similar for all dosages (10 ml-eq = 1:640, 5 ml-eq = 1:480, 1 ml-eq = 1:416), although antibody levels were lower than those elicited by the single *B. bovis* immunogen

Table 7
ESSENTIAL INDICATORS[a] OF PROTECTION INDUCED IN
VACCINATED CATTLE[b] (COMBINED *BABESIA* EXOANTIGENS
IN 10, 5, AND 1 ml-EQ/DOSE) TO SIMULTANEOUS CHALLENGE
EXPOSURE TO HOMOLOGOUS *B. BOVIS* AND *B. BIGEMINA*
ORGANISMS

Immunogen	Vaccinated[c]	Duration of parasitemia (days)	Max % PCV reduction	Weight gain 3 weeks post challenge (kg)
	10 ml-eq	0.5 ± 0.5 (*B. bovis*) 2.5 ± 0.5 (*B. bigemina*)	27.7 ± 3.5	4.5 ± 0.6
Combined *B. bovis* and *B. bigemina* ex-oantigens	5 ml-eq	0.8 ± 0.3 (*B. bovis*) 5.3 ± 0.9[d] (*B. bigemina*)	36.0 ± 2.0[d]	1.7 ± 0.6[d]
	1 ml-eq	1.4 ± 0.4 (*B. bovis*) 4.0 ± 0.2[d] (*B. bigemina*)	36.0 ± 3.0[d]	1.4 ± 0.5[d]
Controls B. bovis		7.0 ± 0.6	53.0 ± 5.0	Loss
B. bigemina		10.0 ± 1.3	50.0 ± 4.0	Loss

[a] Mean ± standard error.
[b] Purebred Holstein animals.
[c] The vaccinated group showed significant differences from the values recorded in each of the control groups.
[d] Significantly different from standard 10-ml-eq dose, $p < 0.05$.

(10 ml-eq = 1:2560, 5 ml-eq = 1:1122, 1 ml-eq = 1:501). The anti-*B. bigemina* responses to the combined *Babesia* immunogen also did not show marked differences in titers between the various dosages (10 ml-eq = 1:2432, 5 ml-eq = 1:1408, 1 ml-eq = 1:1280). However, all responses were greater than those elicited by the single *B. bigemina* immunogen (10 ml-eq = 1:1071, 5 ml-eq = 1:501, 1 ml-eq = 1:562).

Higher and more persistent titers to *B. bigemina* antigen with *B. bovis* immune serum have been reported,[95] and our data support these findings. Additionally, quantitation of cross-reacting antigens using the two-site EIA has further corroborated the avidity of *B. bovis* antibodies for *B. bigemina* antigens (Figure 6). Serologic cross-reactivity between *Babesia* parasites has not been conclusively correlated with significant protection against challenge exposure. Until the common protective antigen(s) are identified, monospecific vaccines against *B. bovis* or *B. bigemina* cannot be expected to provide adequate protection from disease caused by the heterologous species. Therefore, a combined *B. bovis-B. bigemina* immunogen appears to be the most rational.

Nonspecific mechanisms of host immunity, such as innate resistance, affect overall vaccine performance and need to be analyzed. In cattle, it is believed that a long evolutionary association between tropical breeds of zebu or buffalo, *B. bovis* and *B. bigemina* parasites, and the tick vector *Boophilus microplus* has resulted in relatively high resistance to pathogenic effects in these hosts.[110,111] Disease occurs when susceptible cattle, not well adapted to a *Babesia* species, are subjected to fluctuating levels of infection by tick vectors.[112] Animal hosts also show variable susceptibility to the vector, with *Bos indicus* cattle demonstrating

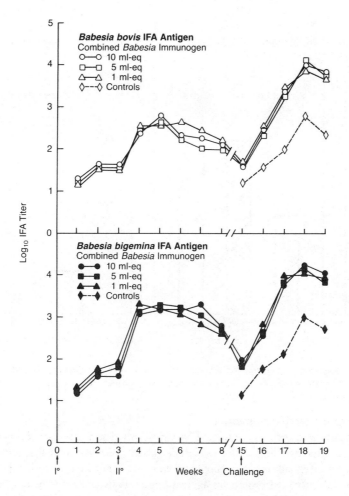

FIGURE 11. Mean log anti-*B. bovis* and anti-*B. bigemina* IFA titers following vaccination with a combined *Babesia* immunogen at 10-, 5-, and 1-ml-eq doses.

greater resistance to infestation with *Boophilus microplus* than European breeds *(Bos taurus)*.[112] Crossbreeding with zebu animals helps to conserve this characteristic, although occasionally *Babesia* may cause serious disease in previously unexposed hybrid cattle.[113]

In our immunization studies, comparable groups of crossbred *(Bos indicus × Bos taurus)* and purebred Holstein *(Bos taurus)* cattle have been utilized during trials with the culture-derived *Babesia* immunogen.[11,12] The most consistent indicator of protection against clinical disease is the gain in body weight after challenge exposure. Figure 12 depicts the influence of cattle breeds on the overall efficiency of culture-derived *Babesia* exoantigens. Vaccinated crossbred cattle showed greater weight gains after virulent heterologous challenge than did vaccinated purebred cattle after homologous challenge. The severity of *Babesia* infections was more accentuated in unvaccinated Holstein animals than in crossbred controls, as demonstrated by the greater weight loss in the former (Holstein unvaccinated cattle: weight loss of *B. bovis* = 15.0 kg, of *B. bigemina* = 9.0 kg; crossbred unvaccinated cattle: weight loss of *B. bovis* = 3.5 kg, of *B. bigemina* = 3.3 kg).

Culture-derived *Babesia* exoantigens have proved useful and effective in protecting highly susceptible cattle against babesiosis. Vaccination of crossbred cattle which exhibit a degree of innate resistance is still beneficial and recommended because active immunization will further enhance the protective immune response against this economically important disease.

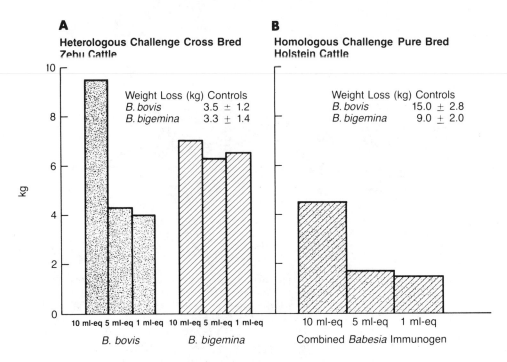

FIGURE 12.   Average weight gain at 3 weeks postchallenge in crossbred (A) and purebred (B) cattle vaccinated with single (A) and combined (B) *B. bovis* and *B. bigemina* immunogens, respectively.

E. Duration of Protective Immunity

One important characteristic of an ideal babesiosis vaccine is the capacity to induce long-term protection after two inoculations. Kuttler et al.[10] demonstrated that two doses of *B. bovis* exoantigen (Mexican isolate), supplemented with 1 mg saponin adjuvant and given at a 3-week interval, provided protection against homologous challenge 6 months thereafter.

Immunogenicity is highly dependent on the mode of vaccination. Antigen composition, immunogen preparation, antigen dosage, and immunization schedule are all factors that influence the immunogenicity and the type of immune response.[114] Closely spaced inoculations are often associated with a rapid decline in antibody levels and a poor booster effect. Vaccination at 2-week intervals with *B. bovis* exoantigens gave suboptimal results.[7,39] Intervals of 3 to 4 weeks have produced consistently better results based on the persistence of high antibody levels after two vaccinal doses.[9,11,12]

In order to optimize the immunization procedure for a *B. bovis-B. bigemina* exoantigen-containing immunogen, two 10-ml-eq doses supplemented with 3 mg Quil A saponin as adjuvant and given at a 4-week interval are recommended.[41] In the same immunization trial Montenegro-James et al.[41] determined the duration of protective immunity conferred by a combined *Babesia* exoantigen after sequential heterologous challenge at 13 and 18 months after vaccination. The antibody kinetics observed during an 18-month period are shown in Figure 13. A 30-d interval between immunizations with a combined immunogen supplemented with 3 mg of saponin (Quil A) produced better results in terms of magnitude of antibody levels after two vaccinal doses. Peak antibody titers (*B. bigemina* = 1:4100; *B. bovis* = 1:1536) occurred 2 weeks after the second vaccination, then slowly declined to baseline levels 13 months thereafter. The combined immunogen was capable of inducing protection against challenge with virulent heterologous *B. bovis* and *B. bigemina* parasites 13 months postvaccination. The mean percent PCV reduction, maximum percent, and duration of parasitemia in vaccinated cattle were significantly different from the values of the

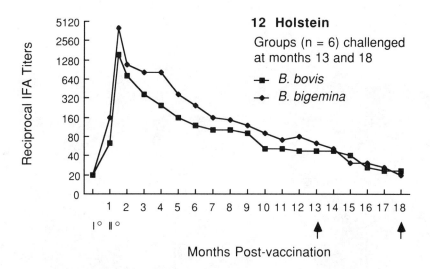

FIGURE 13. Antibody kinetics as expressed by the mean homologous IFA titers of Holstein cattle vaccinated with the combined *Babesia* immunogen and simultaneously needle challenged with virulent *B. bovis* and *B. bigemina* organisms at months 13 or 18 of the experiments.

Table 8
DURATION OF IMMUNITY WITH A *B. BOVIS-B. BIGEMINA* IMMUNOGEN: SEQUENTIAL HETEROLOGOUS CHALLENGE AT 13 AND 18 MONTHS POSTVACCINATION

Months after vaccination	Experimental group	Essential Parameters[a]			
		Max % PCV reduction	Max % parasitemia	Duration of parasitemia (days)	Treatment
13	Vaccinated (n = 6)	35.5 ± 6.4[b]	0.38 ± 0.1[b]	5.4 ± 1.0[b]	0/6
	Control (n = 4)	55.3 ± 4.5	0.73 ± 0.4	8.6 ± 0.5	2/4 (1 death)
18	Vaccinated (n = 6)	49.6 ± 4.2	0.4 ± 0.2[c]	3.6 ± 0.5	4/6
	Control (n = 4)	57.0 ± 4.2	1.0 ± 0.3	5.0 ± 0.4	3/4

[a] Group mean ± standard error.
[b] Significantly different from control values, $p < 0.05$.
[c] Significantly different from control values, $p < 0.01$.

control group (Table 8). All nonvaccinated animals suffered acute clinical disease with central nervous system signs, one animal dying 8 d after challenge. Post-mortem examination revealed brain capillaries clogged with infected erythrocytes (Figure 14). The protective immune status persisted for 18 months. This experiment demonstrated that a combined immunogen derived from the two *Babesia* spp. is capable of fulfilling the important criteria in long-term protection.

Experimental, nonviable immunogens for *B. bovis* and *B. bigemina* satisfy most of the requirements for an effective *Babesia* vaccine. In the near future, the cattle industry should benefit from the availability of such vaccines.

The use of this type of product offers distinct advantages over currently used live vaccines. Besides the demonstrated protective efficacy, exoantigens are antigenically stable for at least 3 years when stored lyophilized at 4°C. Furthermore, the safety of the soluble *Babesia* immunogens has been verified in that (1) they are free of erythrocytic stromal contaminants,[39]

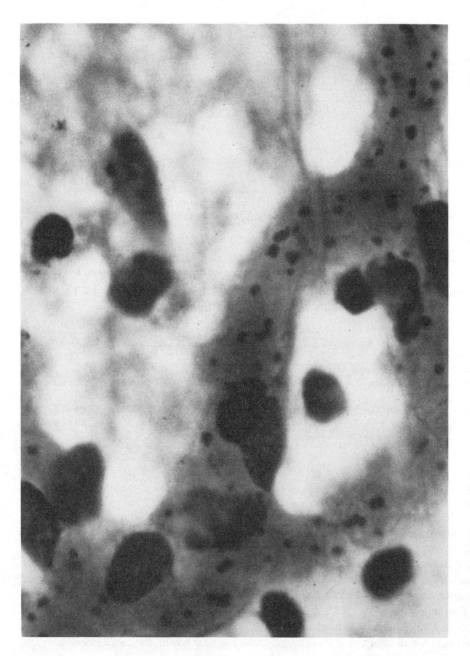

FIGURE 14.   Post-mortem findings in *B. bovis*-infected animal. Sequestration of parasitized erythrocytes within cerebral blood capillary. (Magnification × 3000.)

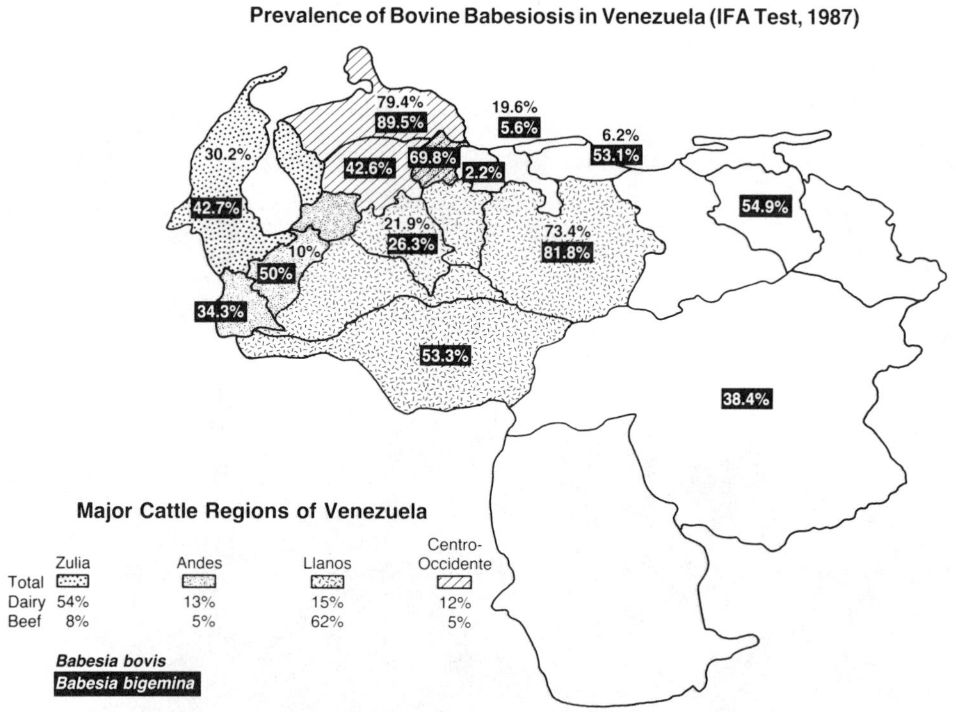

Prevalence of Bovine Babesiosis in Venezuela (IFA Test, 1987)

FIGURE 15. Serological prevalence of *B. bovis* and *B. bigemina* in major cattle regions of Venezuela (1987).

and (2) saponin combines the potency of stimulating humoral[8-12] and cell-mediated immune responses[39] and the unconditional acceptability for use in cattle.[115]

V. FIELD EVALUATION OF CULTURE-DERIVED BABESIAL IMMUNOGENS

A. Vaccination Trials with *Babesia canis* Exoantigen-Containing Immunogens in France

B. canis immunogens derived from *in vitro* cultures are being used on a large scale in France.[14] Approximately 90% protection of naive dogs has been observed without any significant side effects. However, vaccination failures have been recorded in hyperendemic areas, particularly in carrier dogs or those with a long history of the disease. A vaccination regime of two doses, followed by an annual booster has afforded the best protection.[116]

B. Vaccination Trials with a Combined *Babesia bovis-Babesia bigemina* Exoantigen-Containing Immunogen in Venezuela

Vaccination trials conducted under field conditions provide the final evaluation of the efficacy of culture-derived *B. bovis-B. bigemina* immunogens. Seroepidemiological surveys conducted in Venezuela[41,109,117] have indicated that bovine babesiosis and anaplasmosis are endemic hemoparasitic diseases with a prevalence rate of approximately 50% (Figure 15). Venezuela represents an ideal epidemiological setting for immunization studies because (1) highly susceptible, purebred cattle are regularly imported from the U.S. and Canada, and (2) native crossbred animals affect varying degrees of enzootic stability for these infections.

Since the initiation of field vaccination experiments in July 1984, five such trials have been completed. As of August 1987, 16 trials were being conducted in 6 states of Venezuela (Figure 16). Dairy cattle were selected for evaluation of the exoantigen vaccine because most importation involved this type of animal. Furthermore, diversity in management practices, geographic characteristics, and farm size offered the opportunity for analysis within

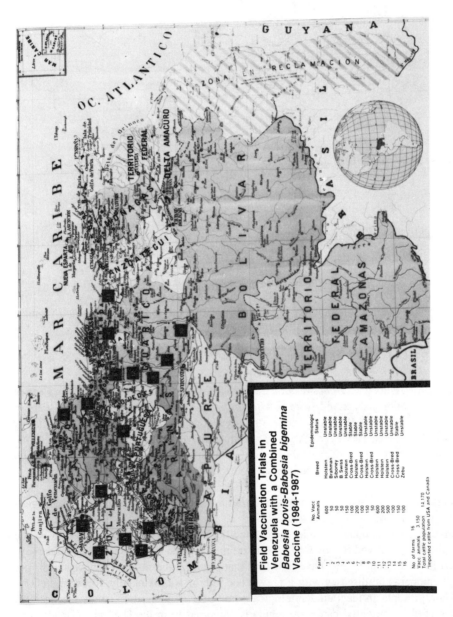

Field Vaccination Trials in Venezuela with a Combined *Babesia bovis-Babesia bigemina* Vaccine (1984-1987)

Farm	No Vacc Animals	Breed	Epidemiologic Status
1	600	Holstein	Unstable
2	50	Brahman	Unstable
3	50	Siboney	Unstable
4	50	B Swiss	Unstable
5	150	Holstein	Stable
6	100	Cross-Bred	Unstable
7	200	Holstein	Stable
8	100	Cross-Bred	Stable
9	150	Holstein	Unstable
10	50	Cross-Bred	Unstable
11	600	Holstein	Unstable
12	200	Holstein	Unstable
13	500	Holstein	Unstable
14	100	Cross-Bred	Unstable
15	150	Stable	Stable
16	100	Zebu	Unstable

No. of farms   16
Vacc. animals   3 150
Total cattle population   14 170
ᵃImported cattle from USA and Canada

FIGURE 16.   Geographic distribution of field vaccination trials (years 1984 to 1987) in Venezuela.

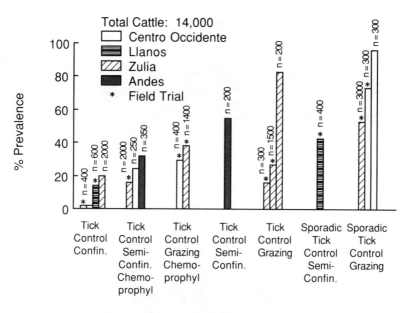

1: IFA test to *B. bovis* and *B. bigemina* performed in 10% of cattle

FIGURE 17. Serological prevalence of bovine babesiosis in 17 dairy farms (various locations throughout Venezuela) managed under varying conditions for parasite and tick control.

a wide range of epidemiologic conditions. During the last 2 years, 17 dairy farms were used for seroepidemiological studies. Prevalence rates of bovine babesiosis ranged from 2.2 to 14.0% on farms with tick control and various degrees of cattle confinement to 96% when sporadic dipping and grazing were practiced (Figure 17). For field evaluation of the *Babesia* exoantigen vaccine, private- and government-owned sites were selected in areas where large-scale dairy and beef production is crucial for the livestock industry. These ranches encompass a total cattle population of 14,000 head, of which 3000 were vaccinated with the combined *B. bovis-B. bigemina* immunogen in a regimen of 2 subcutaneous inoculations at a 4-week interval. With the collaboration of local veterinarians, serological, parasitological, and hematological data are being collected on a monthly basis from 10% of the animals. Data are also collected from a similar number of unvaccinated control cattle. Herd size, breed, and age at the time of vaccination are indicated in Figure 16. Most animals also received other standard vaccinations for dairy cattle in Venezuela, e.g., immunizations against foot-and-mouth disease, brucellosis, and a bacterin containing a combination of *Clostridium* spp. and *Pasteurella* spp. In 75% of the trials, tick exposure occurred between 0 and 4 months postvaccination. The most frequent concurrent infection was *Anaplasma marginale* (69% of the farms). Sporadic outbreaks of rabies and trypanosomiasis (*Trypanosoma vivax*) were also recorded.

Results obtained in the first five trials during a 1-year monitoring period following vaccination and natural tick exposure are presented in Table 9. Preliminary indications are encouraging and suggest that control of babesiosis can be accomplished. Early data show good seroconversion among vaccinated cattle and an absence of mortality due to babesiosis (confirmed upon necropsy). The greater weight gains in vaccinated animals indicate that the use of the culture-derived *Babesia* vaccine can effectively control the clinical manifestation of infection. With these promising results, the Venezuelan Institute for Veterinary Research, under the auspices of the Ministry of Agriculture, will begin production and distribution of the vaccine for use with government-owned cattle. The first priority is for immunization of valuable livestock imported from the U.S. and Canada.

Table 9
DATA[a] FROM FIELD VACCINATION TRIALS IN VENEZUELA

Farm	Breed (age)	Group	Seroconversion		% PCV change	gain(KG)	% mortality
			B. bovis	*B. bigemina*			
1[b]	Holst.	Vacc.	1:1810	1:2152	−22	ND[c]	0
	(2—3 years)	Cont.	1:355	1:597	−31	ND	10
2	Brahm.	Vacc.	1:320	1:640	−5	12.2	0
	(5—6 months)	Cont.	1:80	1:70	−23	4.5	0
3[b]	Sibon.	Vacc.	1.2560	1:5187	−18	20.0	0
	(1 year)	Cont.	1:320	1:508	−31	12.3	10
4[b]	Br. Sw.	Vacc.	1:178	1:94	−11	ND	0
	(1 year)	Cont.	1:160	1:65	−18	ND	40
5[d]	Holst.	Vacc.	1:80	1:180	ND	12.5	0
	(4—8 months)	Cont.	1:40	1:62	ND	6.9	0

[a] 3—5 months after field exposure.
[b] Vaccinated against anaplasmosis.
[c] ND—not done.
[d] Outbreak of anaplasmosis at the beginning of trial.

Until optimal molecular vaccines are developed, the culture-derived soluble *Babesia* immunogens offer the best combination of potency, efficacy, and safety in fulfilling the critical need for effective control measures against babesiosis.

VI. SUMMARY

During the last decade, significant progress has been made in the development and improvement of *in vitro* systems for propagation of *Babesia* parasites. These advances have facilitated numerous investigations which have contributed to our knowledge on the biology of the parasite and, more importantly, on the characterization of exoantigens and their immunoprophylactic potential. Antigenic analyses of culture-derived exoantigens have revealed 9 protein bands in the range of 164 to 36 kDa for *B. bovis,* 14 polypeptides ranging from 92 to 20 kDa for *B. bigemina,* and major protein bands of 55 to 35 kDa for *B. canis.* A greater degree of heterogeneity is observed between *B. bovis* isolates, whereas *B. bigemina* isolates appear to be antigenically stable. A single protein band (89 kDa) has been identified as species-specific for *B. bigemina* exoantigens, whereas three polypeptides (125, 109, and 75 kDa) appear to be unique to *B. bovis* exoantigens. Successful adaptation of *B. equi* to propagation in erythrocyte cultures has facilitated the partial characterization and investigation of these exoantigens and their immunogenic potential.

The development of a soluble, exoantigen-containing immunogen, free of erythrocytic antigens, has been proposed as a practical and realistic solution for the immunoprophylactic control of babesiosis. The protective capacity of such immunogens has been exhaustively tested for factors such as the immunogenicity of different strains, minimal protective dose, and protective efficacy of a combined *B. bovis-B. bigemina* immunogen. During the last 6 years, most of these aspects have been studied as a result of a cooperative research program between the University of Illinois and the Veterinary Research Institute of Venezuela. The *Babesia* exoantigen vaccine prevents clinical disease, is effective against different parasite strains, induces protective immunity for at least 13 months, does not immunize against host blood groups, is antigenically stable for 3 years at 4°C, and is supplied in large quantities. Continued optimization of currently available culture-derived immunogens will best guarantee the successful production of food-producing animals in the tropics. Current research efforts are directed toward isolation of protective antigens and the production of synthetic vaccines. Until optimal molecular vaccines are developed, a combined inactivated *B. bovis*-*B. bigemina* immunogen may offer the best combination of potency, efficacy, and safety in fulfilling the critical need for effective control measures against bovine babesiosis.

ACKNOWLEDGMENTS

These studies were conducted as part of a cooperative research program on hemotropic diseases between the Venezuelan Institute for Veterinary Research and the University of Illinois and was supported in part by funds from the Fondo Nacional de Investigaciones Agropecuarias, Venezuela. The authors are grateful to Dr. M. A. James for his expert and invaluable critique of the manuscript.

REFERENCES

1. **Babes, V.,** Sur l'hemoglobinuria bacterienne du boeuf, *C.R. Acad. Sci.,* 107, 692, 1888.
2. **Smith, T. and Kilbourne, F. L.,** Investigation into the Nature Causation and Prevention of Texas or Southern Cattle Fever, Bull. No. 1, Bureau of Animal Industries, U.S. Department of Agriculture, Washington, D.C., 1893.
3. **Levy, M. G. and Ristic, M.,** *Babesia bovis:* continuous cultivation in a microaerophilous stationary phase culture, *Science,* 207, 1218, 1980.
4. **Aikawa, M., Rabbege, J., Ristic, M., and Miller, L. H.,** Structural alterations of the membranes of erythrocytes infected with *Babesia bovis, Am. J. Trop. Med. Hyg.,* 35, 45, 1985.
5. **James, M. A., Levy, M. G., and Ristic, M.,** Isolation and partial characterization of culture-derived soluble *Babesia bovis* antigens, *Infect. Immun.,* 31, 358, 1981.
6. **Levy, M. G., Kakoma, I., Clabaugh, G. W., and Ristic, M.,** Complement does not facilitate invasion of erythrocytes by *Babesia bovis, Am. J. Trop. Med. Parasitol.,* 80, 377, 1986.
7. **Smith, R. D., Carpenter, J., Cabrera, A., Gravely, S. D., Erp, E. E., Osorno, M., and Ristic, M.,** Bovine babesiosis: vaccination against tick-borne challenge exposure with culture-derived *Babesia bovis* immunogens, *Am. J. Vet. Res.,* 40, 1078, 1979.
8. **Smith, R. D., James, M. A., Ristic, M., Aikawa, M., and Vega, C. A.,** Bovine babesiosis: protection of cattle with culture-derived soluble *Babesia bovis* antigen, *Science,* 212, 335, 1981.
9. **Kuttler, K. L., Levy, M. G., James, M. A., and Ristic, M.,** Efficacy of a nonviable culture-derived *Babesia bovis* vaccine, *Am. J. Vet. Res.,* 43, 281, 1982.
10. **Kuttler, K. L., Levy, M. G., and Ristic, M.,** Cell culture-derived *Babesia bovis* vaccine. Sequential challenge exposure of protective immunity during a 6-month post-vaccination period, *Am. J. Vet. Res.,* 44, 1456, 1983.
11. **Montenegro-James, S., Toro, M., Leon, E., Lopez, R., and Ristic, M.,** Heterologous strain immunity in bovine babesiosis using a culture-derived soluble *Babesia bovis* immunogen, *Vet. Parasitol.,* 18, 321, 1985.
12. **Montenegro-James, S., Toro, M., Leon, E., Lopez, R., and Ristic, M.,** Bovine babesiosis: induction of protective immunity with culture-derived *Babesia bovis* and *Babesia bigemina* immunogens, *Parasitol. Res.,* 74, 142, 1987.
13. **Molinar, E., James, M. A., Kakoma, I., Holland, C. J., and Ristic, M.,** Antigenic and immunogenic status on cell culture-derived *Babesia canis, Vet. Parasitol.,* 10, 29, 1982.
14. **Moreau, Y. and Laurent, N.,** Antibabesial vaccination using antigens from cell cultures: industrial requirements, in *Malaria and Babesiosis,* Ristic, M., Ambroise-Thomas, P., and Kreier, J. P., Eds., Martinus Nijhoff, Dordrecht, Netherlands, 1984, 129.
15. **Ristic, M. and Levy, M. G.,** A new era of research toward solution of bovine babesiosis, in *Babesiosis,* Ristic, M. and Kreier, J. P., Eds., Academic Press, New York, 1981, 509.
16. **Callow, L. L.,** Vaccination against bovine babesiosis, *Adv. Exp. Med. Biol.,* 93, 121, 1977.
17. **Sibinovic, K., Ristic, M., Sibinovic, S., and Phillips, T. N.,** Equine babesiosis: isolation and serologic characterization of a blood serum antigen from acutely infected horses, *Am. J. Vet. Res.,* 26, 147, 1965.
18. **Sibinovic, K., MacLeod, R., Ristic, M., Sibinovic, S., and Cox, H. W.,** A study of some of the physical, chemical, and serologic properties of antigens from sera of dogs and rats with acute babesiosis, *J. Parasitol.,* 53, 919, 1967.

19. **Ristic, M., Lykins, J. D., Smith, A. R., Huxsoll, D. L., and Groves, M. G.,** *Babesia canis* and *Babesia gibsoni:* soluble antigens isolated from blood of dogs, *Exp. Parasitol.,* 30, 385, 1971.

20. **Mahoney, D. F.,** Circulating antigens in cattle infected with *Babesia bigemina* or *Babesia argentina, Nature (London),* 211, 422, 1966.

21. **Mahoney, D. F.,** *Babesia argentina:* immunogenicity of plasma from infected animals, *Exp. Parasitol.,* 32, 71, 1972.

22. **McCosker, P. J.,** The global importance of babesiosis, in *Babesiosis,* Ristic, M. and Kreier, J. P., Eds., Academic Press, New York, 1981, 1.

23. *Animal Health Yearbook,* FAO-WHO-OIE, Rome, 1981.

24. **Lombardo, R. A.,** Socioeconomic importance of the tick problem in the Americas, *Pan Am Health Organ. Sci. Publ.,* 316, 79, 1976.

25. **Anders, R. F., Howard, R. T., and Mitchell, G. F.,** Parasite antigens and method of analysis, in *Immunology of Parasitic Infections,* 2nd ed., Cohen, S. and Warren, K. S., Eds., Blackwell Scientific, Oxford, 1982, 28.

26. **Ristic, M., Smith, R. D., and Kakoma, I.,** Characterization of *Babesia* antigens derived from cell cultures and ticks, in *Babesiosis,* Ristic, M. and Kreier, J. P., Eds., Academic Press, New York, 1981, 337.

27. **Mahoney, D. F., Wright, I. G., and Goodger, B. V.,** Immunization against babesiosis: current studies and future outlook, *Prev. Vet. Med.,* 2, 401, 1984.

28. **Mahoney, D. F. and Goodger, B. V.,** The isolation of *Babesia* parasites and their products from the blood, in *Babesiosis,* Ristic, M. and Kreier, J. P., Eds., Academic Press, New York, 1981, 323.

29. **Mahoney, D. F., Wright, I. G., and Goodger, B. V.,** Bovine babesiosis: the immunization of cattle with fractions of erythrocytes infected with *Babesia bovis* (Syn. *B. argentina*), *Vet. Immunol. Immunopathol.,* 2, 145, 1981.

30. **Mahoney, G. F. and Wright, I. A.,** *Babesia argentina:* immunization of cattle with a killed antigen against infection with a heterologous strain, *Vet. Parasitol.,* 2, 273, 1976.

31. **Goodger, B. V., Wright, I. G., and Waltisbuhl, D. J.,** The lysate from bovine erythrocyte infected with *Babesia bovis, Z. Parasitenkd.,* 69, 473, 1983.

32. **Goodger, B. V., Wright, I. G., Waltisbuhl, D. J., and Mirre, G. B.,** *Babesia bovis:* successful vaccination against homologous challenge in splectomized calves using a fraction of haemagglutinating antigen, *Int. J. Parasitol.,* 15, 175, 1985.

33. **Wright, I. G., Mirre, G. B., Rode-Bramanis, K., Chamberlain, M., Goodger, B. V., and Waltisbuhl, D. J.,** Protective vaccination against virulent *Babesia bovis* with a low molecular weight antigen, *Infect. Immun.,* 48, 109, 1985.

34. **Goodger, B. V., Wright, I. G., Mahoney, D. F., and McKenna, R. V.,** *Babesia bovis:* studies on the composition and location of an antigen associated with infected erythrocytes, *Int. J. Parasitol.,* 10, 33, 1980.

35. **Wright, I. G., White, M., Tracey-Patte, P. D., Donaldson, R. A., Goodger, B. V., Waltisbuhl, D. J., and Mahoney, D. F.,** *Babesia bovis:* isolation of a protective antigen by using monoclonal antibodies, *Infect. Immun.,* 41, 244, 1983.

36. **Wright, I. G., Goodger, B. V., Rode-Bramanis, K., Mattick, J. S., Mahoney, D. F., and Waltisbuhl, D. J.,** The characterization of an esterase derived from *Babesia bovis* and its use as a vaccine, *Z. Parasitenkd.,* 69, 703, 1983.

37. **Goodger, B. V.,** Babesia bovis (= *argentina*): changes in erythrophilic and associated proteins during acute infection of splenectomized and intact calves, *Z. Parasitenkd.,* 55, 1, 1978.

38. **Goodger, B. V., Commins, M. A., Wright, I. G., Waltisbuhl, D. J., and Mirre, G. B.,** Successful homologous vaccination against *Babesia bovis* using a heparin-binding fraction of infected erythrocytes, *Int. J. Parasitol.,* 17, 935, 1987.

39. **Timms, P., Dalgliesh, R. J., Barry, D. N., Dimmock, C. K., Rodwell, B. J.,** *Babesia bovis:* comparison of culture-derived vaccine, non-living antigen and conventional vaccine in the protection of cattle against heterologous challenge, *Aust. Vet. J.,* 60, 75, 1983.

40. **Osorno, M.,** Immunologic Characterization of Soluble Antigen Derived from Cell Culture of *Babesia bovis,* Ph.D. thesis, University of Illinois, Urbana, 1980.

41. **Montenegro-James, S., Toro Benitez, M., Leon, A., Guillen, A. T., Lopez, R., and Ristic, M.,** Manuscript in preparation, 1988.

42. **Trager, W. and Jensen, B.,** Human malaria parasites in continuous culture, *Science,* 193, 673, 1976.

43. **Timms, P.,** Short term cultivation of *Babesia* species, *Res. Vet. Sci.,* 29, 102, 1980.

44. **Erp, E. E., Gravely, S. M., Smith, R. D., Ristic, M., Osorno, B. M., and Carson, C. A.,** Growth of *Babesia bovis* in bovine erythrocyte cultures, *Am. J. Trop. Med. Hyg.,* 27, 1061, 1978.

45. **Erp, E. E., Smith, R. D., Ristic, M., and Osorno, B. M.,** Continuous *in vitro* cultivation of *Babesia bovis, Am. J. Vet. Res.,* 41, 1141, 1980.

46. **Palmer, D. A., Buening, G. M., Carson, C. A.,** Cryopreservation of *Babesia bovis* for *in vitro* cultivation, *Parasitology,* 62, 221, 1982.

47. **Rodriguez, S. D., Buening, G. M., Green, T. J., and Carson, C. A.,** Cloning of *Babesia bovis* by *in vitro* cultivation, *Infect. Immun.,* 42, 15, 1983.

48. **Irvin, A. D. and Young, E. R.,** Introduction and multiplication of bovine *Babesia* in human cells, *Res. Vet. Sci.,* 27, 241, 1979.

49. **Vega, C. A., Buening, G. M., Green, T. J., and Carson, C. A.,** *In vitro* cultivation of *Babesia bigemina, Am. J. Vet. Res.,* 46, 416, 1985.

50. **Rehbein, G., Zweygarth, E., Voigt, W. P., and Schein, E.,** Establishment of *Babesia equi*-infected lymphoblastoid cell lines, *Z. Parasitenkd.,* 67, 125, 1982.

51. **Montenegro-James, S., Toro, M., Leon, E., Baek, B. K., and Ristic, M.,** Manuscript in preparation, 1988.

52. **Vajrynene, R. and Tuomi, J.,** Continuous *in vitro* cultivation of *Babesia divergens, Acta Vet. Scand.,* 23, 471, 1982.

53. **Bautista, C. R. and Kreier, J. P.,** Effect of immune serum on the growth of *Babesia microti* in hamster erythrocytes in short-term culture, *Infect. Immun.,* 25, 470, 1979.

54. **Jack, R. M. and Ward, P.,** *Babesia rodhaini* interactions with complement: relationship to parasitic entry into red cells, *J. Immunol.,* 124, 1566, 1980.

55. **Palmer, K., Hiu, G. S. N., Siddiqui, W. A., and Palmer, E. L.,** A large scale *in vitro* production system for *Plasmodium falciparum, J. Parasitol.,* 68, 1180, 1982.

56. **Leon, E., Toro, M., Lopez, R., and Montenegro-James, S.,** *In vivo* and *in vitro* studies of 3 *Babesia bovis* isolates from Venezuela, IV Congr. Venezolano de Microbiologia, Maracay, Venezuela, November 4 to 8, 1986, 66.

57. **Montenegro-James, S.,** Unpublished observations.

58. **Ristic, M.,** Research on babesiosis vaccines, in *Malaria and Babesiosis,* Ristic, M., Ambroise-Thomas, P., and Kreier, J. P., Eds., Martinus Nijhoff, Dordrecht, Netherlands, 1984, 103.

59. **Yunker, C. E., Kuttler, K. L., and Johnson, L. W.,** Attenuation of *Babesia bovis* by *in vitro* cultivation, *Vet. Parasitol.,* 24, 7, 1987.

60. **Stewart, N. P. and Timms, P.,** Unpublished observations, 1984; as cited by **DeVos, A. J., Dalgliesh, R. J., and Callow, W.,** *Babesia,* in *Immune Responses in Parasitic Infections,* Vol. 3, Soulsby, J. L., Ed., CRC Press, Boca Raton, FL., 1987, 200.

61. **Levy, M. G., Clabaugh, G., and Ristic, M.,** Age resistance in bovine babesiosis: role of blood factors in resistance to *Babesia bovis, Infect. Immun.,* 37, 1127, 1982.

62. **Ristic, M.,** Comments on prospects for antibabesial vaccine, in *Babesiosis,* Ristic, M. and Kreier, J. P., Eds., Academic Press, New York, 1981, 563.

63. **Montenegro-James, S.,** Unpublished observations, 1987.

64. **Fairlamb, A. H., Warhurst, D. C., and Peters, W.,** An improved technique for the cultivation of *Plasmodium falciparum in vitro* without daily medium change, *Ann. Trop. Med. Parasitol.,* 79, 379, 1985.

65. **Schein, E., Rehbein, G., Voigt, W. P., and Zweygarth, E.,** *Babesia equi* (Laveran, 1901). I. Development in horses and in lymphocyte culture, *Tropenmed. Parasitol.,* 32, 223, 1981.

66. **Holbrook, A. A., Johnson, A. J., and Madden, P. A.,** Equine piroplasmosis: intraerythrocytic development of *Babesia caballi* (Nutall) and *Babesia equi* (Laveran), *Am. J. Vet. Res.,* 24, 297, 1968.

67. **Selim, M. K. and Abdel-Gawad, A. M. H.,** Observations on *Babesia equi* (Laveran 1901) in latently infected donkeys after splenectomy, *Egypt. J. Vet. Sci.,* 19, 47, 1982.

68. **Montealegre, F., Montenegro-James, S., Kakoma, I., and Ristic, M.,** Detection of culture-derived *Babesia bovis* exoantigens using a two-site immunoassay, *J. Clin. Immunol.,* 25, 1648, 1987.

69. **Burkot, T. R., Williams, J. L., and Schneider, I.,** Identification of *Plasmodium falciparum*-infected mosquitoes by a double antibody enzyme-linked immunosorbent assay, *Am. J. Trop. Med. Hyg.,* 33, 783, 1984.

70. **Steinbuch, M. and Audran, R.,** The isolation of IgG from mammalian sera with the aid of caprylic acid, *Arch. Biochem. Biophys.,* 134, 279, 1969.

71. **Harboe, N. M. G. and Ingild, A.,** Immunization, isolation of immunoglobulins and antibody titre determination, *Scand. J. Immunol.,* 17(Suppl. 10), 345, 1983.

72. **Wilson, M. B. and Nakane, K. P.,** Recent developments in the periodate method of conjugating horseradish peroxidase (HRPO) to antibodies, in *Immunofluorescence and Related Staining Techniques,* Knapp, W., Holubar, K., and Wicks, G., Eds., Elsevier/North-Holland, Amsterdam, 1978, 215.

73. **Avrameas, S.,** Coupling of enzymes to proteins with glutaraldehyde. Use of the conjugate for the detection of antigens and antibodies, *Immunochemistry,* 6, 43, 1969.

74. **Montealegre, F.,** *Babesia bovis:* Detection and Characterization of Culture-Derived Exoantigens, Ph.D. thesis, University of Illinois, Urbana, 1986.

75. **Wanduragala, L.,** Manuscript in preparation, 1988.

76. **Taylor, D. W., Evans, C. B., Hennkessy, G. W., and Alay, S. B.,** Use of two-sited monoclonal antibody assay to detect a heat-stable malarial antigen in the sera of mice infected with *Plasmodium yoelii, Infect. Immun.,* 51, 884, 1986.

77. **Burkot, T. R., Lavala, F., Gwadz, R. W., Collins, F. H., Nussenzweig, R. S., and Roberto, D. R.,** Identification of malaria infected mosquitoes by a two-site enzyme linked immunosorbent assay, *Am. J. Trop. Med. Hyg.,* 33, 227, 1984.

78. **Eaton, M. D.,** The soluble malarial antigen in the serum of monkeys infected with *Plasmodium knowlesi, J. Exp. Med.,* 69, 517, 1939.

79. **Aikawa, M., Miller, L. H., Johnson, J., Rabbege, J.,** Erythrocyte entry by malaria parasite. A moving junction between erythrocyte and parasite, *J. Cell. Biol.,* 77, 72, 1978.

80. **Ward, P. A. and Jack, R. M.,** The entry process of *Babesia* merozoites into red cells, *Am. J. Pathol.,* 102, 109, 1981.

81. **Miller, L. H., Aikawa, M., and Dvorak, J. A.,** Malaria *(Plasmodium knowlesi)* merozoites: immunity and the surface coat, *J. Immunol.,* 114, 1237, 1975.

82. **Rudzinska, M. A., Trager, W., Lewengrub, S. J., and Gubert, E.,** An electron microscopy study of *Babesia microti* invading erythrocytes, *Cell Tissue Res.,* 169, 323, 1970.

83. **Potgieter, F. T. and Els, H. J.,** The fine structure of intraerythrocytic stages of *Babesia bigemina, Onderstepoort J. Vet. Res.,* 44, 157, 1977.

84. **Brooks, C. B. and Kreier, J. P.,** Role of surface coat on *in vitro* attachment and phagocytosis of *Plasmodium berghei* by peritoneal macrophages, *Infect. Immun.,* 37, 1227, 1982.

85. **Montenegro-James, S., James, M. A., Ristic, M.,** Localization of culture-derived soluble *Babesia bovis* antigens in the infected erythrocyte, *Vet. Parasitol.,* 13, 311, 1983.

86. **Montenegro, S.,** Studies of Culture-Derived Soluble *Babesia bovis* Antigens: Purification, Characterization and Application in Serodiagnosis, Ph.D. thesis, University of Illinois, Urbana, 1981.

87. **Montenegro-James, S. et al.,** Manuscript in preparation, 1988.

88. **Laemmli, U. K.,** Cleavage of structural proteins during the assembly of head of bacteriophage T4, *Nature (London),* 227, 680, 1970.

89. **Tsang, V. C. M., Peralta, J. M., Simons, A. R.,** Enzyme-linked immunoelectron transfer blot techniques (EITB) for studying the specificities of antigens and antibodies separated by gel electrophoresis, *Methods Enzymol.,* 92, 377, 1983.

90. **Callow, L. L.,** Strain immunity in babesiosis, *Nature (London),* 204, 1213, 1964.

91. **Callow, L. L.,** A note on homologous strain immunity in *Babesia argentina* infections, *Aust. Vet. J.,* 44, 208, 1968.

92. **Curnow, J. A.,** In vitro agglutination of bovine erythrocytes infected with *Babesia argentina, Nature (London),* 217, 207, 1968.

93. **Curnow, J. A.,** Studies on antigenic changes and strain differences in *Babesia argentina* infections, *Aust. Vet. J.,* 49, 279, 1973.

94. **Mahoney, D. F., Wright, I. G., and Mirre, G. B.,** Bovine babesiosis: the persistence of immunity to *Babesia argentina* and *Babesia bigemina* in calves *(Bos taurus)* after naturally acquired infection, *Ann. Trop. Med. Parasitol.,* 67, 197, 1973.

95. **Smith, R. D., Molinar, E., Larios, F., Monroy, J., Trigo, F., and Ristic, M.,** Bovine babesiosis: pathogenicity and heterologous species immunity of tick-borne *Babesia bovis* and *Babesia bigemina* infections, *Am. J. Vet. Res.,* 41, 1957, 1980.

96. **Mahoney, D. F.,** Bovine babesiosis: diagnosis of infections by a complement fixation test, *Aust. Vet. J.,* 38, 48, 1962.

97. **Goodger, B. V.,** Preparation and preliminary assessment of purified antigens in the passive haemagglutination test for bovine babesiosis, *Aust. Vet. J.,* 47, 251, 1971.

98. **Todorovic, R. A. and Long, R. F.,** Comparison of indirect fluorescent antibody (IFA) with complement fixation (CF) tests for diagnosis of *Babesia* spp. infections in Colombian cattle, *Tropenmed. Parasitol.,* 27, 169, 1976.

99. **Montenegro-James, S., James, M. A., Levy, M. G., Preston, M. D., Esparza, H., and Ristic, M.,** Utilization of culture-derived soluble antigen in the latex agglutination test for bovine babesiosis and anaplasmosis, *Vet. Parasitol.,* 8, 291, 1981.

100. **Wright, I. G., Goodger, B. V., Leatch, G., Aylward, J. H., Rode-Bramanis, K., and Waltisbuhl, D. J.,** Protection of *Babesia bigemina*-immune animals against subsequent challenge with virulent *Babesia bovis, Infect. Immun.,* 55, 364, 1987.

101. **Bissuel, G., Laurent, N., Moreau, Y., and Vidor, E.,** Identification of culture-derived exoantigens of *Babesia canis,* in Proc. 3rd Int. Congr. on Malaria and Babesiosis, Annecy, France, September 7 to 11, 1987, 190.

102. **Singh, B., Gautam, O. P., and Banerjee, D. P.,** Immunization of donkeys against *Babesia equi* infection using killed vaccine, *Vet. Parasitol.,* 8, 133, 1981.

103. **Pipano, E. and Hadani, A.,** Control of bovine babesiosis, in *Malaria and Babesiosis,* Ristic, M., Ambroise-Thomas, P., and Kreier, J. P., Eds., Martinus Nijhoff, Dordrecht, Netherlands, 1984, 263.

104. **Banerjee, D. P. and Prasad, K. D.,** The use of killed vaccine in immunization of cattle against *Babesia bigemina* infection, *Indian J. Vet. Med.,* 5, 1, 1985.

105. **DeVos, A. J., Combrink, M. P., and Bessenger, R.,** *Babesia bigemina* vaccine: comparison of the efficacy and safety of Australian and South African strains under experimental conditions in South Africa, *Onderstepoort J. Vet. Res.,* 49, 155, 1982.

106. **McCosker, P. J.,** in *Control of Piroplasmosis and Anaplasmosis in Cattle, A Practical Manual,* FAO, Santa Cruz, Bolivia, 1975.

107. **Corrier, D. E., Gonzalez, E. F., and Betancourt, A.,** Current information on the epidemiology of bovine anaplasmosis and babesiosis in Colombia, in *Tick-Borne Diseases and Their Vectors,* Wilde, J. K. H., Ed., University of Edinburgh Press, Edinburgh, 1978, 114.

108. **James, M. A., Coronado, A., Lopez, W., Melendez, R., and Ristic, M.,** Seroepidemiology of bovine anaplasmosis and babesiosis in Venezuela, *Trop. Anim. Health Prod.,* 17, 9, 1985.

109. **Callow, L. L. and Dalgliesh, R. J.,** Immunity and immunopathology in babesiosis, in *Immunology of Parasitic Infections,* 2nd ed., Cohen, S. and Warren, K. S., Eds., Blackwell Scientific, Oxford, 1982, 307.

110. **Daly, G. D. and Hall, W. T. K.,** Note on the susceptibility of British and some Zebu-type cattle to tick fever (babesiosis), *Aust. Vet. J.,* 31, 152, 1955.

111. **Johnston, L. A. Y. and Sinclair, D. F.,** Differences in response to experimental primary infection with *Babesia bovis* in Hereford, Droughtmaster and Brahman cattle, in *Proc. Symp. at the 56th Annu. Conf. of the Australian Veterinary Association,* Johnston, L. A. Y. and Cooper, M. G., Eds., Australian Veterinary Association, Sydney, 1980, 14.

112. **Joyner, L. P. and Donnelly, J.,** The epidemiology of babesial infections, *Adv. Parasitol.,* 17, 115, 1979.

113. **Davidson, K. B.,** An unusual outbreak of babesiosis, *Vet. Rec.,* 85, 391, 1969.

114. **Kaeberle, M. L.,** Function of carriers and adjuvants in induction of immune responses, in *Advances in Carriers and Adjuvants for Veterinary Biologics,* Nervig, R. M., Gough, P. M., Kaeberle, M. L., and Whetstone, C. A., Eds., Iowa State University Press, Ames, 1986, 11.

115. **Morvat, G. M.,** A comparison of the adjuvant effects of saponin and oil emulsions in foot-and-mouth vaccines for cattle, *Bull. Off. Int. Epizool.,* 81, 1319, 1974.

116. **Moreau, Y.,** Practical results obtained with an industrial *Babesia canis* vaccine, Proc. 3rd Int. Conf. on Malaria and Babesiosis, Annecy, France, September 7 to 11, 1987, 236.

117. **Toro, M., Lopez, R., Leon, E., Ruiz, A., and Garcia, J. A.,** Epidemiologic and critical features of bovine babesiosis in Venezuela, *Vet. Trop.,* 2, 91, 1982.

Chapter 5

BABESIAL VACCINATION WITH DEAD ANTIGEN

B. V. Goodger

TABLE OF CONTENTS

I. INTRODUCTION

The manuscript reviews published accounts of vaccination trials against *Babesia* species using killed parasites or fractions thereof as inocula. It is confined to those *Babesia* species which infect animals of veterinary importance. As a generalization, recovery from a single infection with a *Babesia* species confers lifelong protection against that species to the infected animal provided it remains in a healthy state. This fact was exploited early in the century in Australia as an effective form of cattle vaccination, firstly against *B. bigemina*, and then against *B. bovis*.[1,2] Since then, the Queensland Department of Primary Industries, mainly under the guidance of Dr. L. L. Callow, has produced attenuated strains, particularly with *B. bovis*, which are efficient in eliciting protective immunity and have received worldwide adaption.[3] With the development of monoclonal antibody techniques and the sanguine prospects offered by genetic engineering, the time is ripe for the identification and the commercial production of protective antigens *in vitro*. That this concept of "dead" vaccination now appears likely is due mainly to the dogmatic studies initiated by Dr. D. F. Mahoney who demonstrated that effective immunity to *B. bovis* could be induced by vaccination with crude extracts of infected erythrocytes.[4,5]

While live vaccination with attenuated parasites has shown itself to be a reliable and faithful servant, it has several disadvantages which optimistically would not occur with a genetically engineered dead vaccine. The disadvantages include (1) the maintenance of infection in the field, (2) the return to virulence, (3) the possibility of coinfection with contaminating organisms, particularly viruses, (4) the production of erythrocytic isoantibodies in breeding animals with resultant hemolytic disease of the newborn, (5) temperature liability affecting storage and transportation, (6) limited shelf life, and (7) unsuitability for freeze drying.

However, one must not be too complacent about the ease of producing a dead vaccine as features such as ethically acceptable adjuvants, field longevity of the immune response, possibility of booster vaccination(s), and parasite adaptation at the molecular level must be realistically considered and anticipated not only prior to, but also after the release of an acceptable vaccine. In addition, the fact that sterile immunity, in *B. bovis* at least, does not occur in naturally immune animals suggests that effective immunity might be directed more so at the pathophysiological effects caused by infection rather than the parasite itself.[6] Conversely, sterile immunity can occur with *B. bigemina*, and in this species the immunity may be more parasite specific.[7]

II. SELECTION OF VACCINATION ANTIGENS

A concentration of infected erythrocytes is initially essential not only to maximize the amount of babesial antigen, but also to decrease contamination from uninfected erythrocytes. Preferential lysis of uninfected erythrocytes, leaving infected erythrocytes intact, can be achieved with hypotonic salt solutions, and this technique works extremely well for both *B. bovis* and *B. ovis*, moderately well for *B. divergens*, and to some degree with *B. bigemina*.[8-10] As well, concentration has been achieved by centrifugation of infected blood in gradients of Percoll® and similar chemicals[11,12] and by affinity chromotography with chemical ligands.[13] A major problem in isolating antigens from *Babesia*-infected erythrocytes is the difficulty in obtaining merozoites totally free of erythrocytic stroma. Merozoites, seemingly free of stroma, have been produced by continuous sonication,[7,14] by manipulations in media conditions during *in vitro* cultivation, and by differential centrifugation.[12,15] However, studies at our laboratory have shown that merozoites, seemingly free of erythrocyte stroma when stained by Geimsa or similar Romanoswky stains, may be still enmeshed in stroma when examined by electron microscopy or when stained with fluorescent antibodies to normal erythrocytes. Hence, it seems essential that one or both techniques be used to determine if merozoites are, first, extracellular and, second, are not contaminated by stromal residues of both infected and noninfected erythrocytes. Conversely, stroma of infected erythrocytes, especially in *B. bovis* and *B. ovis* and to some degree in *B. bigemina*, contains significant amounts of babesial antigen, and one should not be biased towards antigen located in the merozoite, as protective antigens can be located in the infected erythrocyte itself.[16-18]

Following the initial studies by Mahoney,[4] there are now numerous examples among the babesial species of veterinary importance which demonstrate that immunity can be elicited with crude extracts of infected erythrocytes and with plasma of acutely infected animals. It is realistic to assume that the plasma antigens are excretory/secretory products of the parasite and not unique to plasma alone. Regardless of the species, the most pertinent question to be addressed initially is, how does one select for a protective antigen in a crude parasite extract? Once a fraction, enriched for babesial antigen and shown to elicit protective immunity, is obtained, several factors must be considered before addressing the previous question. One factor is the type of immunity required for protection. Evidence indicates that both humoral and cellular responses occur following infection or vaccination, but controversy exists as to which response is the more important in eliciting protection. The evidence for the humoral response being dominant seems more convincing. This includes protection by passive immunity, protection by acquisition of colostrum, the demonstration of antibody mediated cell toxicity, and the rapid decay in lymphocyte transformation response comparative to antibody longevity following live vaccination.[19-22] Hence, antibody alone should be sufficient to target for protective antigens. However, high-titered, early antisera are not protective in passive immune studies, whereas long-term antisera and antisera from super-infection are protective, even though they may be of low titer.[23] Thus, further care is required in selection of antisera for probing crude antigen as nonprotective dominant antigens may be easily highlighted. In addition, immunosuppression occurs at peak parasitemia, caused possibly by a combination of an acute-phase inflammatory response, circulating immune complexes, and parasite modulation.[24-26] Paradoxically, therefore, immunosuppression and peak antigen concentration occur together, which could result in poor memory induction of important immunogens. Results of vaccination studies at our laboratories support this concept, as a number of antigens, although producing antibody of high titer and of high serological avidity, have failed to elicit a protetive response, but more importantly, the antigens have been sufficiently immunodominant to suppress the immune response against known protective immunogens when used in combination.[27] Compounding the problem are decoy host antigens, mimicry of host antigen by the parasite, and the presence of spurious antigens induced by the host during the inflammatory response against infection.[28] Also, autoantigens of erythrocytic origin exist and must further compromise the immune system.[29,30] Collectively these factors enhance the possibility of protective antigen being effectively camouflaged by irrelevant antigens of both parasite and host origin. Figure 1 lists a probable interplay of such factors relevant to detection of protective immunogens.

Immunoblotting, metabolic labeling, and immunoprecipitation are widely used for detection of antigen. While being precise and elegant techniques, they do not necessarily target for protective antigens. It is far too easy to assume that serological avidity *in vitro* will correlate with protection *in vivo*, and caution is required. The major antigen(s) detected in immunoblotting of *B. bovis* is not detected in extracts of uninfected erythrocytes, and one would initially consider it to be of babesial origin.[28] However, it reacts with some naive bovine serum and avidly with serum from uninfected calves undergoing an inflammatory response, and its origin and potential are therefore suspect (Figure 2). Conversely, one of the more protective *B. bovis* antigens binds very poorly to nitrocellulose in immunoblots. Collectively both these results suggest some caution in interpretation of serological studies when selecting potential target antigens and an awareness that "biggest is not necessarily best".[31]

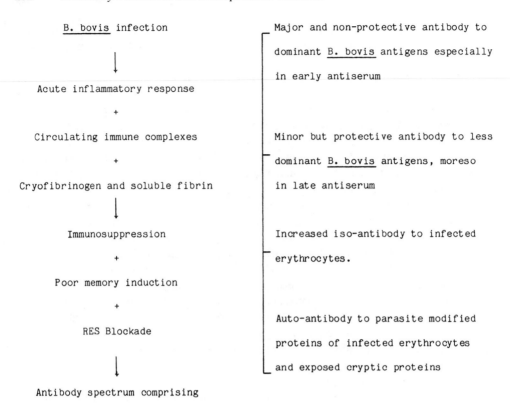

FIGURE 1. Possible interplay of factors at peak infections determining the spectrum of antibodies produced in response to infection with *B. bovis*.

A logical and methodical approach, albeit slow, is that of controlled chemical fractionation and subsequent vaccinations ultimately resulting in fractions enriched for protective antigens. Such an approach has been adapted at this laboratory for *B. bovis* and by Taylor and co-workers for *B. divergens*.[32] Once an enriched protective fraction is obtained, a more meaningful bank of antibodies, both poly- and monoclonal, can be produced and utilized for definitive fractionation. Another approach is to select antigens by biological and pathophysiological parameters. One biological approach at this laboratory was the selection of antigens that were proteolytic, and of two tested, one elicited heterologous protection, while the other unfortunately elicited only homologous protection.[31,33] A pathophysiological selection utilized the knowledge that *B. bovis* causes microvascular stasis. Accordingly, antigens involved in infected erythrocyte adhesion and aggregation were selected, and preliminary studies indicate such antigens elicit a strong protective immunity.[34]

III. SPECIES VACCINATION

A. Bovine Species
1. Babesia bovis
Mahoney initiated dead vaccination studies with *B. bovis* and demonstrated that heterologous immunity could be obtained by vaccinating with crude extracts of infected erythrocytes.[4] Concentration of infected erythrocytes was obtained from serial passage in splenectomized calves and by *in vitro* preferential lysis of uninfected erythrocytes. In contrast, U.S. researchers in recent years have obtained material for vaccination from *in vitro* cultures of *B. bovis*, first with spinner flasks and then with a microaerophilic stationary-phase tech-

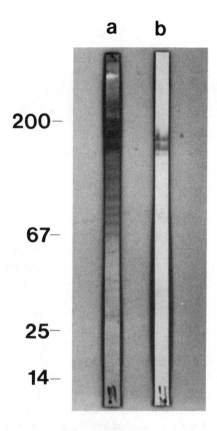

FIGURE 2. Immunoblot following sodium dodecyl sulfate (SDS) electrophoresis and subsequent probing of extracts from *B. bovis*-infected erythrocytes with (a) post- and (b) prevaccination sera, both previously adsorbed with normal erythrocyte stroma. Major bands are detected by prevaccination sera, but, most importantly, are not detected in analagous samples of uninfected erythrocytes.

nique.[15,35] The *in vitro* culturing has initiated a series of dead vaccination studies by groups in the U.S. and Australia, in which culture supernatant was used as vaccination material. All groups have demonstrated that some protective immunity can be induced, but there is disagreement as to its efficacy and longevity, particularly against heterologous parasites.[22,36-38] Timms et al. demonstrated only 50% survival against heterologous challenge comparative to vaccination with attenuated parasites and suggested that the immunity was waning at 6 months.[22] Interestingly, the cell-mediated immune response was enhanced with dead antigen comparative to live antigen, indicating the increased cellular response was not reflected by an increased protection. The results augment Mahoney's passive immune studies and indicate the major immune protection may be humoral rather than cellular. Indeed, colostrum from immune cows confers protection to naive calves, further enhancing the importance of antibody in protection against babesial infection.[20] In contrast, Timms et al. theorized that immunity was both humoral and cellular.[22] Although individual antigens in culture supernatant have been identified and partly characterized, they have as yet not been tested in vaccination trials.[39] Interestingly one of the major antigens is ∼40 kDa, similar in size to that of a known protective *B. bovis* antigen which was detected and purified by monoclonal antibody. Immunity has also been reported following vaccination with a saline eluate from free merozoites obtained from *in vitro* culture.[40] The eluate contains surface coat antigen, and its simple method of preparation suggests such antigens are loosely bound to the outer membrane of the merozoite and, thus, are probably a source of exoantigen *in vivo*.

The decision as whether to obtain *B. bovis* antigens and/or infected erythrocytes from culture or from serial passage in splenectomized calves requires some comment. The latter will normally give a higher parasitemia and certainly will produce larger volumes, but this must be balanced by the simple harvesting of culture supernatant and by the prolonged washing and large-scale centrifugation required for preparation of blood samples from infected calves. Also, some antigen is lost during the concentration step of hypotonic lysis and its potential as a vaccine is not known. Culture would have a tremendous advantage if syncronous life cycles were available, but these have not yet been obtained. Thus, both preparations contain a mixed age group of parasites, and neither is free of erythrocytic isoantigens.

Studies in defined fractionation-vaccination in our laboratory have shown that both a distilled water lysate of infected erythrocytes and soluble extracts of parasite-infected erythrocyte stroma are protective.[41,42] The lysate fraction was of interest in that it contained a dominant antigen which could be readily detected by the relatively insensitive technique of immunodiffusion. Unfortunately, when fractionated to immunological purity it did not elicit a protective immune response, thus presenting as a fine example of serological avidity not reflecting vaccination efficacy.[29] By contrast, another antigen in lysate was extremely effective in eliciting protective immunity against both homologous and heterologous parasites even though it had low serological reactivity and was in low concentration. However, it possessed proteolytic activity which obviously has an important biological role in parasite integrity.[31] Further protective antigens of lysate are currently being purified and characterized. A group of native antigens in lysate, of mean size ~230 kDa, was protective, whereas other groups of antigens of higher molecular weight and stronger serological reactivity were not protective.[43] Once again, *in vitro* serological avidity did not reflect protection *in vivo*.

Fractionation of soluble antigens, released from parasite-infected erythrocyte stoma by sonic disruption and ultracentrifugation, has led to the detection of a highly protective fraction, designated β from its electrophoretic mobility in agarose.[42] Current studies show the β-fraction in the native state contains about 25% lipid which is precipitable by trichloracetic acid. This suggests that the lipid is present as a lipoprotein or proteolipid complex. Enzyme-linked immunosorbent assays show organic solvent fractions of β contain antigenic activity and that 20 to 30% of the serological activity of β is lost following lipase digestion. This strongly suggests that part of the antigencity resides in a lipid moiety. Whether the latter contains protective antigen is currently being tested, but if a lipid-associated antigen is protective, it may prove a daunting task for genetic engineering. A small number of monoclonal antigens against β-fraction has been produced. One of the monoclonals, designated 11C5, avidly stains the infected erythrocyte and has been selected for vaccination studies (Figure 3A). In immunoblotting studies it detects a polymeric ladder (Figure 3B). The antigen has been isolated by affinity chromatography and in a preliminary vaccination trial elicited protective homologous immunity. Currently a genetically prepared sample of the antigen is being tested in a vaccination trial. The question as to whether protective immunogens are more likely to be located in the infected erythrocyte more so than the parasite requires addressing. The parasite per se should be vunerable to the immune system only in the the fleeting period from release to invasion of an uninfected erythrocyte, whereas the infected erythrocyte is theoretically vulnerable for the duration of a life cycle. This may be an oversimplification of antigen selection, but it does serve as another example of biological selection of target antigen rather than *in vitro* serological avidity. Conversely though, the presence of natural or induced erythrocytic isoantibodies may complicate the task of precisely identifying babesial antigens. In indirect fluorescent antibody (IFA) studies, sera from many cattle not infected with *B. bovis* will stain infected erythrocytes, albeit at

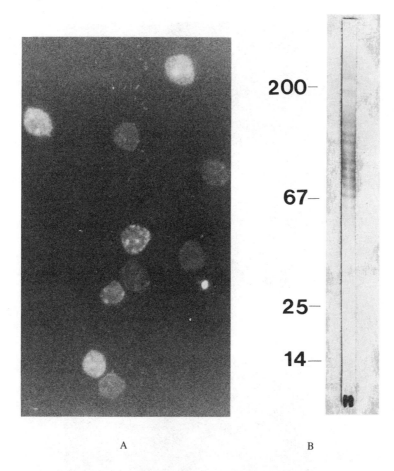

A   B

FIGURE 3.   (A) IFA staining of infected erythrocytes by monoclonal antibody (11C5) to a protective β-antigen. (B) Immunoblot following SDS electrophoresis of *B. bovis* β-antigen and reaction with 11C5 monoclonal.

low dilutions.[44] Likewise, immunologically pure antigen, when injected into cattle, may produce antibody directed against the infected erythrocyte and parasite, whereas the same antigen, when injected into rabbits or used to produce murine monoclonal antibodies, elicits antibody reactive only against the parasite[29] (Figure 4). Presumably there is a complicated cross-reaction between isoantibody/autoantibody and host erythrocytic antigens modified in some manner by the parasite. As a parallel, erythrocytic autoantibodies have been detected in humans infected with *Babesia* species, and the host erythrocyte modifications are surely a result of parasite interaction.[30]

Another highly protective group of *B. bovis* antigens was obtained from intact infected erythrocytes by passing soluble components through a heparin-Sepharose® column and eluting bound components with hypertonic saline.[34] The rationalization for this procedure was that infected erythrocytes bind to heparin *in vitro* and to endothelium *in vivo* and that the inflammation associated with acute infection should cause a massive localization of heparin in endothelium. It was speculated that antibody-to-heparin binding antigens might be protective, and this was demonstrated by subsequent vaccination. *B. bovis* infection produces copious quantities of cryoprecipitable immune complexes immediately after peak parasitemia. The complexes contain babesial antigen, and it was hypothesized that the complexes should be highly immunogenic and perhaps capable of eliciting a protective immunity. Subsequent vaccination did produce an immune response, but unfortunately it was not protective.[45]

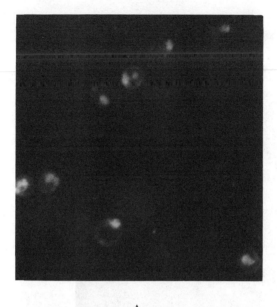

A

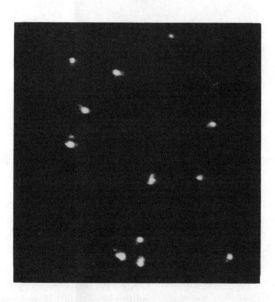

B

FIGURE 4.   IFA staining due to (A) bovine and (B) rabbit
polyclonal antibodies to a dominant *B. bovis* antigen. Only
the bovine antibodies stain the infected erythrocyte.

To date, vaccination experiments have been performed on six different antigens prepared
by affinity chromatography with monoclonal antibodies. Only two have elicited protection,
these being the β one referred to previously and one designated 15B1, which is parasite
specific and was selected on its unique parasite staining by IFA. Surprisingly only microgram
amounts of antigen were required to induce immunity. One of the other monoclonal-derived
antigens was a major immunoblotting one, but it did not elicit a protective response and,
more importantly, was sufficiently dominant to suppress 15B1 in a mixed vaccination.[46]

Once again, "biggest was not best"! The sequence of the 15B1 antigen has been obtained, and genetic engineering is in progress. Evidence suggests that *B. bovis*, like *Plasmodium falciparum*, may contain antigens with repetitive sequences.[47] Such antigens should be immunologically dominant, and once isolated, it will be interesting to see whether they have a role in protective immunity or will be part of a immunological decoy system of the parasite. In similar vein, DNA polymorphism has been detected, and this might indicate potential mutation of a protective antigen following challenge or immune system pressure.[48,49] If so, a "cocktail" of protective antigens capable of a synergistic effect may be required.

2. Babesia bigemina

Only a few studies have been initiated into vaccinating against *B. bigemina* with dead antigen, and all have utilized relatively crude fractions. Kuttler and Johnson[50] used washed infected stroma for vaccination, while Banerjee and Prasad[51] used infected plasma and an infected stromal extract similar to that of Kuttler and Johnson. Banerjee and Prasad claimed the immunity induced from stromal vaccination was of less duration than that induced by plasma antigens. Presumably the major protective antigen was an exoantigen. Todorovic had earlier claimed plasma contained protective antigens.[53] More recently, a soluble extract from infected erythrocytes was first cleared of oxyhemoglobin by chromatography on Celite® columns, and when used as a vaccine, protective homologous immunity was demonstrated.[18] Antigenic variation occurs in *B. bigemina*, and this may confuse future identification of protective antigens.[53] Interestingly, natural *B. bigemina* infections induce a transient cross-protective immunity to *B. bovis*, but the reverse does not occur.[54] Hence, a common protective antigen for both species might exist in *B. bigemina*. Current studies at our laboratories using immunoblotting and IFA show antigen commonality to both species, and concurrent vaccination trials have confirmed the transient, one-way cross-species protective immunity.[18] *B. bigemina* has been successfully cultured *in vitro*, and density gradient studies can now produce an 80% infected erythrocyte concentration.[11,55] Thus, the way is open for detailed studies on protective antigens. In this context, an American group has used monoclonal antibodies to target several merozoite proteins as potential immunogens, but no vaccination studies have been reported.[56]

3. Babesia divergens

Hinaidy demonstrated in 1981 that immunity to *B. divergens* could be produced in cattle by vaccinating with infected erythrocytes pretreated with formalin.[57] Since then, vaccination studies on *B. divergens* have been aided by the fact that the organism can be sustained in a small laboratory rodent, namely the Mongolian gerbil, and also in splenectomized rats.[58,59] In addition, it has also been successfully adapted to *in vitro* culture, while hypotonic lysis of infected blood also produces enriched suspensions of infected erythrocytes. Culture and laboratory animal studies, in combination with monoclonal antibodies, have identified a 55-kDa antigen, which is obtained as a saline "wash-off" protein from merozoites. Subsequent vaccination and homologous challenge have produced encouraging results.[60] This antigen was selected by the ability of its monoclonal antibody to inhibit *in vitro* invasion by merozoites and presents as a fine example of biological selection of target antigen.

A methodical approach of chemical fractionation and subsequent vaccinations has been adapted by Irish workers. Blood from infected cattle was enriched for infected erythrocytes by hypotonic lysis and a soluble extract obtained by sonication and centrifugation. Fractionation by preparative isoelectric focusing and Concanavalin A chromatography has produced an acidic fraction, free of carbohydrate and capable of eliciting protective heterologous immunity.[61] Current studies have shown the protective moiety to have an isoelectric point between pH 6 and 7 and an antigen of size 49 kDa has been selected as the likely immunogen.[62,63] The group has also investigated routes of inoculation, showing that subcutaneous is the best, and have also tested several adjuvants, the best being muramyl dipeptide (MDP) in combination with Freund's incomplete adjuvant.[64,65] The group advocates selecting an adjuvant which will give a persistent humoral response and has demonstrated that injection of *Corynebacterium* spp. prior to challenge does not enhance protection. This confirms earlier studies by Brocklesby and Purnell, which also suggested that, unlike rodent babesial infections, nonspecific immunity may not play an important role in bovine babesial infection.[66]

Monoclonal antibodies have been produced to the rat-adapted strain of *B. divergens*, and two of these have been protective in passive immune studies in rats. Disturbingly though, both monoclonals showed individual variable reactivity in IFA studies of field isolates, indicating antigenic diversity.[67] The relevance of such diversity towards the elicitation of protective immunity from vaccination with the relevant antigens requires resolution.

B. Ovine Species

Little work appears in the literature on ovine species and such work is confined to *B. ovis*. Kyurtov in 1980 demonstrated that a precipitate from infected plasma or infected erythrocyte cytoplasm, obtained at 50 to 70% saturated ammonium sulfate, would induce a protective immune response. Indeed, no deaths were reported in a field trial employing 360 vaccinated sheep.[68] Kyurtov also demonstrated that a soluble extract, obtained from infected erythrocytes by a series of freeze-thawings followed by sonic disruption, had antigenic activity, but no vaccination studies have apparently been performed with the extract.[69] Most recent studies from cojoint research by Australian and Turkish workers indicate that freeze-thawed soluble extracts of *B. ovis*-infected erythrocytes produce protective immunity against *B. ovis* in sheep and that an analogous extract from *B. bovis* also induces protection in sheep against *B. ovis*.[10] Currently a protective antigen common to both *B. ovis* and *B. bovis* is being sought.

C. Equine Species

Singh and co-workers in 1980 vaccinated donkeys with a lysate prepared by freeze-thawing *B. equi*-infected erythrocytes and obtained homologous protection.[70] It has been suggested that humoral immunity is not important in equine babesiosis.[71]

D. Canine Species

B. canis has also been successfully cultured *in vitro*, and vaccination studies with culture supernatant and crude particulate material, using saponin as an adjuvant, have shown that a considerable degree of homologous protection can be obtained in susceptible dogs. It was claimed that effective immunity coincided with high antibody titer. Some antigenic fractionation has been achieved by differential centrifugation and by gel filtration, but no subsequent vaccinations have been reported.[72] As with other *Babesia* spp. cultured *in vitro*, merozoites can be liberated via appropriate medium adaptations and a wash-off "surface coat" antigen can be obtained by saline elution.[73] A commercial vaccine consisting of culture supernatant fluid is available in France.[73]

IV. EVALUATION OF VACCINATION EFFICACY

Temperature fluctuations, packed cell volume decreases, parasite numbers, and survival rates are the time-honored criteria for evaluating vaccination efficacy against erythrocytic protozoa. However, studies on *B. bovis* have indicated that various pathophysiological parameters may also be employed as sensitive indicators of efficacy.[74] Most of these parameters belong to the interrelated kinin-coagulation-firbrinolysis-complement systems, and two of major importance are soluble fibrin and cryofibrinogen.[75] Both are implicated in microvascular stasis and are generally absent or in low concentration in immune animals following challenge. In contrast, control infected animals contain considerable quantities of these proteins. Two other indicators of immunity are fibronectin and conglutinin.[76] The former is the major nonimmune opsonin, and its concentration is an indicator of reticulo endothelial system (RES) efficacy.[77] One of its important functions is the removal of effete debris from the circulation, and during acute *B. bovis* infection, it undergoes massive consumption.[78] In vaccinated animals comparative to susceptible, fibronectin is consumed more so initially, but near the crisis period it rapidly returns to normal levels in immune animals, whereas in nonimmunes it remains at low levels. The same sequence occurs with conglutinin, a protein unique to the Bovidae, the function of which is to remove C3-containing complexes from the circulation.[79] It has been speculated that conglutinin removes infected erythrocytes from the circulation by complexing with C3 which has attached to the erythrocytic membrane either by antigen-antibody complexes or by membrane modifications.[77,78] It is possible that fibronectin is also involved in the removal of infected erythrocytes, as subtle membrane alterations induced by antigen-antibody complexes could present as ligands for fibronectin. Whatever the mechanisms, an increase in these proteins during the later part of acute infection is seemingly a good prognostic indicator. Likewise, the detection of immunoconglutinin, an autoantibody, to fixed C3 is also a good indication of survival, as cattle likely to recover from infection contain immunoconglutinin in their plasma, whereas none can be detected in susceptible cattle. Presumably the immunoconglutinin is augmenting the reactions of fibronectin and conglutinin.[76,79] Figure 5 depicts the changes in fibronectin and conglutinin following challenge of control and vaccinated calves.

A question which needs urgent addressing is the duration of immunity following dead vaccination. The immunity induced by viable parasites, at least in *B. bovis* and *B. bigemina*, lasts at least 4 years, and antibody, albeit at low levels, is still detectable at this time.[6,7] Most vaccination studies have employed challenges shortly after vaccination, and only a few have been performed at 6 months or longer. In parallel, antibody kinetic studies suggest circulating antibody wanes quickly.[37] Thus, field animals in marginal tick areas could be at risk if vaccination-induced immunity is transient and disappears prior to the completion of one tick season.

As *B. bovis* antigens have become more defined, a worrying phenomenon sometimes occurs 1 to 3 d after challenge and before peripheral parasites are detectable.[76] An acute, nonspecific inflammation, probably due to immune complexes depositing in vascular beds, occurs and is detectable as a temperature increase comparative to nonimmunes and by the presence of *in vitro* precipitable immune complexes. Terminal *B. bovis* is characterized by acute inflammation and hypotensive shock, which could be exacerbated by an earlier inflammatory episode. The presenting dilemma is that on one hand, antibodies are required for protection, whereas on the other, by interaction with antigen, they might induce an early inflammatory response upon parasite challenge. A resolution is needed as an early adverse reaction to challenge in vaccinated cattle could result in endogenous suicide due to uncontrolled inflammation.[80]

V. ADJUVANTS

It would be ironical if effective vaccines were developed from genetic engineering studies, but an acceptable adjuvant was not available. At present, Freund's complete adjuvant and saponin produce the most efficient immunity, but are not ethically acceptable. Encouraging results have been obtained wtih glutaraldehyde-cross-linked antigen and dextran sulfate in *B. bovis* vaccination[41,81] and with a mixture of MDP and Freund's incomplete adjuvant in *B. divergens* vaccination.[65] However, antibodies to MDP have been recently detected, and this might circumvent their clinical usage.[82] Also, the present cost of MDP would preclude

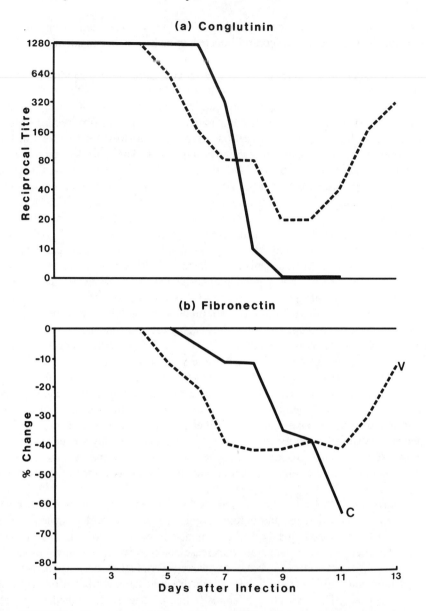

FIGURE 5.   Changes in conglutinin and fibronectin following *B. bovis* challenge of a group of control (————) and a group of vaccinated cattle (- - - -). The terminal increase in conglutinin levels in the vaccinates is due in part to the production of immunoconglutinin, an IgM antibody to C3.

its commercial use. If the premise that the effective immunity is mainly humoral is correct, then a long-lasting depot-type adjuvant should be required. Obviously much research is required before the problem of suitable adjuvants can be resolved.

VI. SUMMARY

Vaccination with crude extracts of *Babesia*-infected erythrocytes elicits the production of protective immunity in animals of veterinary importance. Fractionation of crude extracts and subsequent vaccinations have been performed with *B. bovis* and *B. divergens*. Antigens have been extracted from both species using monoclonal antibodies and used successfully in vaccination studies. In the near future, antigens prepared from genetic engineering should be available. However, problems such as acceptable adjuvants and longevity of immunity have yet to be addressed. It is strongly suggested that biological parameters be used in tandem with serological studies in an attempt to detect potential protective antigens.

ACKNOWLEDGMENTS

I wish to thank fellow staff members, in particular Dr. I. G. Wright and Mr. D. J. Waltisbuhl, for helpful advice and access to unpublished material.

REFERENCES

1. **Pound, C. J.**, Tick fever. Notes on the inoculation of bulls as a preventative against fever at Rathdowney and Rosedale, *Queensl. Agric. J.*, 18, 281, 1907.
2. **Legg, J.**, Recent observations on the preimmunization of cattle against tick-fevers in Queensland, *Aust. Vet. J.*, 15, 46, 1939.
3. **Callow, L. L.**, The control of babesiosis with a highly infective attenuated vaccine, *World Vet. Congr. Proc.*, 1, 357, 1971.
4. **Mahoney, D. F.**, Bovine babesiosis: the immunization of cattle with killed *Babesia argentina*, *Exp. Parasitol.*, 20, 125, 1967.
5. **Mahoney, D. F. and Wright, I. G.**, *Babesia argentina:* immunization of cattle with a killed antigen against infection with a heterologous strain, *Vet. Parasitol.*, 2, 273, 1976.
6. **Mahoney, D. F., Wright, I. G., and Goodger, B. V.**, Immunity in cattle to *Babesia bovis* after single infections with parasites of various origin, *Aust. Vet. J.*, 55, 10, 1979.
7. **Mahoney, D. F.**, Unpublished data.
8. **Mahoney, D. F.**, Bovine babesiosis: preparation and assessment of complement fixing antigens, *Exp. Parasitol.*, 20, 232, 1967.
9. **Taylor, S. M., Kenny, J., Mallon, T., and Elliot, C. T.**, The immunization of cattle against *Babesia divergens* with fractions of parasitized erythrocytes, *Vet. Parasitol.*, 16, 235, 1984.
10. **Alabay, M., Duzgun, A., Cerci, H., Wright, I. G., Waltisbuhl, D. J., and Goodger, B. V.**, Ovine babesiosis: induction of a protective immune response with crude extracts of either *Babesia bovis* or *Babesia ovis*, *Res. Vet. Sci.*, 43, 401, 1987.
11. **Vega, C. A., Buening, G. M., Rodriguez, S. D., and Carson, C. A.**, Concentration of and enzyme content of *in vitro* cultured *Babesia bigemina*-infected erythrocytes, *J. Protozool.*, 33, 514, 1986.
12. **Rodriguez, S. D., Buening, G. M., Vega, C. A., and Carson, C. A.**, *Babesia bovis:* purification and concentration of merozoites and infected bovine erythrocytes, *Exp. Parasitol.*, 61, 236, 1986.
13. **Goodger, B. V., Mahoney, D. F., and Wright, I. G.**, *Babesia bovis:* attachment of infected erythrocytes to heparin sepharose columns, *J. Parasitol.*, 69, 248, 1983.
14. **Gravely, S. M. and Kreier, J. P.**, *Babesia microti* (Gray strain): removal from infected hamster erythrocytes by continuous-flow sonication, *Tropenmed. Parasitol.*, 25, 198, 1974.
15. **Levy, M. G. and Ristic, M.**, *Babesia bovis:* continuous cultivation in a microaerophilous stationary phase culture, *Science*, 207, 1218, 1980.
16. **Ludford, C. G.**, Fluorescent antibody staining of four *Babesia* species, *Exp. Parasitol.*, 24, 327, 1969.
17. **Waltisbuhl, D. J.**, Unpublished data.
18. **Wright I. G., Goodger, B. V., Leatch, G., Aylward, J. H., Rode-Bramanis, K., and Waltisbuhl, D. J.**, Protection of *Babesia bigemina*-immune animals against subsequent challenge with virulent *Babesia bovis*, *Infect. Immun.*, 55, 364, 1987.
19. **Mahoney, D. F., Kerr, J. D., Goodger, B. V., and Wright, I. G.**, The immune response of cattle to *Babesia bovis* (syn. *argentina*). Studies on the nature and specificity of protection, *Int. J. Parasitol.*, 9, 297, 1979.
20. **Hall, W. T. K.**, The immunity of calves to tick-transmitted *Babesia argentina* infection, *Aust. Vet. J.*, 39, 386, 1963.
21. **Goff, W. L., Wagner, G. G., and Craig, T. M.**, Increased activity of bovine ADCC effector cells during acute *Babesia bovis* infection, *Vet. Parasitol.*, 16, 5, 1984.
22. **Timms, P., Stewart, N. P., Rodwell, B. J., and Barry, D. N.**, Immune responses of cattle following vaccination with living and non-living *Babesia bovis* antigens, *Vet. Parasitol.*, 16, 243, 1984.

23. **Mahoney, D. F.,** Bovine babesiosis: the passive immunization of calves against *Babesia argentina* with special reference to the role of complement-fixing antibodies, *Exp. Parasitol.,* 20, 119, 1967.

24. **Callow, L. L. and Stewart, N. P.,** Immunosuppressive effect of *Babesia bovis* on immunity to *Boophilus microplus, Nature (London),* 272, 818, 1978.

25. **Israel, L., Samak, R., Edelstein, R., Bogucki, D., and Breau, J. L.,** Mise en evidence du role immunodepresseur des proteines de l'inflammation. Leur role physiopathologique chez les cancereaux, *Ann. Med. Interne,* 132, 26, 1981.

26. **Goodger, B. V., Wright, I. G., and Mahoney, D. F.,** Initial characterization of cryoprecipitates in cattle recovering from acute *Babesia bovis (argentina)* infection, *Aust. J. Exp. Biol. Med.,* 59, 531, 1981.

27. **Wright, I. G., Mirre, G. B., Rode-Bramanis, K., Chamberlain, M., Goodger, B. V., and Waltisbuhl, D. J.,** Protective vaccination against virulent *Babesia bovis* with a low-molecular-weight antigen, *Infect. Immun.,* 48, 109, 1985.

28. **Goodger, B. V., Wright, I. G., and Waltisbuhl, D. J.,** *Babesia bovis:* the effect of acute inflammation and iso-antibody production in the detection of babesial antigens, *Experientia,* 41, 1577, 1985.

29. **Goodger, B. V., Commins, M. A., Wright, I. G., Mirre, G. B., Waltisbuhl, D. J., and White, M.,** *Babesia bovis:* vaccination trial with a dominant immunodiffusion antigen in splenectomised calves, *Z. Parasitenkd.,* 72, 715, 1986.

30. **Wolf, C. F., Resnick, G., Marsh, W. L., Benach, J., and Habicht, G.,** Autoimmunity to red blood cells in babesiasis, *Transfusion,* 22, 533, 1982.

31. **Commins, M. A., Goodger, B. V., and Wright, I. G.,** Proteinases in the lysate of bovine erythrocytes infected with *Babesia bovis:* initial vaccination studies, *Int. J. Parasitol.,* 15, 491, 1985.

32. **Taylor, S. M., Elliott, C. T., and Kenny, J.,** Isolation of antigenic proteins from erythrocytes parasitised by *Babesia divergens* and a comparison of their immunising potential, *Vet. Parasitol.,* 21, 99, 1986.

33. **Wright, I. G., Goodger, B. V., Rode-Bramanis, K., Mattick, J. S., Mahoney, D. F., and Waltisbuhl, D. J.,** The characterisation of an esterase derived from *Babesia bovis* and its use as a vaccine, *Z. Parasitenkd.,* 69, 703, 1983.

34. **Goodger, B. V., Commins, M. A., Wright, I. G., Waltisbuhl, D. J., and Mirre, G. B.,** Successful homologous vaccination against *Babesia bovis* using a heparin-binding fraction of infected erythrocytes, *Int. J. Parasitol.,* 17, 935, 1987.

35. **Erp, E. E., Smith, R. D., Ristic, M., and Osorno, B. M.,** Continuous *in vitro* cultivation of *Babesia bovis, Am. J. Vet. Res.,* 41, 1141, 1980.

36. **Smith, R. D., James, M. A., Ristic, M., Aikawa, M., and Vega, C. A.,** Bovine babesiosis: protection of cattle with culture-derived *Babesia bovis* immunogens, *Science,* 212, 335, 1981.

37. **Timms, P., Dalgliesh, R. J., Barry, D. N., Dimmock, C. K., and Rodwell, B. J.,** *Babesia bovis:* comparison of culture-derived parasites, non-living antigen and conventional vaccine in the protection of cattle against heterologous challenge, *Aust. Vet. J.,* 60, 75, 1983.

38. **Montenegro-James, S. and Ristic, M.,** Heterologous strain immunity in bovine babesiosis using a culture derived soluble *Babesia bovis* immunogen, *Vet. Parasitol.,* 18, 321, 1985.

39. **James, M. A.,** An update on the isolation and characterization of culture-derived soluble antigens of *Babesia bovis, Vet. Parasitol.,* 14, 231, 1984.

40. **Ristic, M., Smith, R. D., and Kakoma, I.,** Characterization of *Babesia* antigens derived from cell culture and ticks, in *Babesiosis,* Ristic, M. and Kreier, J. P., Eds., Academic Press, New York, 1981, 337.

41. **Goodger, B. V., Wright, I. G., and Waltisbuhl, D. J.,** The lysate from bovine erythrocytes infected with *Babesia bovis.* Analysis of antigens and a report on their immunogenicity when polymerized with glutaraldehyde, *Z. Parasitenkd.,* 69, 473, 1983.

42. **Goodger, B. V., Wright, I. G., Waltisbuhl, D. J., and Mirre, G. B.,** *Babesia bovis:* successful vaccination against homologous challenge in splenectomised calves using a fraction of haemagglutinating antigen, *Int. J. Parasitol.,* 15, 175, 1985.

43. **Waltisbuhl, D. J., Goodger, B. V., Wright, I. G., Mirre, G. B., and Commins, M. A.,** *Babesia bovis:* vaccination studies with three groups of high molecular weight antigens from lysate of infected erythrocytes, *Parasitol. Res.,* 73, 319, 1987.

44. **Johnston, L. A. Y., Pearson, R. D., and Leatch, G.,** Evaluation of an indirect fluorescent antibody test for detecting *Babesia argentina* infection in cattle, *Aust. Vet. J.,* 49, 373, 1973.

45. **Goodger, B. V., Waltisbuhl, D. J., Wright, I. G., Mahoney, D. F., and Commins, M. A.,** *Babesia bovis:* analysis and vaccination trial with the cryoprecipitable immune complex, *Vet. Immunol. Immunopathol.,* 14, 57, 1987.

46. **Wright, I. G., White, M., Tracey-Patte, P. D., Donaldson, R. A., Goodger, B. V., Waltisbuhl, D. J., and Mahoney, D. F.,** *Babesia bovis:* isolation of a protective antigen by using monoclonal antibodies, *Infect. Immun.,* 41, 244, 1983.

47. **McLaughlin, G. L., Edlind, T. D., and Ihler, G. M.,** Detection of *Babesia bovis* using DNA hybridization, *J. Protozool.,* 33, 125, 1986.

48. **Cowman, A. F., Timms, P., and Kemp, D. J.,** DNA polymorphisms and subpopulations in *Babesia bovis, Mol. & Biochem. Parasitol.,* 11, 91, 1984.

49. **Cowman, A. F., Bernard, O., Stewart, N., and Kemp, D. J.,** Genes of the protozoon parasite *Babesia bovis* that rearrange to produce RNA species with different sequences, *Cell,* 37, 653, 1984.

50. **Kuttler, K. L. and Johnson, L. W.,** Immunization of cattle with a *Babesia bigemina* antigen in Fruend's complete adjuvant, *Am. J. Vet. Res.,* 412, 536, 1980.

51. **Banerjee, D. P. and Prasad, K. D.,** The use of killed vaccine in immunization of cattle against *Babesia bigemina* infection, *Indian J. Vet. Med.,* 51, 1985.

52. **Todorovic, R. A., Gonzalez, E. F., and Adams, L. G.,** *Babesia babesiosis:* sterile immunity to *Babesia bigemina* and *Babesia argentina* infection, *Trop. Anim. Health Prod.,* 5, 234, 1973.

53. **Thompson, K. C., Todorovic, R. A., and Hidalgo, R. J.,** Antigenic variation of *Babesia bigemina, Res. Vet. Sci.,* 23, 51, 1977.

54. **Smith, R. D., Molinar, E., Larios, F., Monroy, J., Trigo, F., and Ristic, M.,** Bovine babesiosis: pathogenicity and heterologous species immunity of tick-borne *Babesia bovis* and *Babesia bigemina, Am. J. Vet. Res.,* 41, 1957, 1980.

55. **Vega, C. A., Buening, G. M., Green, T. J., and Carson, C. A.,** *In vitro* cultivation of *Babesia bigemina, Am. J. Vet. Res.,* 46, 416, 1985.

56. **McElwain, T. F., Perryman, L. E., Davis, W. C., and McGuire, T. C.,** Antibodies define multiple proteins with epitopes exposed on the surface of live *Babesia bigemina* merozoites, *J. Immunol.,* 138, 2298, 1987.

57. **Hinaidy, H. K.,** Die babesiose des rindes in Osterreich. IV. Versuche mit totimpfstoffen, *Berl. Muench. Tieraerztl. Wochenschr.,* 94, 121, 1981.

58. **Lewis, D. and Williams, H.,** Infection of the Mongolian gerbil with the cattle piroplasm *Babesia divergens, Nature (London),* 278, 170, 1979.

59. **Phillips, R. S.,** *Babesia divergens* in splenectomised rats, *Res. Vet. Sci.,* 36, 251, 1984.

60. **Winger, C. M., Canning, E. U., and Culverhouse, J. D.,** A monoclonal antibody to *Babesia divergens* which inhibits merozoite invasion, *Parasitology,* 94, 17, 1987.

61. **Taylor, S. M., Elliott, C. T., and Kenny, J.,** Isolation of antigenic proteins from erythrocytes parasitised with *Babesia divergens* and a comparison of their immunising potential, *Vet. Parasitol.,* 21, 99, 1986.

62. **Taylor, S. M., Elliott, C. T., and Edgar, H. W. J.,** Further identification of *Babesia divergens* antigens by immunoprecipitation from parasitised erythrocytes and an examination of their effectiveness as immunogens in cattle, *Vet. Parasitol.,* 1987, in press.

63. **Taylor, S. M., Elliott, C. T., and Davidson, W. B.,** Isolation of a protective antigen capable of inducing protective immunity to *Babesia divergens, Parasitol. Res.,* 1987, submitted.

64. **Taylor, S. M., Kenny, J., and Mallon, T. R.,** The effect of route of administration of a *Babesia divergens* inactivated vaccine on protection against homologous challenge, *J. Comp. Pathol.,* 93, 423, 1983.

65. **Taylor, S. M., Kenny, J., Mallon, T. R., and Elliott, C. T.,** The immunisation of cattle against *Babesia divergens* with fractions of parasitised erythrocytes, *Vet. Parasitol.,* 16, 235, 1984.

66. **Brocklesby, D. W. and Purnell, R. E.,** Failure of BCG to protect calves against *Babesia divergens* infection, *Nature (London),* 265, 343, 1977.

67. **Phillips, R. S., Reid, G. M., McLean, S. A., and Pearson, C. D.,** Antigenic diversity in *Babesia divergens:* preliminary results with three monoclonal antibodies to the rat-adapted strain, *Res. Vet. Sci.,* 42, 96, 1987.

68. **Kyurtov, N.,** A study of soluble *Babesia ovis* antigen, *Vet. Med. Nauki,* 16, 15, 1979.

69. **Kyurtov, N.,** Preparation of an extract of *Babesia ovis* and tests of its antigenic properties, *Vet. Med. Nauki,* 16, 84, 1979.

70. **Singh, B., Gautum, O. P., and Banerjee, D. P.,** Immunization of donkeys against *Babesia equi* infection using killed vaccine, *Vet. Parasitol.,* 8, 133, 1981.

71. **Banerjee, D. P., Singh, B., Gautum, D. P., and Sarup, S.,** Cell-mediated immune response in equine babesiosis, *Trop. Anim. Health Prod.,* 9, 153, 1977.

72. **Molinar, E., James, M. A., Kakoma, I., Holland, C., and Ristic, M.,** Antigenic and immunogenic studies on cell culture derived *Babesia canis, Vet. Parasitol.,* 10, 29, 1982.

73. **Moreau, Y. and Laurent, N.,** Antibabesial vaccination using antigens from cell culture fluids: industrial requirements, in *Malaria and Babesiosis,* Ristic, M., Ambroise-Thomas, P., and Kreier, J. P., Eds., Martinus Nijhoff, Dodrecht, Netherlands, 1984, 129.

74. **Goodger, B. V., Wright, I. G., and Mahoney, D. F.,** The use of pathophysiological reactions to assess the efficacy of the immune response to *Babesia bovis* in cattle, *Z. Parasitenkd.,* 66, 41, 1981.

75. **Goodger, B. V.,** A cold precipitable fibrinogen complex in the plasma of cattle dying from infection with *Babesia argentina, Z. Parasitenkd.,* 48, 1, 1975.

76. **Goodger, B. V., Wright, I. G., Waltisbuhl, D. J., and Mirre, G. B.,** *Babesia bovis*: successful vaccination against homologous challenge in splenectomised calves using a fraction of haemagglutinating antigen, *Int. J. Parasitol.,* 15, 175, 1985.

77. **Saba, T. A. and Jaffe, E.,** Plasma fibronectin (opsonic glycoprotein) its synthesis by vascular endothelial cells and role in cardiopulmonary integrity after trauma as related to reticuloendothelial function, *Am. J. Med.,* 68, 577, 1980.
78. **Goodger, B. V., Wright, I. G., and Mahoney, D. F.,** Changes in conglutinin, immunoconglutinin, complement C3 and fibronectin concentrations in cattle acutely infected with *Babesia bovis, Aust. J. Exp. Biol. Med. Sci.,* 59, 531, 1981.
79. **Lachmann, P. J.,** Conglutinin and immunoconglutinins, *Adv. Immunol.,* 6, 479, 1967.
80. **Kalter, E. S.,** Inflammatory mediators and acute infections, *Resuscitation,* 11, 133, 1984.
81. **Goodger, B. V.,** Unpublished work.
82. **Bahr, G. M., Majeed, H. A., Yousof, A. M., Chedid, L., and Behbehani, K.,** Detection of antibodies to muramyl dipeptide, the adjuvant moiety of streptococcal cell wall in patients with rheumatic fever, *J. Infect. Dis.,* 154, 1012, 1986.

Chapter 6

VACCINES AGAINST *THEILERIA PARVA*

A. D. Irvin and W. I. Morrison

TABLE OF CONTENTS

I. INTRODUCTION

Theileria parva is a tick-borne protozoan parasite of cattle and African buffalo (*Syncerus caffer*). Buffalo invariably are asymptomatic carriers of the infection, whereas infection of cattle results in an acute, often fatal disease known as East Coast fever (ECF). The disease is currently known to exist in Burundi, Kenya, Malawi, Mozambique, Rwanda, Southern Sudan, Tanzania, Uganda, Zaire, and Zambia.[1] A closely related form of theileriosis (January disease) occurs in Zimbabwe.[2,3]

The economic losses caused by ECF are hard to ascertain, but it is estimated that some 25 million cattle may be at risk from the disease and that some half million deaths occur each year.[1] Production losses may also occur in animals chronically infected with ECF, a condition which appears to be common in indigenous cattle in endemic areas.[4,5]

In addition to direct losses from ECF, the amount of money spent on trying to control the disease constitutes a major drain on the economy of African farmers and national veterinary services; losses result from cost of acaricides to control ticks, cost of therapeutic drugs, loss of grazing time of animals going for dipping, and increased management costs. Some of these costs have to be met with foreign exchange, and this can further cripple the ailing economy of poor countries. The situation has been reached in some countries where these costs cannot be met and ECF has to be left to run its course; this in turn virtually rules out the prospects of introducing improved livestock to meet the increasing demands of meat and milk for human consumption. In order for African countries to overcome ECF, ways of controlling the disease must be sought which do not need to rely on payment of foreign exchange and similar costly measures. In this respect, vaccination could provide a solution. The current review, therefore, focuses on the possible approaches to vaccination against ECF. Details of other and wider aspects of the disease are given in a number of previous reviews.[6-9]

II. LIFE CYCLE

The most important vector of *T. parva* is the ixodid tick *Rhipicephalus appendiculatus*. A number of other tick species can transmit the parasite but, of these, only *R. evertsi*[10] and *R. zambeziensis*[11,12] are likely to be natural vectors, and then only on a very limited scale. The parasite is transmitted transstadially by the tick vector (i.e., from larva to nymph or nymph to adult), and there is no transovarial transmission.

An outline of the life cycle of *T. parva* is shown in Figure 1. Herein, we will focus on the developmental stages of the parasite within the mammalian host. More detailed accounts of the life cycle are given elsewhere.[8,13]

When an infected tick feeds on a susceptible host, sporozoites are ejected in the saliva of the tick and thus enter the host at the site of the tick bite. Sporozoites very rapidly enter host lymphocytes,[14] causing them to transform to lymphoblasts,[15] while at the same time, the parasite itself changes to a schizont which divides in synchrony with the host cell.[16,17] Rapid lymphoproliferation of parasitized cells, followed by their subsequent destruction (and possibly also that of nonparasitized lymphoid cells), constitute the main pathological features of ECF.[18,19] The kinetics of lymphoid cell changes and the resultant pathology of the disease have been discussed in detail elsewhere.[8,19,20]

During the course of the infection, a proportion of the schizonts undergo merogeny to produce large numbers of merozoites which, when they are released from the lymphoid cells, enter erythrocytes and develop to piroplasms.[13] Despite the fact that a high percentage of cells may become infected with piroplasms, this stage does not appear to be pathogenic in that there is no evidence of hemolytic anemia. This is in contrast to some of the other theilerial parasites such as *T. mutans* and *T. annulata*, which can cause severe anemia. It

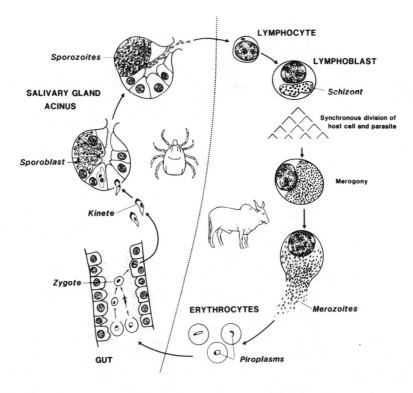

FIGURE 1. Life cycle of *T. parva*.

is generally believed that *T. parva* piroplasms are not pathogenic because they undergo minimal division. However, the recent finding by Conrad and others[21] that the intraery-throcytic stage multiplies to form merozoites *in vitro*, indicates that parasite replication may contribute to piroplasm parasitemia, although the extent to which this occurs *in vivo* remains to be determined.

Ticks which feed on infected cattle imbibe parasitemic blood, which initiates the tick cycle of *T. parva*.

III. THE DISEASE

In addition to classical ECF, there are two other forms of the disease: buffalo theileriosis or Corridor disease and Zimbabwean theileriosis or January disease. Formerly, the three syndromes were regarded as specific disease entities, for which the causative agents were respectively designated *T. parva*, *T. lawrencei*, and *T. bovis*.[22] But since it has been shown that the parasites cannot readily be differentiated, except on the basis of the clinical syndromes they cause, the subspecific names of *T. parva parva*, *T. p. lawrencei*, and *T. p. bovis* have been proposed[23] and are now generally accepted, at least for convenience, even if not taxonomically fully justified. Herein, the term ECF will be used to encompass all three forms of the disease. The clinical and pathological features of the diseases have recently been described in detail.[1,8]

In susceptible cattle, classical ECF caused by *T. p. parva* is normally characterized by high morbidity and mortality rates, the presence of large numbers of schizonts throughout the lymphoid system, and readily detectable levels of piroplasm parasitemia. Infected animals usually die within 3 to 4 weeks of infection as a result of widespread lymphocytolysis within the lymphoid tissues and pulmonary edema associated with invasion of the lungs by parasitized lymphocytes.

T. p. lawrencei is a widespread parasite of the African buffalo (*Syncerus caffer*), in which it causes inapparent disease. However, when ticks which have fed on infected buffalo feed in a subsequent instar on cattle, fatal theileriosis can be transmitted. Although the disease is highly fatal, the number of circulating parasites is low and, unlike *T. p. parva* and *T. p. bovis*, the parasite is not readily transmitted by ticks; thus, the disease tends to be self-limiting. However, if *T. p. lawrencei* is passaged experimentally in cattle using artificially applied ticks, the parasite appears to change after only four or five passages and then produces a disease in cattle which is indistinguishable from classical ECF.[20-24] Conrad and colleagues[27] have suggested that this apparent change may be due to selection of parasites, present in buffalo in low numbers, which have a propensity for growing in cattle, rather than a change in pathogenicity of the starting parasite population.

T. p. bovis is a parasite of cattle which appears to be restricted to Zimbabwe and parts of Zambia. It causes a generally less severe disease than *T. p. parva*, and mild reactions with recovery commonly occur.[2,28] Rapid passage of this parasite through ticks and cattle experimentally has so far failed to result in reversion to classical *T. p. parva*.[29]

To date, it has not been possible to distinguish the three forms of *T. parva* by serological methods. Current studies on their antigenic composition and DNA, using monoclonal antibodies (MAb)[27,30,31] and DNA probes,[32] will hopefully resolve whether they are true subspecies or merely variations within the same species.

Cattle which recover from ECF either naturally[4] or with the aid of chemotherapy[33] commonly become carriers of the infection. In some instances this carrier state can only readily be detected by feeding ticks on the animals and testing the ticks for infectivity.

In animals that have been raised in regions where ECF is not endemic, there is little evidence for differences in susceptibility to infection between different breeds of cattle. However, in some locations where the disease prevails, herds of indigenous zebu cattle (*Bos indicus*) survive in the absence of disease control measures.[4,5,34] The basis of the resistance of these animals has not been studied in detail, but the available evidence suggest that it is at least partly due to genetic factors.[34]

IV. NATURE OF IMMUNITY

The nature of immunity in theileriosis is extremely complex, involving both humoral and cellular immune responses by the host to all the different parasite stages (sporozoite, schizont, merozoite, and piroplasm). The nature and mechanisms of immunity have been reviewed extensively in recent years,[8,9,35,36] and only a brief summary is given here to provide a background against which immunization methods have been or are being developed.

A. Humoral Immunity

Animals which recover from ECF develop antibodies against all parasite stages.[37] Antibodies against schizonts and piroplasms can readily be detected by screening sera against the appropriate antigen in an indirect fluorescent antibody (IFA) test.[38-40] However, these antibodies appear to play very little role in protecting the host against subsequent challenge. Thus, although the levels of antibody tend to wane below detectable limits after about 6 months,[41] the host is often still resistant to challenge (even in the absence of reinfection) for several years.[22,42,43] Attempts to transfer immunity passively with serum from immune cattle have been unsuccessful.[44] Moreover, although isolated schizont and piroplasm antigens,[45] or killed schizont-infected cells,[46] elicit a humoral response in the host, they provide no effective protection.

Recent work, however, has shown that antisporozoite antibodies can neutralize the infectivity of sporozoites,[47,48] at least *in vitro*. These findings indicate a possible approach to vaccination, which is discussed in further detail below (Section V.B).

Although antischizont and antipiroplasm antibodies appear to play a negligible role in host protection, their presence in host sera is a good indicator of previous exposure to disease or of effective vaccination. These antibodies are therefore extensively screened in routine diagnostic serology of ECF.[38-40,49]

B. Cell-Mediated Immunity

There is clear evidence that in cattle which recover from infection with *T. parva*, a major component of the protective immune response is directed against the schizont-infected cell.[9,35] The finding that immunity can be transferred between immune and naive cattle twins with thoracic duct lymphocytes,[50] together with the evidence that antischizont antibody does not play a significant role in immunity, indicate that cell-mediated immune responses are important in immunity. Several studies have demonstrated that when immunity is established, cytotoxic T lymphocytes are generated which kill autologous parasitized cells.[51-53] These T cells appear to recognize a specific parasite-induced antigen on the cell surface in conjunction with host major histocompatibility complex (MHC) molecules.[35,53,54] Since the latter are highly polymorphic between individuals of a species, the cytotoxic T cells recognize and kill autologous parasitized cells, but in most instances, not parasitized cells from allogeneic animals. This phenomenon, known as MHC-restriction, probably accounts for the difficulty in inducing immunity with nonviable preparations of schizonts. Another important feature of the cytoxic T cells is that, in some instances, they are parasite strain specific.[53,54] Recent evidence indicates that noncytotoxic helper T cells, which are similarly MHC restricted and parasite specific, are also involved in the generation of these immune responses.[55]

Initial priming of this cellular mechanism results in generation of memory cells which respond rapidly to reinfection of the host by a homologous parasite and stimulate reactivation of specific cytotoxic T lymphocytes and elimination of infection.

In order, however, for this immune response to be initiated following challenge with sporozoites, infection must develop to the schizont level,[8] and this of course can jeopardize the health of the host unless the schizont reaction can be contained. Much of the effort towards developing an effective immunization strategy with viable parasites is, therefore, directed towards controlling the level of schizont reaction within the host.

V. APPROACHES TO VACCINATION

A. Vaccination Against the Tick

A number of studies (recently reviewed by Willadsen[56] and Wickel[57]) have demonstrated the feasibility of immunizing mammalian hosts against tick antigens and of inducing a level of immunity in them which can give a substantial reduction in tick infestation levels, in tick feeding, and in tick reproductive capacity. In preliminary experiments with *R. appendiculatus*, reduced transmission rates of *T. parva* were also demonstrated.[58]

Although vaccination against ticks is unlikely ever to result in total inhibition of tick feeding, reduction in tick burdens and of parasite transmission can be expected, and the method could be used as an adjunct to a vaccine directed more specifically towards one or more stages of *T. parva*.

B. Vaccination Against Sporozoites

Following feeding of infected ticks on immune cattle or injections of purified sporozoites into the cattle, antisporozoite antibodies are produced which neutralize, *in vitro*, the infectivity of sporozoites for bovine lymphocytes.[47,48] Repeated sporozoite challenge substantially boosts antibody levels in the host.[48]

Subsequent work has shown that antisporozoite MAbs can also neutralize sporozoites, not only those of a strain homologous to that against which the MAb was raised, but also sporozoites of heterologous strains,[59,60] suggesting that a common antigenic determinant exists on sporozoites of different strains of *T. parva*. This antigen appears to be synthesized during sporogony in the tick salivary gland.[61] Electron microscopy of sporozoites labeled with protein A colloidal gold bound to specific MAb, indicates that the antigen is located in the surface coat of mature sporozoites.[62,63] This coat is shed when sporozoites invade host lymphocytes and appears to remain attached to the lymphocyte surface.[61,63]

These findings indicate that it may be possible to immunize cattle against infection with sporozoite antigen. The possibility of using intact sporozoites, however, is remote because of practical difficulties in obtaining them in sufficient numbers from ticks. Moreover, since sporogony occurs within a highly specialized cell type in the tick salivary gland,[64] and as sporozoites do not replicate, there is little prospect of cultivating the organisms *in vitro*. The alternative approach is to identify and isolate the relevant sporozoite surface antigens for immunization. Several antisporozoite MAbs which exhibit neutralizing activity have been shown to react with a sporozoite protein of approximately 68 kDa.[62,65] Whether or not there are other sporozoite surface molecules which elicit neutralizing antibody responses has yet to be determined. However, the demonstration that at least one surface antigen is protein in nature offers the hope that the antigen can be produced *in vitro* in sufficient quantities for immunization. This may be achieved by identifying and isolating the gene which encodes the molecule, thus allowing production of the antigen in an *in vitro* expression system, or by determining the amino acid sequence of the molecule and constructing synthetic peptides corresponding to segments of the molecule which make up determinants recognized by neutralizing antibody and helper T cells.

Such an approach to vaccination does not involve potentially infective material, and the immunity would theoretically protect against all strains of the parasite. However, in order to be effective high levels of antibody would be needed, and these levels would need to be sustained because any sporozoites which broke through the antibody screen would be potentially infective to the host and fatal disease could result. If this approach can be validated, work would then be justified on developing means of potentiating and prolonging the antibody response, for example, by use of adjuvants.

Antisporozoite antibodies can be found in the sera of indigenous cattle constantly exposed to challenge in endemic areas.[66] Theoretically, such antibodies can provide protection against challenge and, when passed from dams to their calves via colostrum, may protect the neonatal calf. The antisporozoite vaccine approach, therefore, appears to be feasible, although it remains to be determined whether this approach will confer complete or partial protection against the infection.

The possibility that antibodies to sporozoites may also influence the development of sporozoites in the tick when it feeds on an immunized animal has recently been examined.[67] Passive transfer of antisporozoite MAb into rabbits on which infected ticks were fed resulted in a small but significant delay in the maturation of sporozoites in the tick salivary glands. Further studies are required to determine whether or not this effect could significantly reduce sporozoite challenge.

C. Immunization by Infection and Treatment
1. Infection and Treatment with Tetracyclines
It has been known for many years that cattle which recover from ECF are immune to subsequent challenge by the parasite. However, attempts to immunize cattle by titrating the infective dose of sporozoites with the intention of producing mild, self-limiting infections have been unsuccessful.[68,69] Very low doses of parasites did not infect all animals, with the result that some cattle were susceptible to challenge, while higher doses which were uniformly infective produced severe clinical disease in a proportion of animals. Subsequent studies, therefore, have focused on using drugs to attenuate the infection.

Immunization by infection and treatment (or chemoprophylaxis) with tetracyclines, is currently the most successful method of immunizing cattle against ECF. Its successful development has depended on advances in two areas which have made the method, when used under controlled conditions, effective, economical, and safe. The first advance was on the therapeutic front following the initial finding that chlortetracycline or oxytetracycline, if administered during the incubative stage of the disease, would suppress the development of ECF,[70,71] even though the drugs had very little therapeutic activity against clinical disease.[72] Developments in the chemotherapy of ECF have been reviewed by Dolan[73] and McHardy.[74]

The other area of advance which has made immunization by infection and treatment feasible is that infective sporozoites can be harvested from ticks, cryopreserved in stabilate form, and then used to give a quantified level of challenge by inoculation.[75] The earlier work leading to development of stabilates is reviewed elsewhere.[7,8]

Immunization by infection and treatment involves simultaneous inoculation of animals with infective tick stabilate and a long-acting formulation of oxytetracycline given at 20 mg/kg body weight.[76] This preparation of oxytetracycline provides significant drug activity in the animal for a period of 4 to 5 d. However, other formulations of the drug have been shown to be equally effective,[77-79] and some appear to offer economic advantages,[80] although they need to be administered on more than one occasion. The mechanism of action of oxytetracycline appears to be in reducing protein synthesis in parasitized cells and in parasite mitochondria;[81] this slows down cell and parasite development and allows time for host immune mechanisms to be engaged.

The work which has led to the successful development of immunization by infection and treatment with tetracyclines has been reviewed elsewhere,[7,8,79,82] and the practical aspects of preparing stabilates and immunizing cattle are given in a recent FAO manual.[83]

The strength of immune protection afforded by infection and treatment immunization is unrelated to the level of clinical response generated during immunization.[82] Thus, cattle which undergo mild or even inapparent reactions are fully resistant to lethal homologous challenge. This finding has recently been endorsed by work in which immunized cattle, which were given up to a 1000-fold lethal challenge, underwent only mild reactions to this challenge, whereas controls died very rapidly.[84] However, the major shortcoming of the immunity engendered by infection and treatment is that cattle may be susceptible to challenge with a parasite strain heterologous to that which initiated immunity.[82]

2. Infection and Treatment with Other Compounds

Recently, two theilericidal drugs, parvaquone[85] and halofuginone,[86] have been produced which are effective in the therapy of ECF. Because some isolates of *T. parva*, particularly those of buffalo origin, are more difficult to control than others with oxytetracycline,[87] an alternative strategy of immunization using parvaquone has been advocated. If parvaquone is used to suppress the initial parasite reaction, it either kills the parasites, with the result that immunity does not develop, or delays the onset of clinical disease.[88] However, if treatment is delayed until 8 d after inoculation of cattle with sporozoites, when clinical symptoms are first apparent, animals can successfully be immunized.[89] As with prophylactic treatment with tetracyclines, cattle immunized with one stock of the parasite may be susceptible to challenge with heterologous stocks.[88] Although the method works well and appears safe, it does involve handling animals on at least two occasions, and there is a danger that an infected animal could be overlooked at the time of treatment.

Recent work has indicated that a parvaquone analogue, buparvaquone, although initially developed as a therapeutic,[90] can be used prophylactically.[91] This drug, therefore, appears to have potential for use in infection and treatment immunization for ECF.

3. The Use of Parasite Strains of Low Pathogenicity

In the case of *T. p. bovis* from Zimbabwe, a number of mild strains have been isolated;[92]

these have been used for immunization on the basis of infection alone, with no chemotherapy. Animals immunized in this way have been shown to resist challenge not only with more virulent *T. p. bovis* stocks, but also with virulent *T. p. parva* and *T. p. lawrencei* stocks.[93]

The use of a mild *T. parva* strain to immunize against more virulent strains is attractive as a field strategy because it is cheap and easy to administer, but stocks for immunization should be well characterized and the possibility of reversion to virulence should be considered. This latter problem should be avoidable if a reference stabilate is maintained and working stabilates are always derived from such early passage material. Changes in antigenicity are more likely to occur if parasites are sequentially passaged.

4. Parasite Strain Heterogeneity

The major obstacle to widespread application of the infection and treatment method of immunization is the problem of parasite strain heterogeneity. However, current evidence indicates that the number of antigenically different strains of *T. parva* may be limited. In the case of *T. p. parva*, immunization with one or two key stocks can give good protection against field challenge.[94] Where *T. p. lawrencei* stocks are involved, the antigenic spectrum is broader and breakthroughs occur more commonly.[95-98] This problem can usually be overcome by using a cocktail of stocks to provide a wide spectrum of cover.[76,97-99] However, because animals immunized by infection and treatment may become carriers of the infection and thus may transmit the parasites to the resident tick population,[33,94] there is an understandable reluctance to implement immunization programs with parasite strains which may not be present in the region where immunization is to be applied. Thus, the strategy which has been adopted is to isolate and characterize local parasite stocks for use in infection and treatment immunization in each region.[94,98]

At present, the characterization and selection of parasite stocks for use in immunization programs relies largely on immunization and cross-challenge experiments in cattle. However, a number of methods of characterizing parasite stocks *in vitro* are now being explored.[100]

A series of monoclonal antibodies specific for schizont antigens has been produced, and these have been shown to detect antigenic differences between parasite stocks, so that different stocks exhibit distinct MAb profiles.[27,30,31] Although preliminary studies have shown that among parasite stocks which differ in MAb profile, some combinations do not cross-protect, those stocks showing similar profiles give good cross-protection.[101] It remains to be determined to what extent the heterogeneity detected by MAbs correlates with differences in the capacity to cross-protect.

Recently, DNA sequences which appear as multiple copies in the *Theileria* genome have been isolated, and when used as DNA probes on restriction enzyme digests of parasite DNA, they detect heterogeneity between different parasite stocks.[32] This work is still in its early stages, and further studies will be required to determine whether or not the restriction fragment length polymorphisms detected with DNA probes correlate with differences observed in cross-protection.

The third method which has been used to detect parasite strain differences has utilized T cell clones specific for parasitized lymphoblasts. Cytotoxic T cell clones generated from cattle immunized with either *T. parva* (Muguga) or *T. parva* (Marikebuni) have been tested on target cells infected with either of these two stocks.[54,102] Those generated against *T. parva* (Muguga) were specific for target cells infected with the Muguga stock, whereas those generated against *T. parva* (Marikebuni) killed target cells infected with either of the two stocks. These results appear to correlate with the finding that cattle immunized with *T. parva* (Marikebuni) are resistant to challenge with either of the two stocks,[103] whereas a proportion of animals immunized against *T. parva* (Muguga) are susceptible to challenge with the Marikebuni stock.[101] Since T cell responses are believed to be important in mediating immunity against *T. parva*, it is indeed likely that the antigenic heterogeneity detected by

T cell clones will be of relevance to cross protection. However, it is unlikely that such T cells could be widely applied as laboratory reagents for detection of parasite strain heterogeneity, because their production and maintenance requires a high degree of expertise and they require to be tested on target cells of defined MHC phenotype.

5. Field Application of Infection and Treatment

Infection and treatment immunization has now reached the stage of extended field trials, and provided supervision is good, the success rate has been high.[79,80,87,99] Problems of quality control and mode of delivery still need to be explored, but the method is practically and economically feasible on a large scale.

Parasite strain selection is probably the most important criterion to consider in field strategies using infection and treatment immunization, but other factors also need to be considered, of which age and breed of cattle may be important. Young calves below 1 month of age can be successfully immunized,[104] and there appears to be no age difference in response to immunization, although the possible influence of age on duration of immunity has not been fully explored. Cattle immunized at 6 months of age and then maintained free of ticks showed protection when exposed to field challenge a year later.[94] Under field conditions, it is unlikely that the challenge interval would be as prolonged, and it would seem that infection and treatment immunization would therefore only be needed on a single occasion; boosting of immunity would occur by natural challenge. *Bos taurus* breeds are more likely to show clinical symptoms on immunization[105] and, in such breeds, more careful selection of stabilate dose may be required than for *Bos indicus* breeds.

Immunization against ECF will clearly not directly affect other tick-borne diseases, and as ECF is brought under control by immunization, their impact can expect to become greater, particularly in situations where rigorous dipping to control ECF is relaxed. Thus, immunization or other control measures against these diseases still need to be considered and implemented where necessary.

D. Vaccination Against Schizonts

Although immunization by infection and treatment involves inoculation of the host with *T. parva* sporozoites, the resultant immunity which develops is largely directed against the infected cell. Thus, it should be feasible to use schizont-infected cells themselves as an immunogen; indeed, many of the early attempts at vaccination (reviewed in Reference 10) utilized crude cell suspensions obtained from *Theileria*-infected cattle. Although many thousands of animals were immunized in this way in the early days,[106] the crudity of material and the difficulty of quantifying infectivity meant that the method was both hazardous and unpredictable. When, however, schizont-infected cells were grown *in vitro* and continuous lines of *T. parva*-infected bovine lymphoid cells obtained,[107,108] the prospect of developing a tissue culture vaccine seemed very real, particularly as such an approach had already been shown to be feasible as a means of vaccination against *T. annulata*.[109]

The work with *T. annulata* progressed to the development of a successful vaccination procedure for field use, which involves inoculating animals with 10^6 parasitized cells.[110,111] Unfortunately, this approach proved to be less successful for *T. parva* because the minimum number of cells required to give protection in the majority of animals was around 10^8, representing about 100 ml of culture material.[112,113] Moreover, a small percentage of animals developed severe clinical reactions and sometimes mortalities occurred. The induction of immunity was apparently dependent on transfer of the parasite into the cells of the recipient host,[46,112] a process which, in the case of *T. parva*, only occurs consistently when animals receive large numbers of parasitized cells.

If lymphocytes are taken from an uninfected animal and infected *in vitro* with sporozoites of *T. parva*, a schizont-infected cell line can be obtained.[15] If the infected cells are then injected into the autologous host from which they were derived, infection can be established and immunity induced in the host with as few as 10^2 cells.[114] These autologous cells are presumably capable of establishing infection in the animal without the need for transfer of the schizont. When parasitized cells are inoculated into animals with which they are matched for class I MHC A-locus-encoded specificities, 10^3 to 10^5 cells induce immunity in a proportion of animals.[115] In this situation, it is unclear whether the inoculated cells themselves induce immunity or whether there is more efficient transfer of infection into the cells of recipient animals, possibly due to prolonged survival of the inoculated cells. Since allogeneic cells are rejected by the host through host-vs.-graft responses directed principally against MHC antigens, and since cells infected with *T. parva* and *T. annulata* express similar levels of MHC antigens,[116] the differences between the two parasites in cell numbers needed for effective immunization is unlikely to be due to a difference in survival of the cells following inoculation. Rather, it is probable that the phenomenon is due to a difference in the capacity of the free schizont to infect fresh target cells. Moreover, the cell types which the parasites infect may have an influence. In this regard, *T. parva* has been shown *in vitro* to infect T and B lymphocytes as well as non-T and non-B lymphocytes, but not monocytes;[117] by contrast, preliminary experiments indicate that *T. annulata* infects monocytes in addition to B lymphocytes.[118]

Even if the problems of cell numbers and cell "take" can be overcome, the use of schizont-infected cells to immunize against ECF still poses a danger to the host of using live, potentially virulent material. Passage of schizont-infected cells *in vitro* can result in partial attenuation and loss of virulence, but this is a difficult property to monitor, and there is the danger that loss of immunogenicity may also occur.[112,113] In addition, the problem of parasite strain variation also arises since animals immunized against one strain may not be protected if challenged with a heterologous strain. Thus, again, a cocktail of strains would be needed for field use.

In order to circumvent the problem associated with the infectivity of schizont-infected cells, it will be necessary to identify the antigens on the surface of parasitized cells against which cell-mediated immune responses are directed. Evidence supporting the argument that cell surface antigens are important in inducing protective immunity has come from experiments in which cattle inoculated with cell membrane preparations from autologous parasitized cells were resistant to challenge with sporozoites, whereas allogeneic cattle receiving similar inocula were fully susceptible.[119] Although such an approach to immunization is clearly impractical, the work suggests that once the genes coding for the parasite-induced antigens can be cloned, the way towards an alternative molecular vaccine for ECF may be feasible. An important consideration in this approach is that it will be necessary to deliver the antigens to the host in such a way that they are presented to the immune system in conjunction with those MHC molecules of the animals. One way of achieving this may be to exploit recombinant virus vectors such as vaccinia.[120]

E. Vaccination Against Merozoites and Piroplasms

Apart from sporozoites, merozoites are probably the only stage of *T. parva* which has a significant extracellular existence in the bovine host. They are, therefore, potentially vulnerable to antibody-mediated attack, which could prevent or reduce the development of piroplasm parasitemia and thus reduce cyclical transmission. However, because invasion of erythrocytes is not a significant pathogenic feature of infections with *T. parva*, little attention is currently being given to vaccination against the merozoite or piroplasm stages of the parasite. In the longer term, once effective vaccines are available for protection of the host against the pathogenic effects of ECF, a vaccine against these intraerythrocytic stages may be of more value as a means of reducing the level of parasitized erythrocytes in carrier animals and thus reducing transmission. The carrier state is a common sequel to infection and treatment immunization,[94] and although it may not be a sequel to other new methods of immunization utilizing defined parasite antigens, once immunized cattle are exposed to natural challenge, a piroplasm carrier state commonly develops. Such carrier animals provide a source of infection to ticks and a means of perpetuating disease.

There are, on the other hand, arguments against reducing the carrier state by immunization against merozoites. First, in situations where buffalo are prevalent, it is unlikely that prevention of the carrier state in cattle would have much impact on the level of field challenge. Second, new vaccines against ECF using defined parasite antigens may require field challenge to boost immunity and provide long-lasting protection, so that any steps which might eliminate or reduce natural challenge may be undesirable. The same may be true in situations where there is enzootic stability of ECF. Cattle in such locations are exposed to constant low-level challenge and invariably are carriers of the infection.[4] If the carrier state is disturbed, enzootic stability may be disrupted and the epidemiology of ECF could change, with disastrous effects on cattle populations. Thus, should effective means of immunizing against merozoites or piroplasms become available, great care should be taken in deciding where application of immunization will have beneficial rather than deleterious effects on control of ECF.

VI. FUTURE DEVELOPMENTS

It will be some years before synthetically derived vaccines can be used in the field, but in the meantime, infection and treatment immunization can be applied under supervised conditions. It is essential to ensure that such immunization is used in appropriate situations, for example, where improved cattle are to be introduced into an enzootic area. Care should be taken that introduction of immunization does not disturb enzootic stability. Likewise, where high-grade cattle are maintained in a tick-free environment, introduction of infection through use of a live vaccine may be undesirable. Thus, situations for immunization by infection and treatment need to be carefully assessed and monitored, but the method, if properly applied, could have a major impact on controlling ECF and improving cattle productivity. However, some aspects of the method need further attention; these include potency and safety testing of stabilates, improved methods of strain characterization *in vitro*, improved delivery systems, and a better understanding of duration of immunity in animals immunized during the first few months of life.

Immunization alone will not provide the solution to ECF; an integrated approach to control is needed which includes immunization, together with tick control, chemotherapy, and management. The relative emphasis to be placed on each control method will vary in different situations and circumstances, and these differences need to be defined carefully.

The prospects of developing genetically engineered or synthetic peptide vaccines for ECF control are also good in view of the great advances being made in these fields in general and the specific advances made already in identifying immunogenic sporozoite proteins. Further advances can be expected as the genes encoding the antigens on the surface of parasitized lymphoblastoid cells are identified and the prospect of using these gene products or the genes themselves within suitable vectors for immunization can be envisaged.

Other areas which justify attention include use of adjuvants to provide improved presentation to the host of parasite antigens, use of virus vectors, such as vaccinia, to carry theilerial genes for immunization, and modification of the parasite genome to produce attenuated or hybrid parasites with enhanced immunogenicity, but reduced virulence.

It is all too easy to be dazzled by the topicality, concept, and prospects of molecular and synthetic vaccines and overlook the facts that one animal dies approximately every minute from ECF, and that an effective (even if mundane) method of immunization exists in the form of infection and treatment. Until such time as an improved vaccine is developed, there is still every reason to exploit and seek to improve the current methodology available to reduce cattle death or productivity loss associated with ECF, and thus relieve some of the hardship currently faced by African farmers in ECF enzootic areas.

ACKNOWLEDGMENTS

We wish to thank Drs. Patricia Conrad and S. P. Morzaria for helpful discussion during the preparation of this manuscript. This is ILRAD publication number 000.

REFERENCES

1. **Irvin, A. D. and Mwamachi, D. M.,** Clinical and diagnostic features of East Coast fever (*Theileria parva* infection of cattle), *Vet. Rec.,* 113, 192, 1983.
2. **Lawrence, J. A.,** Bovine theileriosis in Zimbabwe, in *Advances in the Control of Theileriosis,* Irvin, A. D., Cunningham, M. P., and Young, A. S., Eds., Martinus Nijhoff, Dordrecht, Netherlands, 1981, 74.
3. **Lawrence, J. A. and Norval, R. A. I.,** A history of ticks and tick-borne diseases of cattle in Rhodesia, *Rhod. Vet. J.,* 10, 28, 1979.
4. **Young, A. S., Leitch, B. L., Newson, R. M., and Cunningham, M. P.,** Maintenance of *Theileria parva* infection in an endemic area of Kenya, *Parasitology,* 93, 9, 1986.
5. **Moll, G., Lohding, A., and Young, A. S.,** The epidemiology of theileriosis in the Trans-Mara Division, Kenya, in *Advances in the Control of Theileriosis,* Irvin, A. D., Cunningham, M. P., and Young, A. S., Eds., Martinus Nijhoff, Dordrecht, Netherlands, 1981, 56.
6. **Neitz, W. O.,** Theileriosis, *Adv. Vet. Sci.,* 5, 241, 1959.
7. **Purnell, R. E.,** East Coast fever: some recent research in East Africa, *Adv. Parasitol.,* 15, 83, 1977.
8. **Irvin, A. D. and Morrison, W. I.,** Immunopathology, immunology and immunoprophylaxis of *Theileria* infections, in *Immune Responses in Parasitic Infections: Immunopathology, Immunology and Immunoprophylaxis,* Soulsby, E. J. L., Ed., CRC Press, Boca Raton, FL, 1987, 223.
9. **Morrison, W. I., Lalor, P. A., Goddeeris, B. M., and Teale, A. J.,** Theileriosis: antigens and host-parasite interactions, in *Parasite Antigens: Towards New Strategies for Vaccines,* Pearson, T., W., Ed., Marcel Dekker, New York, 1986, 167.
10. **Barnett, S. F.,** Theileriosis, in *Infectious Blood Diseases of Man and Animals,* Vol. 2, Weinmann, D. and Ristic, M., Eds., Academic Press, New York, 1968, 269.
11. **Walker, J. B., Norval, R. A. I., and Corwin, M. D.,** *Rhipicephalus zambeziensis* sp. nov., a new tick from eastern and southern Africa, together with a redescription of *Rhipicephalus appendiculatus* Neumann, 1901 (Acarina, Ixodidae), *Onderstepoort J. Vet. Res.,* 48, 87, 1981.
12. **Lawrence, J. A., Norval, R. A. I., and Uilenberg, G.,** *Rhipicephalus zambeziensis* as a vector of bovine *Theileria, Trop. Anim. Health Prod.,* 15, 39, 1983.
13. **Mehlhorn, H. and Schein, E.,** The piroplasms: life cycle and sexual stages, *Adv. Parasitol.,* 23, 37, 1984.
14. **Fawcett, D. W., Doxsey, S., Stagg, D. A., and Young, A. S.,** The entry of sporozoites of *Theileria parva* into bovine lymphocytes *in vitro.* Electron microscopic observations, *Eur. J. Cell Biol.,* 27, 10, 1982.
15. **Brown, C. G. D., Stagg, D. A., Purnell, R. E., Kanhai, G. K., and Payne, R. C.,** Infection and transformation of bovine lymphoid cells *in vitro* by infective particles of *Theileria parva, Nature (London),* 245, 101, 1973.
16. **Hulliger, L., Wilde, J. K. H., Brown, C. G. D., and Turner, L.,** Mode of multiplication of *Theileria* in cultures of bovine lymphocytic cells, *Nature (London),* 203, 728, 1964.
17. **Irvin, A. D., Ocama, J. G. R., and Spooner, P. R.,** Cycle of bovine lymphoblastoid cells parasitized by *Theileria parva, Res. Vet. Sci.,* 33, 298, 1982.
18. **De Martini, J. C. and Moulton, J. E.,** Responses of the bovine lymphatic system to infection by *Theileria parva.* I. Histology and ultrastructure of lymph nodes in experimentally-infected calves, *J. Comp. Pathol.,* 83, 281, 1973.

19. **Morrison, W. I., Buscher, G., Murray, M., Emery, D. L., Masake, R., Cook, R. H., and Wells, P. W.,** *Theileria parva*: kinetics of infection in the lymphoid system of cattle, *Exp. Parasitol.*, 52, 248, 1981.

20. **Emery, D. L.,** Kinetics of infection with *Theileria parva* (East Coast fever) in the central lymph of cattle, *Vet. Parasitol.*, 9, 1, 1981.

21. **Conrad, P. A., Denham, D., and Brown, C. G. D.,** Intraerythrocytic multiplication of *Theileria parva in vitro*: an ultrastructural study, *Int. J. Parasitol.*, 16, 223, 1986.

22. **Neitz, W. W.,** Theileriosis, gonderiosis and cytauxzoonosis: a review, *Onderstepoort J. Vet. Res.*, 27, 275, 1957.

23. **Uilenberg, G.,** Tick-borne livestock diseases and their vectors. II. Epizootiology of tick-borne diseases, *World Anim. Rev.*, 17, 8, 1976.

24. **Barnett, S. F. and Brocklesby, D. W.,** The passage of *"Theileria lawrencei"* (Kenya) through cattle, *Br. Vet. J.*, 122, 396, 1966.

25. **Young, A. S. and Purnell, R. E.,** Transmission of *Theileria lawrencei* (Serengeti) by the ixodid tick *Rhipicephalus appendiculatus*, *Trop. Anim. Health Prod.*, 5, 146, 1973.

26. **Young, A. S., Branagan, D., Brown, C. G. D., Burridge, M. J., Cunningham, M. P., and Purnell, R. E.,** Preliminary observations on a theilerial species pathogenic to cattle isolated from buffalo (*Syncerus caffer*) in Tanzania, *Br. Vet. J.*, 129, 382, 1973.

27. **Conrad, P. A., Stagg, D. A., Grootenhuis, J. G., Irvin, A. D., Newson, J., Njamunggeh, R. E. G., Rossiter, P. B., and Young, A. S.,** Isolation of *Theileria* parasites from African buffalo (*Syncerus caffer*) and characterization with anti-schizont monoclonal antibodies, *Parasitology*, 94, 413, 1987.

28. **Lawrence, J. A.,** The differential diagnosis of the bovine theilerias in Southern Africa, *J. S. Afr. Vet. Assoc.*, 50, 311, 1979.

29. **Uilenberg, G., Perie, N. M., Lawrence, J. A., de Vos, A. J., Paling, R. W., and Spanjer, A. A. M.,** Causal agents of bovine theileriosis in Southern Africa, *Trop. Anim. Health Prod.*, 14, 127, 1982.

30. **Pinder, M. and Hewett, R. S.,** Monoclonal antibodies detect diversity in *Theileria parva* parasites, *J. Immunol.*, 124, 1000, 1980.

31. **Minami, T., Spooner, P. R., Irvin, A. D., Ocama, J. G., Dobbelaere, D. A. E., and Fujinaga, T.,** Characterization of stocks of *Theileria parva* by monoclonal antibody profiles, *Res. Vet. Sci.*, 35, 334, 1983.

32. **Conrad, P. A., Iams, K., Brown, W. C., Sohanpal, B., and ole-MoiYoi, O. K.,** DNA probes detect genomic diversity in *Theileria parva* stocks, *Mol. Biochem. Parasitol.*, 25, 000, 1987.

33. **Dolan, T. T.,** Chemotherapy of East Coast fever: the long term weight changes, carrier state and disease manifestations of parvaquone treated cattle, *J. Comp. Pathol.*, 96, 137, 1986.

34. **Barnett, S. F.,** Theileriosis control, *Bull. Epizoot. Dis. Afr.*, 5, 343, 1957.

35. **Morrison, W. I., Goddeeris, B. M., Teale, A. J., Baldwin, C. L., Bensaid, A., and Ellis, J.,** Cell-mediated immune responses of cattle to *Theileria parva*, *Immunol. Today*, 7, 211, 1986.

36. **Irvin, A. D.,** Immunity in theileriosis, *Parasitol. Today*, 1, 124, 1985.

37. **Cowan, K. M.,** The humoral responses in *Theileria* in *Advances in the Control of Theileriosis*, Irvin, A. D., Cunningham, M. P., and Young, A. S., Eds., Martinus Nijhoff, Dordrecht, Netherlands, 1981, 368.

38. **Burridge, M. J.,** Application of the indirect fluorescent antibody test in experimental East Coast fever (*Theileria parva* infection of cattle), *Res. Vet. Sci.*, 12, 338, 1971.

39. **Burridge, M. J. and Kimber, C. D.,** The indirect fluorescent antibody test for experimental East Coast fever (*Theileria parva*) infection of cattle: evaluation of a cell culture schizont antigen, *Res. Vet. Sci.*, 13, 451, 1972.

40. **Goddeeris, B. M., Katende, J. M., Irvin, A. D., and Chumo, R. S. C.,** Indirect fluorescent antibody test for experimental and epizootiological studies on East Coast fever (*Theileria parva* infection in cattle). Evaluation of a cell culture antigen fixed and stored in suspension, *Res. Vet. Sci.*, 33, 360, 1982.

41. **Burridge, M. J. and Kimber, C. D.,** Duration of serological response to the indirect fluorescent antibody test of cattle recovered from *Theileria parva* infection, *Res. Vet. Sci.*, 44, 270, 1973.

42. **Barnett, S. F.,** Theileriosis control, *Bull. Epizoot. Dis. Afr.*, 5, 343, 1957.

43. **Burridge, M. J., Morzaria, S. P., and Cunningham, M. P.,** Duration of immunity to East Coast fever (*Theileria parva* infection of cattle), *Parasitology*, 64, 511, 1972.

44. **Muhammed, S. I., Lauerman, L. H., and Johnson, L. W.,** Effect of humoral antibodies on the course of *Theileria parva* infection (East Coast fever) of cattle, *Am. J. Vet. Res.*, 36, 399, 1975.

45. **Wagner, G. G., Duffus, W. P. H., and Burridge, M. J.,** The specific immunoglobulin response in cattle immunized with isolated *Theileria parva* antigens, *Parasitology*, 69, 43, 1974.

46. **Emery, D. L., Morrison, W. I., Nelson, R. T., and Murray, M.,** The induction of cell-mediated immunity in cattle inoculated with cell lines parasitized with *Theileria parva*, in *Advances in the Control of Theileriosis*, Irvin, A. D., Cunningham, M. P., and Young, A. S., Eds., Martinus Nijhoff, Dordrecht, Netherlands, 1981, 295.

47. **Gray, M. A. and Brown, C. G. D.,** *In vitro* neutralization of theilerial sporozoite infectivity with immune serum, in *Advances in the Control of Theileriosis,* Irvin, A. D., Cunningham, M. P., and Young, A. S., Eds., Martinus Nijhoff, Dordrecht, Netherlands, 1981, 127.

48. **Musoke, A. J., Nantulya, V. M., Buscher, G., Masake, R. I., and Otim, B.,** Bovine immune response to *Theileria parva*: neutralizing antibodies to sporozoites, *Immunology,* 45, 663, 1982.

49. **Duffus, W. P. H and Wagner, G. G.,** Comparison between certain serological tests for diagnosis of East Coast fever, *Vet. Parasitol.,* 6, 313, 1980.

50. **Emery, D. L.,** Adoptive transfer of immunity to infection with *Theileria parva* (East Coast fever) between cattle twins, *Res. Vet. Sci.,* 30, 364, 1981.

51. **Emery, D. L., Eugui, E. M., Nelson, R. T., and Tenywa, T.,** Cell-mediated immune responses to *Theileria parva* (East Coast fever) during immunization and lethal infections in cattle, *Immunology,* 43, 323, 1981.

52. **Eugui, E. M. and Emery, D. L.** Genetically restricted cell-mediated cytotoxicity in cattle immune to *Theileria parva, Nature (London),* 290, 251, 1981.

53. **Morrison, W. I., Goddeeris, B. M., Teale, A. J., Groocock, C. M., Kemp, S. J., and Stagg, D. A.,** Cytotoxic T cells elicited in cattle challenged with *Theileria parva* (Muguga): evidence for restriction by class I MHC determinants and parasite strain specificity, *Parasite Immunol.,* in press.

54. **Goddeeris, B. M., Morrison, W. I., Teale, A. J., Bensaid, A., and Baldwin, C. L.,** Bovine cytotoxic T-cell clones specific for cells infected with the protozoan parasite *Theileria parva*: parasite strain specificity and class I major histocompatibility complex restriction, *Proc. Natl. Acad. Sci. U.S.A.,* 83, 5238, 1986.

55. **Baldwin, C. L., Goddeeris, B. M., and Morrison, W. I.,** Bovine helper T-cell clones specific for lymphocytes infected with *Theileria parva* (Muguga), *Parasite Immunol.,* in press.

56. **Willadsen, P.,** Immunity to ticks, *Adv. Parasitol.,* 18, 293, 1980.

57. **Wikel, S. K.,** Immunomodulation of host response to ectoparasite infestation — an overview, *Vet. Parasitol.,* 14, 321, 1984.

58. **Cunningham, M. P.,** Biological control of ticks with particular reference to *Rhipicephalus appendiculatus,* in *Advances in the Control of Theileriosis,* Irvin, A. D., Cunningham, M. P., and Young, A. S., Eds., Martinus Nijhoff, Dordrecht, Netherlands, 1981, 160.

59. **Musoke, A. J., Nantulya, V. M., Rurangirwa, F. R., and Buscher, G.,** Evidence for a common protective antigenic determinant on sporozoites of several *Theileria parva* strains, *Immunology,* 52, 231, 1984.

60. **Dobbelaere, D. A. E., Spooner, P. R., Barry, W. C., and Irvin, A. D.,** Monoclonal antibody neutralizes the sporozoite stage of different *Theileria parva* stocks, *Parasite Immunol.,* 6, 361, 1984.

61. **Dobbelaere, D. A. E., Webster, P., Leitch, B. L., Voigt, W. P., and Irvin, A. D.,** *Theileria parva*: expression of a sporozoite surface coat antigen, *Exp. Parasitol.,* 60, 90, 1985.

62. **Dobbelaere, D. A. E., Shapiro, S. Z., and Webster, P.,** Identification of a surface antigen on *Theileria parva* sporozoites by monoclonal antibody, *Proc. Natl. Acad. Sci. U.S.A.,* 82, 1771, 1985.

63. **Webster, P., Dobbelaere, D. A. E., and Fawcett, D. W.,** The entry of sporozoites of *Theileria parva* into bovine lymphocytes *in vitro*. Immunoelectron microscopic observations, *Eur. J. Cell Biol.,* 36, 157, 1985.

64. **Fawcett, D. W., Buscher, G., and Doxsey, S.,** Salivary gland of the tick vector of East Coast fever. III. The ultrastructure of sporogony in *Theileria parva, Tissue Cell,* 14, 183, 1982.

65. **Musoke, A. J.,** Unpublished data.

66. **Musoke, A. J. and Morzaria, S. P.,** Unpublished data.

67. **Fujisaki, K., Irvin, A. D., Dobbelaere, D. A. E., Voigt, W. P., Musoke, A. J., Morzaria, S. P., and Gettinby, G.,** Anti-sporozoite antibody passing across the gut of *Rhipicephalus appendiculatus* ticks delays maturation of *Theileria parva* in the salivary glands, in preparation.

68. **Cunningham, M. P., Brown, C. G. D., Burridge, M. J., Musoke, A. J., Purnell, R. E., Radley, D. E., and Sempebwa, C.,** East Coast fever: titration in cattle of suspensions of *Theileria parva* derived from ticks, *Br. Vet. J.,* 130, 336, 1974.

69. **Dolan, T. T., Young, A. S., Losos, G. J., McMillan, I., Minder, C. H., and Soulsby, K.,** Dose dependent responses of cattle to *Theileria parva* stabilate, *Int. J. Parasitol.,* 14, 89, 1984.

70. **Neitz, W. O.,** Aureomycin in *Theileria parva* infection, *Nature (London),* 171, 34, 1953.

71. **Brocklesby, D. W. and Bailey, K. P.,** The immunization of cattle against East Coast fever (*Theileria parva* infection) using tetracyclines: a review of the literature and a reappraisal of the method, *Bull. Epizoot. Dis. Afr.,* 13, 161, 1965.

72. **Brown, C. G. D., Radley, D. E., Burridge, M. J., and Cunningham, M. P.,** The use of tetracyclines in the chemotherapy of experimental East Coast fever (*Theileria parva* infection of cattle), *Tropenmed. Parasitol.,* 28, 513, 1977.

73. **Dolan, T. T.,** Progress in the chemotherapy of theileriosis, in *Advances in the Control of Theileriosis,* Irvin, A. D., Cunningham, M. P., and Young, A. S., Eds., Martinus Nijhoff, Dordrecht, Netherlands, 1981, 186.

74. **McHardy, N.,** Recent advances in the chemotherapy of theileriosis, in *Impact of Diseases of Livestock Production in the Tropics,* Riemann, H. P. and Burridge, M. J., Eds., Elsevier, Amsterdam, 1984, 179.

75. **Cunningham, M. P., Brown, C. G. D., Burridge, M. J., and Purnell, R. E.,** Cryopreservation of infective particles of *Theileria parva, Int. J. Parasitol.,* 3, 583, 1973.

76. **Radley, D. E., Brown, C. G. D., Cunningham, M. P., Kimber, C. D., Musisi, F. L., Payne, R. C., Purnell, R. E., Stagg, S. M., and Young, A. S.,** East Coast fever. III. Chemoprophylactic immunization of cattle using oxytetracycline and a combination of theilerial strains, *Vet. Parasitol.,* 1, 51, 1975.

77. **Radley, D. E., Brown, C. G. D., Burridge, M. J., Cunningham, M. P., Kirimi, I. M., Purnell, R. E., and Young, A. S.,** East Coast fever. I. Chemoprophylactic immunization of cattle against *Theileria parva* (Muguga) and five theilerial strains, *Vet. Parasitol.,* 1, 35, 1970.

78. **Radley, D. E., Young, A. S., Brown, C. G. D., Burridge, M. J., Cunningham, M. P., Musisi, F. L., and Purnell, R. E.,** East Coast fever. II. Cross immunity trials with a Kenya strain of *Theileria lawrencei, Vet. Parasitol.,* 1, 43, 1975.

79. **Radley, D. E.,** Infection and treatment method of immunization against theileriosis, in *Advances in the Control of Theileriosis,* Irvin, A. D., Cunningham, M. P., and Young, A. S., Eds., Martinus Nijhoff, Dordrecht, Netherlands, 1981, 227.

80. **Young, A. S.,** Immunisation of cattle against theileriosis in the Trans Mara Division of Kenya, in *Immunization Against Theileriosis in Africa,* Irvin, A. D., Ed., International Laboratory for Research on Animal Diseases, Nairobi, 1985, 64.

81. **Spooner, P. R.,** Unpublished data.

82. **Cunningham, M. P.,** Immunization of cattle against *Theileria parva,* in *Theileriosis,* Henson, J. B. and Campbell, M., Eds., IDRC, Ottawa, 1977, 66.

83. FAO, *Ticks and Tick-Borne Disease Control. A Practical Field Manual,* Food and Agricultural Organization, Rome, 1984.

84. **Morzaria, S. P., Irvin, A. D., Voigt, W. P., and Taracha, E. L. N.,** Effect of timing and intensity of challenge following immunization against East Coast fever, *Vet. Parasitol.,* 26, 29, 1987.

85. **McHardy, N., Hudson, A. T., Morgan, D. W. T., Rae, D. G., and Dolan, T. T.,** Activity of ten naphthoquinones, including parvaquone (993c) and menoctone, in cattle artificially infected with *Theileria parva, Res. Vet. Sci.,* 35, 347, 1983.

86. **Schein, E. and Voigt, W. P.,** Chemotherapy of bovine theileriosis with halofuginone, *Acta Trop.,* 36, 391, 1979.

87. **Robson, J., Pedersen, V., Odeke, G. M., Kamya, E. P., and Brown, C. G. D.,** East Coast fever immunization trials in Uganda: field exposure of zebu cattle immunized with 3 isolates of *Theileria parva, Trop. Anim. Health Prod.,* 9, 219, 1977.

88. **Dolan, T. T.,** The choice of drug for infection and treatment immunisation against East African theileriosis, in *Immunization Against Theileriosis in Africa,* Irvin, A. D., Ed., International Laboratory for Research on Animal Diseases, Nairobi, 1985, 100.

89. **Dolan, T. T., Linyoni, A., Mbogo, S. K., and Young, A. S.,** Comparison of long-acting oxytetracycline and parvaquone in immunization against East Coast fever by infection and treatment, *Res. Vet. Sci.,* 37, 175, 1984.

90. **McHardy, N., Wekesa, L. S., Hudson, A. T., and Randall, A. W.,** Antitheilerial activity of BW720C(buparvaquone): a comparison with parvaquone, *Res. Vet. Sci.,* 39, 29, 1985.

91. **McHardy, N. and Wekesa, L. S.,** Buparvaquone (BW720c), a new anti-theilerial napthoquinone: its role in the therapy and prophylaxis of theileriosis, in *Immunization Against Theileriosis in Africa,* Irvin, A. D., Ed., International Laboratory for Research on Animal Diseases, Nairobi, 1985, 88.

92. **Koch, H. T., Ocama, J. G. R., Munatswa, F. C., Byrom, B., Norval, R. A. I., Spooner, P. R., Conrad, P. A., and Irvin, A. D.,** Isolation and characterization of bovine *Theileria* parasites in Zimbabwe, *Vet. Parasitol.,* 28, 19, 1988.

93. **Irvin, A. D., Morzaria, S. P., Munatswa, F. C., and Koch, H. T.,** Immunization of cattle with a *Theileria parva bovis* stock from Zimbabwe protects against challenge with virulent, *T. p. parva* and *T. p. lawrencei* stocks from Kenya, submitted.

94. **Morzaria, S. P., Irvin, A. D., Taracha, E., Spooner, P. R., Voigt, W. P., Fujinaga, T., and Katende, J.,** Immunization against East Coast fever: the use of selected stocks of *Theileria parva* for immunization of cattle exposed to field challenge, *Vet. Parasitol.,* 23, 23, 1987.

95. **Young, A. S., Radley, D. E., Cunningham, M. P., Musisi, F. L., Payne, R. C., and Purnell, R. E.,** Exposure of immunized cattle to prolonged natural challenge of *Theileria lawrencei* derived from African buffalo (*Syncerus caffer*), *Vet. Parasitol.,* 3, 283, 1977.

96. **Cunningham, M. P., Brown, C. G. D., Burridge, M. J., Irvin, A. D., Kirimi, I. M., Purnell, R. E., Radley, D. E., and Wagner, G. I.,** Theileriosis: the exposure of immunized cattle in a *Theileria lawrencei* enzootic area, *Trop. Anim. Health Prod.,* 6, 39, 1974.

97. **Radley, D. E., Young, A. S., Grootenhuis, J. G., Cunningham, M. P., Dolan, T. T., and Morzaria, S. P.,** Further studies on the immunization of cattle against *Theileria lawrencei* by infection and chemoprophylaxis, *Vet. Parasitol.,* 5, 117, 1979.

98. **Dolan, T. T., Radley, D. E., Brown, C. G. D., Cunningham, M. P., Morzaria, S. P., and Young, A. S.,** East Coast fever. IV. Further studies on the protection of cattle immunized with a combination of theilerial strains, *Vet. Parasitol.,* 6, 325, 1980.

99. **Uilenberg, G., Silayo, R. S., Mpangala, C., Tondeur, W., Tatchell, R. J., and Sanga, H. J. N.,** Studies on Theileriidae (sporozoa) in Tanzania, X. A large-scale field trial on immunization against cattle theileriosis, *Tropenmed. Parasitol.,* 28, 499, 1977.

100. **Irvin, A. D.,** Characterization of species and strains of *Theileria, Adv. Parasitol.,* 26, 145, 1987.

101. **Irvin, A. D., Dobbelaere, D. A. E., Mwamachi, D. M., Minami, T., Spooner, P. R., and Ocama, J. G. R.,** Immunization against East Coast fever: correlation between monoclonal antibody profiles of *Theileria parva* stocks and cross immunity *in vivo, Res. Vet. Sci.,* 35, 341, 1983.

102. **Morrison, W. I., Goddeeris, B. M., and Teale, A. J.,** Bovine cytotoxic T cell clones which recognise lymphoblasts infected with two antigenically different stocks of the protozoan parasite *Theileria parva,* submitted.

103. **Morzaria, S. P.,** Unpublished data.

104. **Irvin, A. D., Dobbelaere, D. A. E., Morzaria, S. P., Spooner, P. R., Goddeeris, B. M., Chumo, R. S., Taracha, E. L. N., Dolan, T. T., Young, A. S., and Gettinby, G.,** East Coast fever: the significance of host age in infection or immunization of cattle with *Theileria parva,* submitted.

105. **Radley, D. E.,** Immunization against East Coast fever by chemoprophylaxis, in *Research on Tick-Borne Diseases and Tick Control Kenya, Tanzania, Uganda,* Tech. Rep. 1, AG: DP/67/077, Food and Agricultural Organization, Rome, 1978, 37.

106. **Spreull, J.,** East Coast fever inoculations in the Transkeian Territories, South Africa, *J. Comp. Pathol. Ther.,* 27, 299, 1914.

107. **Malmquist, W. A., Nyindo, M. B. A., and Brown, C. G. D.,** East Coast fever: cultivation *in vitro* of bovine spleen cell lines infected and transformed by *Theileria parva, Trop. Anim. Health Prod.,* 2, 139, 1970.

108. **Brown, C. G. D.,** Propagation of *Theileria,* in *Practical Tissue Culture Applications,* Maramorosch, K. and Hirumi, H., Eds., Academic Press, New York, 1979, 223.

109. **Pipano, E. and Tsur, I.,** Experimental immunization against *Theileria annulata* with a tissue culture vaccine, *Refu. Vet.,* 23, 133, 1966.

110. **Pipano, E.,** Schizont and tick stages in immunization against *Theileria annulta* infection, in *Advances in the Control of Theileriosis,* Irvin, A. D., Cunningham, M. P., and Young, A. S., Eds., Martinus Nijhoff, Dordrecht, Netherlands, 1981, 242.

111. **Pipano, E.,** Immune response in calves to varying numbers of attenuated schizonts of *Theileria annulata, J. Protozool.,* 17, 31, 19.

112. **Brown, C. G. D., Crawford, J. G., Kanhai, G. K., Njuguna, L. M., and Stagg, D. A.,** Immunization of cattle against East Coast fever with lymphoblastoid cell lines infected and transformed by *Theileria parva,* in *Tick-borne Diseases and Their Vectors,* Wilde, J. K. H., Ed., University of Edinburgh Press, Edinburgh, 1978, 331.

113. **Brown, C. G. D.,** Application of *in vitro* techniques to vaccination against theileriosis, in *Advances in the Control of Theileriosis,* Irvin, A. D., Cunningham, M. P., and Young, A. S., Eds., Martinus Nijhoff, Dordrecht, Netherlands, 1981, 104.

114. **Buscher, G., Morrison, W. I., and Nelson, R. T.,** Titration in cattle of infectivity and immunogenicity of autologous cell lines infected with *Theileria parva, Vet. Parasitol.,* 15, 29, 1984.

115. **Dolan, T. T., Teale, A. J., Stagg, D. A., Kemp, S. J., Cowan, K. M., Young, A. S., Groocock, C. M., Leitch, B. L., Spooner, R. L., and Brown, C. G. D.,** A histocompatibility barrier to immunization against East Coast fever using *Theileria parva*-infected lymphoblastoid cells, *Parasite Immunol.,* 6, 243, 1984.

116. **Spooner, R. L. and Brown, C. G. D.,** Bovine lymphocyte antigens (BoLA) of bovine lymphocytes and derived lymphoblastoid lines transformed by *Theileria parva* and *Theileria annulata, Parasite Immunol.,* 2, 163, 1980.

117. **Baldwin, C. L., Black, S. J., Brown, W. C., Conrad, P. A., Goddeeris, B. M., Kinuthia, S. W., Lalor, P. A., MacHugh, N. D., Morrison, W. I., Morzaria, S. P., Naessens, J., and Newson, J.,** Bovine T-cells, B-cells and null cells are transformed *in vitro* and *in vivo* by the protozoan parasite *Theileria parva,* submitted.

118. **Spooner, R. L. and Brown, C. G. D.,** Unpublished data.

119. **Emery, D. L., Morrison, W. I., and Jack, R. M.,** Induction of immunity against infection with *Theileria parva* (East Coast fever) in cattle using plasma membranes from parasitized lymphoblasts, *Vet. Parasitol.,* 19, 321, 1986.

120. **Mackett, M. and Smith, G. L.,** Vaccinia virus expression vectors, *J. Gen. Virol.,* 67, 2067, 1986.

Chapter 7

THE POTENTIAL FOR *TRYPANOSOMA* VACCINE DEVELOPMENT

Stuart Z. Shapiro

TABLE OF CONTENTS

I. INTRODUCTION

A. Genus *Trypanosoma*

The genus *Trypanosoma* is composed of protozoa which have a single flagellum extending from a dense, DNA-containing organelle called a kinetoplast. Members of this genus are heteroxenous parasites. They live alternately in the bloodstream and tissues of vertebrates and in the gut of leeches or arthropods. The genus is split into two divisions (*Stercoraria* and *Salivaria*) which differ primarily in their course of development in their vectors.[1]

The first division, *Stercoraria*, contains parasites that complete their development in the terminal gut of their vector; they are passed with vector feces. *Trypanosoma cruzi*, the American trypanosome that causes Chagas' disease in humans, is the most well-known species of this division. Although *T. cruzi* can also cause disease in some domestic animals (e.g., the dog), its primary importance is as a pathogen of human beings. The other stercorarian trypanosomes are relatively nonpathogenic in animals.[2]

This chapter will focus on vaccine development against the members of the other section of *Trypanosoma*, the salivarian trypanosomes. These parasites cause diseases with profound economic implications in livestock in the developing world. The salivarian trypanosomes complete their course of development in the anterior part of the digestive tract of tsetse flies. The mature parasites are then transmitted with vector saliva when the fly salivates while taking a blood meal.[3,4] These parasites are also called African trypanosomes because it is on the African continent where the diseases they cause are most prevalent.

The major pathogens among the African trypanosomes fall into three subgenera: *Duttonella* (type species *Trypanosoma [Duttonella] vivax*), *Nannomonas* (type species *Trypanosoma [Nannomonas] congolense*), and *Trypanozoon* (type species *Trypanosoma [Trypanozoon] brucei*). Parasites of all species cause disease in domestic animals, while disease in human beings is caused only by two subspecies of *T. brucei*: *T.[T.] b. gambiense* and *T.[T.] b. rhodesiense*.

B. African Trypanosomes

African trypanosomes are responsible for widespread disease in human beings and domestic animals. In Africa these parasites are usually transmitted by tsetse flies which infest an area of about 10 million km^2.[5] Outside this area these parasites may be transmitted by other biting flies, by vampire bats, or venereally. They cause disease in the rest of the world over an area approximately three times the size of the tsetse belt[5] (Figure 1). It is in the African tsetse belt, however, that the African trypanosome has had its greatest impact.

The African trypanosomes were first discovered in the latter part of the 19th century. *T. evansi* was discovered to be the causative agent of the disease called surra in camels and horses by Evans in 1880, in India.[1] In 1894, Bruce discovered *Trypanosoma brucei* in the blood of cattle suffering from the cattle disease nagana. Bruce demonstrated the parasite's transmission by tsetse flies in 1897.[1] *Trypanosoma brucei gambiense* and *Trypanosoma brucei rhodesiense*, the organisms which cause African sleeping sickness in human beings, were first reported in 1902 and 1903, respectively.[6]

The diseases caused by African trypanosomes have been recognized for a long time. African trypanosomiasis in human beings, African sleeping sickness, was described as early as the 14th century by an Arab traveler, al Qualquashaudi.[7] In the past, this disease has been responsible for millions of deaths in great epidemics.[6] Today, however, it has been reduced to only about 20,000 newly reported cases per year.[8,9] Recently, the disease has been on the increase in many endemic areas. Presently available control measures (drug treatment of detected cases to reduce the parasite reservoir, and vector control) require a continuous commitment. Thus, the breakdown in medical services and vector control programs coupled with the large population movements which have been caused by political,

133

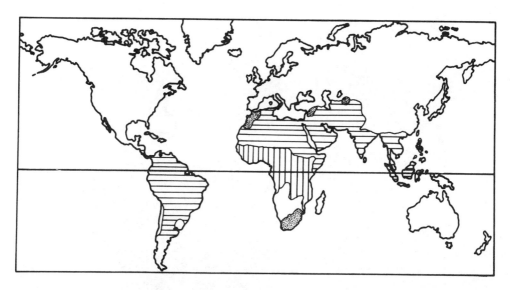

FIGURE 1. Worldwide distribution of disease caused by African trypanosomes. The area of the African tsetse belt is indicated by vertical stripes. The horizontally striped regions are where salivarian trypanosome are transmitted by other biting flies and vampire bats. The stippled regions are where *T. [T.] equiperdum* is transmitted as a venereal disease. (From Woo, P. T. K., in *Parasitic Protozoa*, Vol. 1, Kreier, J. P., Ed., Academic Press, New York, 1977, chap. 7. With permission.)

economic, and social instabilities has led to new epidemics in many areas (e.g., Sudan,[8] Zaire,[10] Ethiopia,[11] and Uganda[12]). An estimated 50 million people in 34 countries in Africa are at risk of getting human trypanosomiasis.[9]

While these parasites certainly cause a significant amount of disease in human beings, their impact on human society has been even greater because of the disease they cause in domestic animals. Many African cattle-owning people have been aware of African trypanosomiasis in livestock from very early times; indeed, the disease may have been known to the Hebrew prophet Isaiah about 3000 years ago.[13] It was probably this disease that limited the southward travel of Arab trading caravans and, thus, the penetration of Islam into sub-Saharan Africa.[7] Animal trypanosomiasis today still presents a serious endemic economic problem because it renders the vast area of the tsetse belt unfit for cattle production[11] (Figure 2). In addition to denying the rapidly expanding human population in this area a needed source of animal protein, the scarcity of cattle restricts crop production because of lack of draft animals and animal manure. This disease also seriously affects most other economically important domestic animals in Africa including sheep, goats, horses, donkeys, and camels. Outside the tsetse belt, across a vast area of North Africa, South and Central America, the Middle East, and parts of Europe, India, and Southeast Asia, the African trypanosome *T. [T.] evansi* causes substantial disease, primarily in camels and horses, but also in water buffaloes and cattle (Figure 2). *T. [T.] equiperdum* causes dourine, a venereal disease of horses, in South Africa, northwest Africa, and parts of the Middle East (Figure 2). *T. vivax* has been reported to cause disease in cattle in the Caribbean and parts of South and Central America (Figure 2). Altogether, human and animal trypanosomiasis present a serious obstacle to health and development, especially in Africa, but also in many other parts of the developing world.

Limited control of the diseases have been achieved by great and continuous effort. Vector control by aerial and ground spraying of insecticides has produced some success in reducing disease prevalence.[8,15,16] However, insecticide application can be expensive and is a potential source of environmental pollution.[8] Also, since total eradication of tsetse is not feasible, these vector control strategies involve a long-term effort requiring trained personnel and

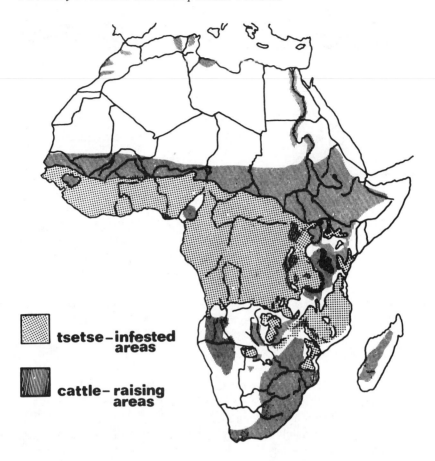

FIGURE 2.   The distribution of tsetse flies and cattle on the continent of Africa. (From Nantulya.[14])

technology not always available in many areas of Africa. Vector control by sterile male release, while showing limited success in small areas,[17,18] does not show major promise for disease control.[8] More recently, vector control by low-cost, locally produced tsetse traps impregnated with insecticide has shown great promise,[9] but this approach may still be subject to disruptions caused by political instabilities. Medical surveillance by mobile teams with subsequent chemotherapeutic treatment of discovered cases has reduced the reservoir of human disease in many areas[15] and remains the backbone of human trypanosomiasis control.[8] However, the trypanocidal drugs are costly and have serious toxic side effects, and the development of drug resistance can be a problem.[19] As a result, mass drug prophylaxis has largely been abandoned.[8] Clearly, additional control measures are needed. Possibilities for the development of an immunization program against African trypanosomiasis should not be overlooked. Disease control by vaccination would probably require less highly trained personnel and certainly would require less continuous effort than the strategies currently employed.

II. ATTEMPTS AT IMMUNIZATION AGAINST AFRICAN TRYPANOSOMIASIS

A. Early Experiments

Early vaccination attempts against African trypanosomiasis followed the standard classical approaches to vaccine development. Indeed, several of the earliest experiments were performed by two of the great men of early modern medicine, Robert Koch and Paul Ehrlich. In 1901, Koch is reported to have attempted to produce attenuated parasites for vaccination by passage of African trypanosomes through other host species.[20] This approach was soon abandoned; repeated passage, which today is a standard procedure to make the parasite more virulent for laboratory hosts,[21] appears to have very little effect on virulence in the natural host.[22] Ehrlich was more successful; it is reported that in 1904 he produced immunity to challenge with a strain of trypanosome by infecting with that strain and then curing the infected animals with a trypanocidal drug.[23] (Perhaps it is a reflection on the slow progress made in chemotherapy against this disease that we still use today two drugs, suramin and tryparsamide, which were derived from drugs developed by Ehrlich in the early days of pharmacology.)

Later investigators tried most of the other tricks of early immunologists: "killed" vaccines, radiation-attenuated parasites, formalinized parasites, and homogenates or partially purified antigens from trypanosomes. They reported producing strain-specific immunity with killed parasites;[24] strain-specific immunity was also produced with irradiated (motile but division-suppressed) parasites,[25-27] formalinized parasites,[28,29] and parasite homogenates or partially purified parasite antigens.[30-34] In fact, it seemed relatively easy to generate strain-specific immunity to African trypanosomes. The failures that did occur were discussed in terms of insufficient adjuvant activity[24] or unsuccessful preservation of functional antigen.[29] However, rarely was immunity against heterologous strains reported.[20]

At the same time, as these studies were being performed, other studies suggested that not only did strains differ antigenically, but also that the antigenic characteristics of African trypanosomes within a strain were unstable. Franke,[35] in 1905, showed that the serologic characteristics of an African trypanosome population appeared to change during the course of infection in a mammalian host. Ritz[36] then showed that this change was not a reflection of heterogeneity in the inoculum, but rather was the inherent property of a single parasite. Subsequently, several investigators were able to derive many serologically different types from single trypanosome lines.[37-40] It is now known that this serologic variation reflects a change in the glycoprotein antigen that composes the coat present on the surface of the bloodstream-form African trypanosome (Figure 3). The protective immune responses induced by early vaccinators were directed at these antigens. Many early investigators passaged their trypanosome lines at 3-d intervals to stabilize their antigenic characteristics.[42] This probably explains why antigenic variation did not interfere with the early vaccination attempts with homologous strains. Also, it seems likely that the occasional reports of protection against heterologous strains were because of incidental antigenic cross-reactivity between the surface glycoproteins.[20]

B. Antigenic Variation

1. Serological and Biochemical Observations

The instability observed in the serologic characteristics of an African trypanosome population during the course of an infection is called "antigenic variation". The potential number of distinct variable surface antigens expressible by any given trypanosome is unknown, but serologic studies indicate that it is in excess of 100,[43] and molecular biological studies suggest it may be up to 1000.[44] Upon infection, the immune response of the host appears to consist largely of production of antibodies against these surface antigens.[20,45,46] While most of the trypanosomes in the parasitemic wave are eliminated by this host immune response, some survive because they have changed their surface antigens. Antibody which binds readily to the early surface antigen does not recognize the new antigens. Parasites with new surface antigens proliferate and repopulate the host bloodstream, stimulating a new antibody response which, in turn, is evaded by parasites further changing their surface antigens.

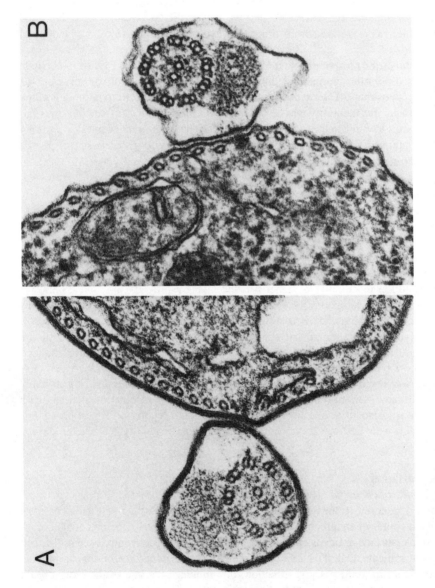

FIGURE 3. The variant surface antigen coat of *T. brucei.* At the left (A) is shown an electron micrograph of a cross section of a bloodstream-form parasite. Note the electron-dense surface coat on the peripheral side of the cell membrane and on the flagellum. At the right (B) is shown a similar cross section of an uncoated tsetse fly midgut-form (procyclic) trypanosome. (Magnification × 132,500.) (From Shapiro, S. Z. and Pearson, T. W., in *Parasite Antigens: Toward New Strategies for Vaccines,* Pearson, T. W., Ed., Marcel Dekker, New York, 1986, chap. 7. With permission.)

Two theories were developed to explain the variation of the surface antigens. As early as 1909, it was suggested that host antibodies might induce the antigenic variation observed with the African trypanosome.[47] Experiments testing this hypothesis were inconclusive until an *in vitro* culture system for the bloodstream form of the parasite was developed. It was then discovered that antigenic variation can occur entirely in the absence of specific antibody.[48] Other early investigators hypothesized that random mutations in the gene encoding the variant surface glycoprotein (VSG) were the cause of antigen variation. This hypothesis became testable when procedures were developed to identify, purify, and biochemically characterize the parasite surface antigens.

Many of the early biochemical studies on the parasite's variant surface antigens concentrated on the soluble "exoantigen" found in the plasma of infected animals. Exoantigen was observed to induce the production of antibodies that could agglutinate living bloodstream-form trypanosomes.[49] Furthermore, these antigens appeared to be variant type specific.[50] It had been observed also that the African trypanosome had an unusual electron-dense surface coat, 12 to 15 nm in thickness, peripheral to its plasma membrane[51] (Figure 3). Specific antibodies were seen to bind to this surface coat on antigenically homologous, but not on heterologous, trypanosomes.[52] It was suggested that the exoantigens were, in fact, released surface coat of the parasite.[53]

In 1975, Cross used a surface protein-reactive reagent to radiolabel the surfaces of antigenically different populations derived from one line of *T. brucei*.[54] He showed that the major component on the surface of the bloodstream parasite was a glycoprotein of about 65,000 Da in molecular mass. The glycoprotein was homogeneous for each antigenic type. Cross developed a general procedure for purification of these VSGs and analyzed them to determine their amino acid content. The glycoproteins varied so considerably in amino acid content among parasite populations that random mutational events could not account for the phenomenon of antigenic variation with this parasite.

Other investigators have confirmed these observations. Figure 4 shows, by surface labeling with radioiodine, the homogeneous protein content of the parasite surface coat. Highly variable amino acid compositions were found in other laboratories comparing VSGs purified from antigenically different parasite populations.[55,56] Comparison of tryptic peptide maps of VSGs purified by Le Page gave more evidence for substantial structural variation in the VSGs;[57] this was confirmed by similar experiments of others.[58,59] Substantial change in protein primary sequence was also indicated when amino acid sequencing of amino-terminal[60] and internal tryptic fragments[61] was performed.

2. Molecular Mechanism

In the past 8 years the effort to understand the molecular basis for antigenic variation in African trypanosomes has benefited greatly from the application of recombinant DNA technology. Clones of variant surface glycoprotein genes have been selected from trypanosome cDNA banks and genomic libraries. These clones have been used in restriction enzyme digest experiments and chromosome separation experiments to study changes in the organization of VSG genes during antigenic variation. VSG gene clones have also been sequenced to produce more information about the basic structure of the genes and their protein products. This work has been the subject of recent reviews[62,63] to which the interested reader should refer for a more detailed discussion. The following is a brief description of trypanosome VSG genes and their rearrangements and the implications of this data for understanding the molecular basis for antigenic variation.

Two different mechanisms were considered in the search for a molecular explanation of antigenic variation after the dismissal of the early hypothesis of random mutation of the VSG gene. It was possible that chromosomal sequences might be rearranged to create VSG genes in a manner similar to that which occurs in the generation of antibody diversity.[64]

T. brucei ETat 1.2

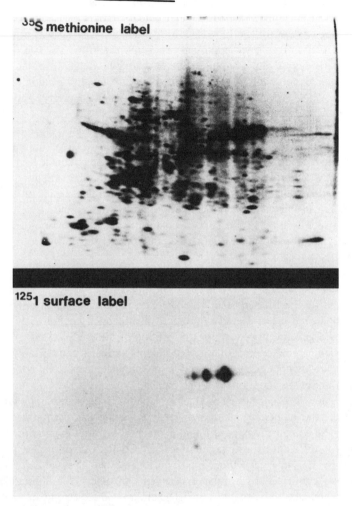

FIGURE 4.   Two-dimensional (2D) gel analysis of the surface antigen of *T. brucei*. In the top frame is shown a 2D gel autoradiogram of ³⁵S-methionine-labeled proteins from *T. b. rhodensiense* isolate ETat 1.2. Beneath it is shown an autoradiogram of a similar 2D gel where only the parasite protein exposed on the surface was labeled, with ¹²⁵I by the lactoperoxidase method; only the trypanosome variant surface antigen is labeled. (The satellite spotting in the horizontal dimension indicates charge heterogeneity in the molecule which is frequently observed with VSGs.[41])

Alternatively, it was considered that all VSGs in the repertoire of surface antigens of a given trypanosome might be encoded as genes in the genome of the parasite, and were expressed only one at a time. According to this model, antigenic variation would involve turning off the expression of one gene while turning on another.

The earliest recombinant DNA data provided clear evidence for the second mechanism. Every cDNA probe made from mRNA encoding a variant surface antigen hybridized with genomic DNA from every variant antigen type tested within the parasite line from which the probe was made.[65-69] This proved that the gene for a VSG was present in the genome whether or not that VSG was being expressed. In Northern blot experiments, it was found that probes for VSG genes hybridized to cellular mRNA only if the particular VSG was being expressed.[66,70-72] Therefore, although the genes for many VSGs are present in the genome of the African trypanosome, only one at a time appears to be transcribed into mRNA in any single trypanosome.

Serologic studies had shown that an African trypanosome could express at least 100 different VSGs.[43] With DNA probes that hybridize to highly conserved sequences at the edges of VSG genes, molecular biologists examined the parasite genome for how many VSG genes it might encode. Van Der Ploeg and co-workers[44] thus estimated that the trypanosome genome may encode about 1000 different VSG genes.

Antigenic variation, then, results from switching in which one VSG gene, of a large repertoire of such genes present in the genome, is selectively transcribed and translated. Molecular biological studies have attempted to elucidate the mechanisms for selective transcription and this switching. Early such studies indicated that an extra copy of the gene for a particular VSG appeared in the genome of parasites expressing that variant surface antigen.[66,67,69,71,73] These extra copies were called ''expression-linked copies'', or ELCs, while the original copies of the VSG genes were called ''basic copies''. These studies also indicated that the ELCs are located in a different place within the genome; that place was discovered to be at the end of a chromosome.[74] The ELCs were also found to be preferentially sensitive to DNAase I digestion, which suggested that they are the copies of the VSG genes which are actually transcribed.[68,75] The earliest model for the molecular mechanism of antigenic variation incorporated these data to propose that antigenic variation was effected by the duplication of a basic copy gene, which is not transcribed, with transposition of the new copy into a unique transcriptionally active site.[76] Transposition of a new ELC into this ''expression site'' would result in the displacement of the VSG gene previously in the site so that only one VSG gene would be expressed at a time. This was called the ''cassette model''. It is similar to the mechanism of mating-type switching in yeast.[77]

Not all VSG genes, however, demonstrated ELCs when they were expressed.[69,71,78,79] The genes that did not have been referred to as ''nonduplication-activated'' genes. The copies of these genes that are expressed are also located at the ends of chromosomes (in telomeres).[80] The cassette model was then modified to account for the expression of these VSG genes by suggesting that a VSG gene need not be duplicated, but could be brought into the expression site by reciprocal recombination of telomeres. This modification of the model then explained the observation that antigenic variation was not always accompanied by the removal of the ELC of the previously expressed VSG (this phenomenon was called a ''lingering ELC'').

Lately, other observations have suggested that there may not be only one potential expression site; VSGs may be expressed from several telomeric regions in the genome.[81] Since the number of such potential expression sites is considerably smaller than the total number of VSG genes encoded in the genome, it simplifies the process of antigenic variation. However, the mechanism whereby these expression sites are coordinated to result in the expression of only one VSG gene at a time in a given parasite is still not understood. It has recently been suggested that trypanosomes may exist that express more than one VSG at a time, but that these organisms are strongly selected against because of the disruption the presence of two different VSGs would have on the packing of the surface coat.[82] Such disruption would make the coat less protective against host immune responses.

3. Biochemistry of VSGs

While the VSGs of the different species of African trypanosomes vary somewhat in size, most appear to be proteins of about 55 to 65 kDa in molecular mass. Differential staining of bands of VSG in isoelectric focusing gels gave the first indication that the VSG might be glycoprotein.[53] This was confirmed by Cross[54] and by Johnson and Cross.[83] Cytochemical staining for carbohydrate localized the carbohydrate in *T. brucei* VSG to the boundary of the surface coat with the plasma membrane; it did not appear to be on the external surface of the parasite.[58,84] Studies with *T. [T.] equiperdum* showing that lectin bound to agarose beads failed to bind to the unaltered parasite surface also indicated that the VSG carbohydrate is not exposed on the surface of the living trypanosomes.[85] Experiments on cleavage fragments of VSG have suggested that, in some cases, there may be more than one carbohydrate moiety attached to the protein.[86,87]

The organization and attachment of VSG molecules on the surface of the trypanosome has generated much interest. The early cytochemical staining results showing all the carbohydrate of the glycoprotein to be on the plasma membrane side of the surface coat suggested the VSG was present as a layer only one molecule deep. De Almeida and Turner showed that all molecules of membrane-attached VSGs possess a hydrophobic carboxy terminal residue, which is lost upon disruption of the parasite.[88] This hydrophobic residue consists of myristic acid[89] on phsophatidyl glycerol,[90] linked via an ethanolamine group[91] to the carboxy terminal amino acid of the VSG. Thus, the VSG appears to be a monotopic membrane protein attached to the parasite plasma membrane by the insertion of covalently bound fatty acid residues into the lipid bilayer. X-ray diffraction and electron microscopy have indicated that the VSG molecule on the parasite surface contains large regions of rod-like alpha helical coils.[92,93] The association of VSG molecules with each other in the array they form on the surface of the parasite has also been studied. Cross linking of VSG molecules with bifunctional reagents has shown that VSGs can associate as dimers in solution.[94,95] Intramolecular disulfide bonding can occur with some of these purified VSGs, although whether such bonding occurs *in situ* is not known. That VSG molecules are tightly packed together on the parasite surface is indicated by the electron-dense nature of the coat,[51] by calculations of surface VSG density based on protein content,[54] and by capping studies.[96]

The data on the tight and regular organization of the VSG on the parasite surface make it unsurprising that only a small portion of the VSG molecule is exposed to the outside in the living parasite. Monoclonal antibodies have been prepared against purified VSGs and reacted with living trypanosomes to show that only a small number of the epitopes of the molecule are exposed on the surface,[97,98] and these epitopes are at the amino terminus of the VSG.[99] This is probably why the noticeable amino acid sequence homologies near the carboxy terminus of VSGs[100] do not produce serological cross-reactivity between living parasites. Such cross-reactivity would reduce the usefulness of variation of the surface antigens during the course of an infection if those epitopes were exposed. Similarly, the antigenic cross-reactivity between all VSGs, based on the antigenicity of the unusual carbohydrate linkage at the carboxy terminus,[87,101-103] does not appear to compromise the parasite because that epitope is also not exposed on the surface of the living parasite.[101]

4. Immunity Based on the Variant Surface Antigens

African trypanosomes are readily killed by variant surface antigen-specific host antibodies. A variant antigen-specific antibody response eliminates most of the parasites from the bloodstream in the first wave of parasitemia[104] and appears to be responsible for parasite killing throughout infection.[105] Immunization with purified VSG can protect animals against infection with parasites bearing that VSG on their surface.[54] Likewise, animals can be protected by transfer of serum[106,107] or B lymphocytes[107] from animals exposed to parasites of the same VSG type. Some investigators have even shown that an anti-idiotype response against parasite VSG-specific antibodies can protect an animal against trypanosomes of that VSG type. So the limitation to protective immunization using the variant surface antigens is not the lack of immunogenicity of these molecules or the inability of antibody against them to function, but rather the enormity of the task of immunizing with thousands of different antigens to protect against a single disease.

However, in some limited situations protective immunity apparently based on a host antibody response to VSGs can be produced and may be useful. Strain-specific immunity has been induced in trypanosome-infected cattle kept under laboratory conditions. These animals were well fed and watered and were protected from secondary infections, so they were better able to resist the pathologic effects of the disease than similar animals in the field. With time, the infected bovids appear to have developed antibodies against all of the parasite's VSGs, and the parasite infection disappeared with exhaustion of its VSG repertoire.[108,109] Also, many workers have reported the development of protective immunity in cattle kept under continuous prophylactic drug treatment while exposed to African trypanosome infection in the field.[20,23] The fact that such immunity appears to be trypanosome strain-specific strongly suggests that the antibody response against the variant surface antigens is the basis of protection.[110] It is hypothesized that animals which are repeatedly infected, but drug-cured to prevent the development of significant pathologic effects from the disease, eventually develop antibodies that react with all of the VSG types produced by a given trypanosome line. These animals are then refractory to infection with that line of parasite. Unfortunately this type of immunity has limited practical usefulness; if such "immune" animals are moved to an area where they are presented with another strain of the parasite they succumb to the disease,[20,23] and studies of different isolates of African trypanosomes suggest that there are very many different strains in the field.[111,112] The repeated drug treatment approach to immunity is further limited by the danger of encouraging development of drug-resistant trypanosomes.

C. Vaccination With Nonvariable Antigens

1. Trypanosome Membrane Antigens

Beneath the variant antigen surface of the African trypanosome lies a lipid bilayer plasma membrane (Figure 3A). This membrane contains proteins and other potentially antigenic molecules like the plasma membranes of other eukaryotic cells. Many of these molecules should be common to trypanosomes of different variant antigen type and different parasite line. Some of these molecules may show some species or subspecies specificity, but others probably will show antigenic cross-reactivity with many different African trypanosomes.

The view is widely held that the presence of the variant antigen surface coat precludes antibody attack on nonvarying parasite membrane antigens under the coat. However, the coat need not be totally impenetrable to provide a good immune defense. The variant surface antigens are highly immunogenic, present in substantial amounts (10% of the total protein of the bloodstream-form parasite is VSG), and each trypanosome presents its animal host with many antigenically different VSGs. It is reasonable that the VSGs might suppress the host immune response to other parasite antigens by immune competition.[113] Indeed, relatively few nonvarying parasite antigens seem to stimulate an antibody response in infected animals.[114] Thus, it could be theorized that the VSG coat acts as a smoke screen, shielding parasite membrane antigens from stimulating an immune response rather than as a barrier to antibody.

The smoke screen hypothesis led this author to attempt to vaccinate against african trypanosomiasis with parasite membrane antigens. In one experiment, rabbits were repeatedly inoculated with live procyclic trypanosomes.[115] This is the uncoated stage of the parasite found in the gut of the tsetse fly; it shares several membrane antigens with the blood stream-form parasite.[116] These parasites were not capable of establishing an infection; nevertheless, they did induce a high-titer antibody response against nonvarying parasite membrane antigens. However, when the inoculated rabbits were challenged with a bloodstream-form trypanosome infection, they showed no evidence of resistance. In another unpublished experiment, vaccination was attempted with trypanosome membrane glycolipids.[115] Such molecules can be very immunogenic.[117] Glycolipid was purified from *T. brucei* by a standard chloroform/methanol extraction.[118] Rabbits were inoculated with this material, and they produced antibodies which clearly reacted with fixed parasites in immunofluorescence tests. However, as in the experimental vaccination with procyclic trypanosomes, these antibodies failed to protect the rabbits from a challenge infection. It therefore appears that parasite membrane antigens beneath the variant antigen surface coat are probably inaccessible to antibody in the living bloodstream-form parasite.

At one time, it was suggested that the metacyclic (tsetse fly salivary gland) form of *T. vivax* might not possess a surface coat.[119] This led Rovis and co-workers[120] to vaccinate animals with purified trypanosome plasma membrane and a purified trypanosome plasma membrane protein (of 83 kDa). Although high antibody titers to these antigens were detected in all immunized animals, the animals became infected readily when challenged with *T. vivax* or *T. brucei* metacyclic forms, and the course of the disease was the same as in unimmunized control animals. This experiment was then considered evidence that *T. vivax* metacyclic forms possess a protective surface antigen coat which has been recently confirmed by electron microscopy.[121]

2. Trypanosome Enzymes

Some enzymes of the African trypanosomes are sufficiently different from comparable host enzymes to be immunogenic. Several of these molecules have been purified and used in immunizations; they are trypanosome hexokinase,[122] aldolase,[123] and phosphohexose isomerase.[124] However, as intracellular molecules, these proteins appear to be inaccessible to host antibodies. Thus, antibodies to these antigens failed to protect mice from challenge infections.[125]

III. THE PARASITE AND DISEASE

Attempts at vaccination against African trypanosomiasis have clearly been frustrated by the surface coat of the parasite and by antigenic variation. The reader will notice that this chapter, unlike other chapters in book, has the title qualified (i.e., it is ''The *Potential* for *Trypanosoma* Vaccine Development''). This is because knowledge of the immune evasion tactics of this parasite has led many to conclude that it will not be possible to develop a vaccine against these parasites. Certainly what could be called the neoclassical approach to vaccine development, that is the identification of major parasite surface antigens with monoclonal antibodies, followed by cloning of the antigen and use of purified cloned antigen to vaccinate, will not work in this case. There are simply too many varying surface antigens. This author believes that it may still be possible to develop immunizations that protect against these important disease-producing parasites. However, this task may require a more sophisticated knowledge of the parasite and host-parasite interactions than is required to tackle other diseases. Thus, this section consists of more detailed background information about the parasites and the disease they cause in infected animals in order to prepare the reader for the speculative approaches to vaccination presented in Section IV.

A. The Parasite and its Life Cycle

During their life cycle in the mammalian host and in the tsetse fly vector, the African trypanosomes proceed through several morphologically distinguishable stages. The life cycle of *T. brucei*, the most widely studied African trypanosome, is presented in Figure 5.

Infection of the mammalian host is initiated by the injection of metacyclic trypanosomes present in the vectors saliva.[1] These organisms differentiate into rapidly proliferating (generation time of 4 to 6 h), long, slender, bloodstream-form parasites.[127,128] This form of the organism is quite wasteful in its metabolism, it appears to rely entirely on glycolysis for energy production.[129] Studies indicate that it has only a rudimentary mitochondrion which appears to be nonfunctional.[130] (The repression of mitochondrial function in *T. congolense* and *T. vivax* is not as complete as it is in *T. brucei*.[126]) It is the long slender bloodstream form of the organism that engages in variation of its surface antigens.[131]

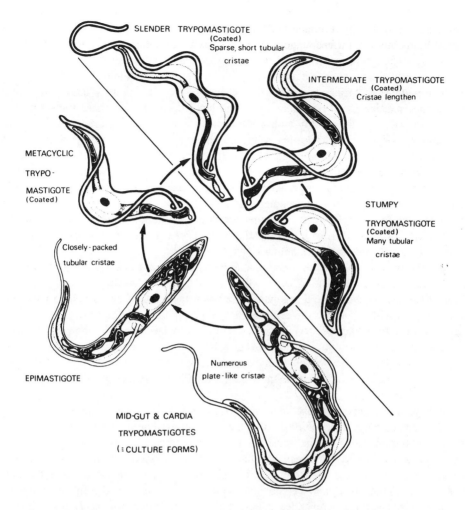

FIGURE 5.   The life cycle of the African trypanosome *T. brucei* in its tsetse fly vector (lower left) and in the mammalian host (upper right). (Modified from Vickerman, K., in *Ecology and Physiology of Parasites*, Fallis, A. M., Ed., University Press, Toronto, 1971, 58. With permission.)

The bloodstream parasite relies on the presence of an abundant supply of host-derived carbohydrates; this allows it to survive despite its inefficient anaerobic metabolism. The parasite, when it lives in the relatively carbohydrate-poorer environment of the tsetse midgut, must be much more efficient.[126] In order to survive in this new environment, it develops a fully functional mitochondrion.[126] This change can be observed both physiologically and morphologically. Several mitochondrial enzymes are seen to be activated,[132] and the mitochondrion is observed to transform from a simple tube into an elaborate network of connecting canals, well supplied with cristae.[130] This differentiation begins in the mammalian host with the production of the *T. brucei* short, stumpy bloodstream form which possesses a semi-developed mitochondrion.[133]

The *T. brucei* short, stumpy bloodstream form appears to be an intermediate stage the parasite has developed for efficient transmission to the tsetse fly vector.[134] With ingestion into the gut of the feeding tsetse fly, most slender forms are probably unable to differentiate rapidly enough and thus die,[135] while the stumpy forms complete differentiation to the procyclic parasite stage and proliferate. Interestingly, the stumpy form of the parasite is a nondividing cell.[134] A similar production of nondividing bloodstream parasites is also observed with both *T. congolense*[136] and *T. vivax*,[134] although these parasites do not transform

from slender into stumpy morphology. It may be that cessation of cell division is a requirement of the process of differentiation from the early bloodstream form into the stage specialized for transmission to the vector. Cessation of cell division before differentiation has
been frequently observed in metazoan systems.[137] This differentiation, however, may be
doubly important for African trypanosomes because of the restriction it places upon parasite
population expansion,[138,139] thus limiting the virulence of infection.[127] That transition to the
stumpy form occurs more readily with some parasite lines in their natural bovid hosts than
in laboratory rodents[22] suggests that a specific host-parasite interaction functions in the
induction of this differentiation event. The molecular nature of this interaction is unknown.

In the midgut of the fly vector the stumpy parasite completes the activation of its mitochondrion,[133] renews DNA synthesis[134] and differentiates into the proliferating procyclic
form of the parasite. In this environment the parasite is no longer confronted with a host
immune response to evade, so it no longer needs to continually vary its surface antigens.
This is reflected in its lack of a peripheral surface coat (Figure 3); this uncoated form of
the parasite is not infective for mammals.[132]

While the parasite establishes a continuous infection in the tsetse fly midgut, some procyclic-form organisms migrate forward to the fly mouth parts and salivary glands. Here they
differentiate into epimastigote forms which are characterized by a change in the location of
the kinetoplast with respect to the cell nucleus[140] (see Figure 5). Epimastigote forms in turn
differentiated into metacyclic trypanosomes. During this last transition the parasite begins
to synthesize a variable antigen surface coat and becomes once again infective for mammals.[126] When inoculated with tsetse saliva into the new environment of the skin of a mammal,
the metacyclic parasites differentiate into proliferating long, slender bloodstream forms and
establish infection of the animal.[141]

The African trypanosomes lose infectivity for mammals when they enter the tsetse fly gut
and must complete their developmental cycle with differentiation to the metacyclic stage
before infective parasites can be transmitted in tsetse saliva. This mode of transmission of
infection is called *cyclical transmission*. It is distinguished from the *mechanical transmission*
that can occasionally occur when bloodstream-form parasites survive unchanged in the mouth
parts of a fly long enough to be inoculated into the next animal the fly feeds upon.[142]
Mechanical transmission is much less efficient than cyclical transmission. Biting flies, other
than tsetse, can transmit the parasite in this way, and this is how infections with the African
trypanosomes *T. evansi* and *T. vivax* are transmitted outside the tsetse-infested areas of
Africa.[140]

Whether mating takes place in the life cycle of African trypanosomes has generated much
controversy. A sexual stage in the life cycle had long been postulated without much evidence
for its existence.[1] In 1980 there were two independent reports that isoenzyme studies of
field isolates of the parasite suggested that recombination of genetic markers had occurred
in nature.[143,144] Recently, recombination of genetic markers between different strains of *T.
brucei* has been accomplished in the laboratory.[145] It is still not known how frequently such
events occur in nature; however, their possibility has significant implications for the ability
of the parasite to reduce the efficacy of certain potential vaccines (e.g., antimetacyclic VSG
vaccines, Section IV.B.1) as well as new trypanocidal drugs.

B. The Disease in the Mammalian Host

The disease in most animals has two stages: the initial stage in which the parasite is
localized at the site of the tsetse fly bite, and the systemic stage in which the trypanosomes
are widely distributed by the bloodstream throughout the body. In human beings and some
domestic animals infected with certain *T. brucei* subspecies, an advanced or neurologic
stage of the disease exists in which the parasite invades the central nervous system.

The initial stage of the disease is characterized by a swelling, called a chancre, at the bite site. This swelling develops 5 to 10 d after metacyclic trypanosomes have been inoculated into the animal's skin with tsetse saliva.[146,147] This chancre is clearly a response to the trypanosome, rather than an inflammatory response to tsetse saliva, as bites from uninfected flies fail to elicit a chancre.[147,148] The lesion is a combination of an acute inflammatory response and an immune reaction to the locally proliferating trypanosomes. It is characterized initially by the infiltration of polymorphonuclear leukocytes and small lymphocytes. The small lymphocytes are replaced by lymphoblasts after a few days. Many plasma cells and macrophages are then found in the lesion as it regresses. The appearance of the chancre is delayed until sufficient parasite numbers have been generated; the size of the chancre appears to be related to the amount of parasite proliferation at the bite site. Far fewer metacyclics are injected in *T. vivax* infections, and a smaller chancre or no chancre develops.[149] The initial leukocyte infiltration probably occurs in response to chemotactic factors[150] and other, as yet uncharacterized factors generated by the parasite that increase vascular permeability.[151] This appears to be followed by an immune response to the parasite surface antigens. The exact nature of the immune response in the chancre is still under investigation.[152] Animals that have been previously exposed to metacyclics of the same parasite strain do not develop a chancre, probably because they have enough circulating antibodies against the variant surface antigens of those parasites to restrict the initial parasite proliferation.[148] Parasites that escape the initial immune response in the skin make their way via the local draining lymph node and the lymphatic system into the bloodstream where they establish the systemic infection.[147] It has been postulated that differences in the host immune response to the parasite during these early events in the skin may account for the resistance of some animals to trypanosomiasis.[153]

In the second stage of the disease, the trypanosomes are widely distributed throughout the body by the bloodstream and they proliferate rapidly. During this stage the animal suffers from an intermittent fever and may show other nonspecific signs of infection such as lymphadenopathy and weakness. It is best to think of this stage as having two stages itself: an early period when a profound host immunoproliferative response occurs and the parasite population expansion is brought under some control, and a later period when, after continued presence of the parasite, the animal becomes progressively anemic, cachectic, and immunodepressed. The parasite-induced anemia and cachexia lead to weakness and inability to forage for food, and most animals that succumb die of malnutrition. The anemia, along with myocardial damage and increased vascular permeability may also lead to death from congestive heart failure.[154] Parasite-induced immune suppression causes some animals to die of secondary infections, frequently pneumonia. Unlike human trypanosomiasis, the infection in animals in not always fatal. After varying periods of infection, cattle, sheep, and goats of certain breeds, as well as some species of wild animals, may completely eliminate the parasite and recover.[155]

In most African trypanosome infections the initial systemic proliferation of the parasite is short lived. The parasitemia reaches a peak within 1 to 2 weeks and then declines. The initial remission appears to occur as a result of two factors. First, near the height of the first parasitemic peak there appears to be a somewhat synchronous transition of many parasites to the nondividing, short, stumpy bloodstream parasite form.[134,156,157] Second, the host's antibody response to the parasite's surface antigens eliminates most of the trypanosomes from the bloodstream by a combination of complement-mediated lysis[20] and antibody-dependent phagocytosis.[158,159] Unfortunately, while most of the parasites are eliminated because of the antibody response, a few manage to survive. These trypanosomes have changed the glycoprotein antigen comprising their surface coat and thus are not recognized by the antibody induced by the initial parasites. Parasites bearing new surface antigens proliferate and repopulate the host's bloodstream. They also stimulate a new antibody response which in turn eliminates most of them, while a few parasites change to yet newer surface coat antigen types. Most subsequent waves of parasitemia are not as well defined as the first peak. It is thought that the transition of trypanosomes to the nondividing stage is not as synchronous or possibly as extensive as in the first peak of parasitemia. Furthermore, the antigenic types present may become more numerous, so that parasite elimination by the immune system is not as efficient. The result of repeated parasite proliferation and elimination is that the host animal develops a chronic fluctuating parasitemia.

Early in the second stage of the disease, when the parasite become widely distributed in the body, a generalized immunoproliferative reaction develops in most lymphoid organs. There is an early but transient T lymphocyte proliferation,[160,161] a null lymphocyte proliferation,[160] and an increase in phagocytic cells.[162] But perhaps the most striking part of this generalized immune cell proliferation is the profound multiplication of B lymphocytes in lymph nodes, bone marrow, and spleen.[160,163-166] This B cell proliferation leads to a significant increase in circulating IgM.[167,168] While some of the newly produced immunoglobulin is clearly parasite specific, some is not. Autoantibodies and antibodies with apparent specificity for neither parasite nor host are also produced.[166,169] The genesis of this immunoproliferation is not certain. However, with respect to B lymphocyte proliferation, much evidence seems to favor the hypothesis that trypanosomes either produce a B cell mitogen directly[170,171] or stimulate host macrophages to produce such a factor.[172]

During the course of lethal African trypanosome infections, a profound anemia develops. In the early phase of the disease this anemia is hemolytic and hemophagocytic in origin.[173-175] Several laboratories have found that immune complexes containing trypanosome antigen and complement adhere to the erythrocyte membrane.[176,177] Also, red blood cells with bound immune complexes have been detected just before each wave of hemolysis.[177] These data are consistent with the hypothesis that it is the binding of immune complexes to the erythrocyte membrane that renders red blood cells more susceptible to phagocytosis and complement-mediated lysis in infected animals. This hypothesis may not, however, account for all of the hemolysis. Some evidence has been presented that autoantibodies to red blood cell antigens[178-181] and trypanosome-released or -induced hemolysins[154,182] may also contribute to the early systemic stage hemolytic anemia. With continued infection, impaired iron recirculation and dyserythropoiesis occurs.[174]

The other major pathologic effect observed in African trypanosomiasis is wasting, or cachexia. Recent studies have suggested that this is the result of parasite-stimulated production, by host macrophages, of a factor known either as cachectin or as tumor necrosis factor.[183] Among other activities, this molecule appears to suppress lipoprotein lipase activity, thereby impairing serum triglyceride utilization which leads to the wasting. Incubation of macrophages *in vitro* with a trypanosome lysate can stimulate the production of this factor.[184] In the course of a trypanosome infection, host macrophages are likely to be stimulated by factors in phagocytosed parasites, although release of a macrophage stimulating factor by the living or dying parasites into the circulation is also possible.

Many animals that succumb to infection with African trypanosomes actually die of a secondary infection with another pathogen. This clinical observation led many investigators to look for evidence of parasites-induced immunosuppression. It was observed that, despite the early immunoproliferation, lymphoid organs become depleted of lymphoid cells late in infection, and a systemic neutropenia develops. Lymph nodes shrink and patchy fibrosis replaces the earlier lymphocytosis.[185] While the level of circulating immunoglobulins remains high,[167] the ability to mount specific antibody responses to new antigens decreases.[105,186] Several mechanisms have been suggested to explain the clinical immunodepression. These include the involvement of suppressor cells,[187,188] soluble suppressor substances,[182,189,190] depressed marcophage antigen-presenting activity,[172] and "clonal exhaustion" of B lymphocytes.[191]

C. Trypanotolerance

In many domestic animals, and in human beings, if African trypanosomiasis is not treated with trypanocidal drugs, the infection usually leads to death.[45,192] However, certain breeds of cattle, sheep, and goats, as well as some species of wild animals, have a significant degree of reduced susceptibility to the pathologic effects of the infection and may even be able to eliminate the parasites completely.[155,193,194] This trait has been termed trypanotolerance. The mechanism(s) of resistance to the disease in these animals is unknown, and factors of both acquired immunity and innate resistance may contribute.

Several different types of innate factors may operate. Some animals may have deleterious substances present in[195-197] or essential nutrients absent from[198] their bloodstream or tissues. Some such deleterious factors may be induced as a nonantigen-specific aspect of the host response to trypanosome infection. For example, it has been shown that if given before trypanosome inoculation, the immunostimulants *Corynebacterium parvum* (now called *Propionibacterium acnes*), BCG, and *Bordetella pertussis* can significantly reduce parasitemias.[199] Perhaps trypanotolerance is a reflection of the varying ability of different animals, in response to trypanosome infection, to activate the nonspecific aspect(s) of resistance that is also activated by these immunostimulants. Another nonantigen-specific factor that might influence susceptibility to trypanosomiasis is the inherent and inducible phagocytic activity of neutrophils. In support of this possibility, it has been shown that there are greater numbers of more active neutrophils in trypanotolerant breeds of cattle than in more susceptible breeds.[200] All of the factors mentioned in this paragraph would reduce parasite numbers. Since resistance to the disease often appears to be related to the magnitude of the first parasitemic wave,[199] these animals would suffer less anemia and immunodepression and thus might be able to survive the waves of new antigenic variants of the parasite until the parasite exhausts its antigenic repertoire.

Alternatively, rather than factors that depress parasite population expansion, there may be innate reasons why some animals are less affected than others by similar levels of parasitemia. The red blood cells of some animals may be less easily lysed by hemolysins, which are hypothesized to be released by the parasite. There is some evidence that erythrocytes of some wild animals are resistant to lysis by trypanosomal phospholipases, while bovine red blood cells are lysed.[201] Also, some resistant strains of mice appear to respond to trypanosome infection with production of higher levels of haptoglobin,[202] which could enhance iron recirculation, thus reducing the chronic anemia. Haptoglobin bound to hemoglobin also has peroxidase activity,[203] so it may be involved in reducing damage from free oxygen radicals released by parasite-stimulated macrophages; this could reduce long-term pathologic effects. Furthermore, some animals may be less responsive to trypanosome-released B lymphocyte mitogens or may be less responsive to trypanosome factors which induce the production of cachectin or other such potentially deleterious host-made molecules.

An alternate explanation for trypanotolerance is that some animals are able to mount or maintain a better immune response to parasite antigens than others. Indeed, there is evidence for a more effective IgM response to the parasite's variant surface antigens in infections of relatively resistant breeds of mice.[204] The fact that most immunity to trypanosomiasis acquired in the field appears to be trypanosome strain specific suggests that the antibody response against the variant surface antigens, the repertoire of which is also strain specific, is an important part of disease resistance. A further possibility is that better recognition of subsurface antigen(s) common to trypanosomes of different variant surface antigen type may contribute to more effective control of trypanosome infection in trypanotolerant animals. Animals infected with African trypanosomes frequently recognize parasite-common antigens.[41] Whether or not these antibodies influence the disease in these animals is not known.

IV. POSSIBLE APPROACHES TO IMMUNOLOGICAL CONTROL OF TRYPANOSOMIASIS

African trypanosomes have been studied as much as any other protozoan pathogen and probably more than most. Experiments performed in attempting to develop a vaccine against these parasites have considerably broadened our knowledge about basic mechanisms in immunology and gene expression. However, the remarkable immune evasion tactic, antigenic variation, continues to frustrate vaccine development. Indeed, it is the widely held view that the presence of the variant antigen surface coat so precludes antibody attack on nonvarying antigens of the parasite that it will not be possible to develop any practical immunization to protect against African trypanosomiasis. Nevertheless, disease control by vaccination holds attraction for the developing world because it would probably require less highly trained personnel, less economic resources, and less continuous and concerted efforts than other disease control measures. Thus, this author believes in the necessity of continuing investigation into vaccine development for African trypanosomiasis. This section of the chapter will detail several possible approaches that investigators are already examining or may examine in the future. The approaches are divided into two types: empirical and theoretical. First will be discussed the empirical approaches, that is, those investigations where nature (or laboratory serendipity) shows that protective immunization is possible, and then it remains for the investigator to identify the protective antigens. Second, it may be theorized that particular categories of molecules could induce some protective immunity; these molecules can be purified according to their known biochemical characteristics and then tested for vaccination potential.

A. Empirical Approaches

1. Antigens Recognized by Self-Curing Animals

Many animals appear to be able to recover from African trypanosomiasis. The mechanism of their disease resistance is not known. It is possible that the immune recognition of some subsurface antigen(s) common to trypanosomes of different variant surface types might contribute to recovery.

Animals infected with African trypanosomes frequently recognize parasite-common antigens.[41] In one study[114] an attempt was made to correlate the trypanosome common antigens recognized with the clinical course of disease in trypanosomiasis-sensitive and -resistant breeds of cattle. Sera from nine N'dama (resistant) and nine Zebu (susceptible) cattle infected with *T. brucei* were examined. The clinical course of disease had varied in these animals. The N'dama cattle developed less severe anemia and suffered no deaths, whereas five of the nine Zebus died of trypanosomiasis. The humoral immune responses also varied. Overall, antibodies from these animals were found to bind protein antigens weighing 20, 40, 45, 100, 110, 150, 180, and 300 kDa, as determined by sodium dodecyl sulfate polyacrylamide gel electrophoresis performed under reducing conditions. None of the animals appeared to have produced antibodies against the cross-reacting antigenic determinant at the carboxy terminus of the VSG molecule. Not all cattle produced detectable antibodies against African trypanosome common antigens, nor did any animal produce antibody to all of the antigens enumerated above. However, all cattle able to control the disease (both N'dama and Zebu) recognized at least one of three specific antigens, i.e., proteins of molecular masses 110, 150, and 300 kDa. The N'damas (the more disease-resistant cattle) generally responded to more of the three identified parasite protein antigens than did the Zebus.

The role of the antibody response to the trypanosome antigens (identified by that study) in host control of the disease is unclear. Probably the recognition of these antigens was only coincidental to some other mechanism of host resistance. However, it is possible that this antibody response contributed to disease resistance. In light of the studies on immunization with parasite membrane antigens and enzymes, it is difficult to imagine that host antibody penetrated the parasite surface coat to reach a target within the cell; however, if the antigens are present in the parasite's flagellar pocket (which may be less densely coated with VSG) they might be accessible to direct antibody effects. Alternatively, an antibody response interfering with the action of any pathogenic molecules possibly generated by trypanosomes[182,205,206] might moderate the disease so that a less debilitated animal could more efficiently eliminate new waves of parasitemia. A third possibility is that these three common antigens act as carrier molecules, promoting a primed immune response to the VSGs; such a mechanism was proposed in immunity to malaria.[207]

Such studies as the one performed with N'dama and Zebu cattle should be performed with other animals that show resistance to African trypanosomiasis such as the many species of resistant wild mammals in Africa. However, in order to assess completely the role of the recognition of trypanosome common antigens in disease resistance, such antigens must be prepared in purified form so that the effects of preimmunization with them upon the course of infection can be studied. These antigens, at least in the case reported above, will probably be present in very small quantity in the parasite and, thus, recombinant DNA technologies may be required for further studies.

2. Nonvariant Surface Antigens

Several studies have suggested the possibility that nonvarying parasite antigens may be present on the surface of African trypanosomes. Beat and co-workers[208] investigated parasite common antigens limited to the bloodstream stage of *T. brucei*. They prepared antitrypanosome antibodies by hyperimmunizing rabbits with *T. brucei* homogenates, and then absorbed the sera with procyclic parasite homogenates to remove antibodies directed at antigens common to both life cycle stages of the parasite. Their absorbed sera revealed by immunoelectrophoresis the presence of an antigen common to blood stream forms of the parasite having different variant surface antigens. Indirect immunofluorescence tests on living trypanosomes suggested that the antigen(s) might be diffusely distributed over the parasite surface. Burgess and Jerrels[209] studied monoclonal antibodies prepared from mice exposed to living bloodstream-form trypanosomes and found two monoclonal antibodies which reacted with trypanosomes of differing variant surface antigen type. The antigen recognized by these monoclonal antibodies appeared by immunofluorescence tests to be on the living parasite's surface. The antibodies recognized a doublet band of approximately 22 kDa in immunoblots from sodium dodecyl sulfate polyacrylamide gels. Burgess and Jerrels were unable, however, to neutralize living trypanosomes with these monoclonal antibodies. Also, the antigen recognized by the monoclonal antibodies did not appear to induce an antibody response in trypanosome-infected humans.

Most investigators have difficulty believing that such nonvariant surface antigens exist. Why should the parasite compromise such a sophisticated immune evasion tactic as variation of the antigens comprising its surface coat by placing in that coat a nonvarying antigen? Both sets of investigators whose studies are described in the preceding paragraph indicate that the antibodies that recognize these antigens were produced in repeatedly immunized animals. Perhaps nonvariant surface antigen does not compromise the parasite because it does not induce an antibody response in the course of a natural trypanosome infection because of antigenic competition[113] with the highly immunogenic VSGs or because of the generalized immunosuppression caused by the disease. However, in the absence of any necessity for a nonvarying antigen to be exposed on the parasite surface, it is easier to argue that some experimental artifact accounts for the interesting results in the two reports. It has recently become apparent that the nonphysiologic conditions under which African trypanosomes are separated from other infected blood components by DEAE chromatography cause significant biochemical changes in the parasites.[210] Changes in internal levels of ATP occur, and several proteins, including enzymes and VSG, are released from the cell. Both groups of investigators purified trypanosomes in this manner before testing for the presence of nonvariant antigen on the surface, so the antigens they found could have been released during parasite isolation or could have been exposed through a surface coat less dense than the intact coat on the parasite in infected animals.

3. Parasite Subcellular Fractions

One investigator has reported an immunization protocol that gives protection against heterologous strains of African trypanosomes.[211,212] Powell produced protection by immunizing with a crude mitochondrial fraction prepared from the parasite. He reported that when laboratory rodents were immunized with such fractions prepared from either *T. b. brucei* or *T. b. rhodesiense* they survived significantly longer than unimmunized control animals when challenged with *T. brucei* parasites of different strains. Unfortunately, these results have proved difficult to repeat in other laboratories.[213,214] More recent studies have led Powell to suggest that the protection is not based on an immune response to a protein antigen, since the activity remains with a small-molecular-weight fraction in attempts to purify it from the crude mitochondrial preparation.[215] What trypanosome antigen or, indeed, whether an antigen or an immunomodulating factor (see Section IV.A.4) is functioning in these immunizations is unclear. It is certainly difficult to imagine how antibodies could affect an intracellular structure like a mitochondrion, while the parasite membrane just under the surface coat is inaccessible. However, it is probable that the simple fractionation procedure used in these experiments included other components with mitochondria in the immunizing preparation. Thus, the functional immunogen could have been some component in the flagellar pocket with which one end of the parasite's single mitochondrion is closely associated.

4. Immunomodulators

Treatment of mice with nonspecific immunomodulators such as *C. parvum*, *B. pertussis*, and BCG,[199] or bacterial endotoxin[216] can significantly increase resistance to African trypanosome infection if given shortly before or at the same time as the parasite. Mice so treated survive with the parasite for longer periods of time and have a greater ability to control the level of parasitemia. The mechanism(s) by which such treatment affects the level of parasitemia is not clearly understood. Much of the effect may result from the activation of host macrophages which are involved in parasite killing.[158] Recent evidence suggests that it may involve, also, an acceleration of the transition to the nondividing bloodstream parasite stage.[217] Other experiments have shown that the nondividing parasite is a much more efficient inducer of a host antibody response against the trypanosome variant surface antigens that the dividing bloodstream parasite.[138] Thus, although some of the immediate effect of the immunomodulators may not be on the immune response, they should ultimately enhance the antibody response to the parasite surface antigens. If agents could be found that would induce a similar long-acting, antitrypanosomal effect, then perhaps parasitemias could be lowered enough in animals in the field to decrease pathologic changes to the point where less debilitated animals could develop disease resistance based on antibodies to all the parasites' VSG types. This requires more basic research on the mechanism of action on trypanosomes of the immunomodulators.

B. Theoretical Targets
1. Metacyclic VSGs

Although each trypanosome line can potentially produce many different variant surface antigens, early investigators noted that each line had a tendency to produce a particular antigen type in the first wave of parasitemia after cyclical transmission by the tsetse fly.[111]

This antigen type was called the "basic" antigen, and it was suggested that there existed a characteristic basic antigen for each parasite strain. Recent studies have explained this early observation; it has been shown that the VSGs synthesized by the metacyclic parasites in the fly salivary glands are represented in the first parasitemic wave.[218,219] Furthermore, the metacyclic surface antigens are constant for a given trypanosome line regardless of the surface antigen type of the bloodstream-form parasites ingested by the fly.[220] Thus, the "basic antigen" of the earlier investigators was probably a composite of the VSGs of the metacyclic population. The repertoire of metacyclic VSGs appears to be much more limited in number than the bloodstream parasite VSG repertoire.[221] One study with monoclonal antibodies showed that the metacyclic VSG repertoire of a *T. congolense* line contained only 12 different VSGs.[222]

Immunization based on the metacyclic stage of the parasite would be advantageous. If infection would be prevented, then the morbidity (usually seen as reduced weight gain which can pose a significant economic problem with meat animals) associated even with infections that do not kill the animals would be avoided. The findings that the metacyclic VSG repertoire was limited in number and constant for a trypanosome line suggested that immunity to tsetse fly challenge might be induced by immunizing with metacyclic surface antigens. Studies in which mice[223] and cattle[224] have been immunized with metacyclic antigens by infection followed with trypanocidal drug treatment have shown that parasite strain-specific immunity can be produced. However, no one knows how many different African trypanosome strains with different metacyclic VSG repertoires are distributed throughout Africa. Recent data showing that at least four different T. *congolense* lines were present in one small, isolated area of low trypanosomiasis incidence suggests that the situation is probably quite complex.[225] Furthermore, if genetic exchange occurs between African trypanosomes in the field, then new metacyclic VSG repertoires may be generated constantly. Thus, a vaccine based on metacyclic VSGs would probably be effective only in a very few places, and even in those places only for limited time. This probably would not justify the effort that would be involved in cloning metacyclic VSG genes for large-scale antigen production.

2. Endocytotic Vesicle Receptors

Despite the general superposition of the VSG antigens on the parasite surface, there is a class of nonvarying molecules that the African trypanosomes might require, exposed on its surface, to maintain its existence. These molecules are receptors for adsorptive endocytosis. Adsorptive or receptor-mediated endocytosis is a process by which materials are taken into cells. In this process, selected extracellular components are first bound to specific cell surface receptors, and then taken into the cell by invagination of the clustered receptors to form intracelluar vesicles. Receptor-mediated endocytosis plays an important role in the growth, nutrition, and differentiation of many eukaryotic cells.[226,227] This type of endocytosis could also be called coated vesicle endocytosis because the pinocytotic vesicles are covered with a lattice-like coat of a protein called clathrin.[226-228] Coated-vesicle endocytosis occurs in African trypanosomes in the mammalian bloodstream,[51,229,230] and serum albumin has been suggested to be one protein selectively endocytosed.[231] Whether there are other selectively endocytosed host molecules is not known. Some selectively endocytosed host molecule could be a growth factor or signal for differentiation of the dividing long, slender, bloodstream-form parasite into the nondividing short, stumpy form (see Figure 5).

If adsorptive endocytosis occurs, then the receptors must be accessible to their intended ligands. If the receptors are sufficiently exposed that they are accessible to large molecules, they should also be accessible to antibodies despite the otherwise overwhelming presence of the variant antigen surface coat. Coated-vesicle receptors are usually present only in very small amounts on any cell surface.[232] In the case of African trypanosome infection, in which the variant surface glycoproteins are such dominant antigens, because of immune compe-

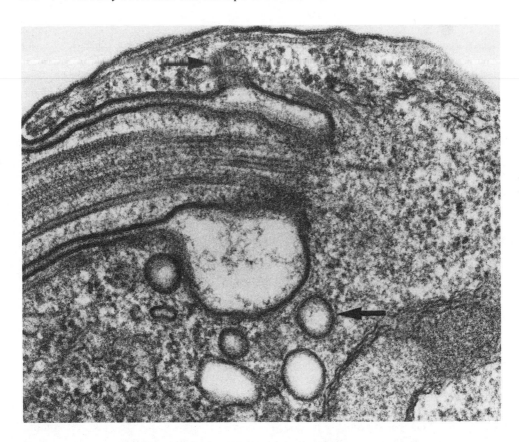

FIGURE 6.   Coated endocytotic vesicles in the African trypanosome. The arrows indicate a coated vesicle forming (above) in the flagellar pocket of *T. brucei* and an internalized coated vesicle (below). (Magnification × 107,000.) (From Shapiro, S. Z. and Pearson, T. W., in *Parasite Antigens: Toward New Strategies for Vaccines,* Pearson, T. W., Ed., Marcel Dekker, New York, 1986, chap. 7. With permission.)

tition,[113] the relatively rare coated vesicle receptor molecules would probably produce at best an ineffective immune response. Although these molecules would thus not normally induce a strong enough response to compromise the highly developed protective strategy of the trypanosome, if the isolated protein(s) were used as immunogen, a stronger immune response should be elicited.

Some observations have been made that support the hypothesis that antibody-accessible endocytotic vesicle receptors exist in African trypanosomes. Coated endocytotic vesicles are formed only in a small region of the parasite, the flagellar pocket[51,229] (Figure 6). Although the variant surface antigen coat appears to cover the surface inside the flagellar pocket, evidence from monoclonal antibody binding studies suggests that the conformation of the coat glycoprotein differs in this region as different epitopes are exposed on the VSG molecule.[97] If receptor molecules extend through the VSG coat in order to bind ligands, the resulting conformational disruption could account for the exposure of these additional epitopes of the VSG. Secondly, the parasites appear to require some growth-promoting factor from host serum which has a molecular mass of at least 100 kDa[233] and may be larger than 250 kDa.[234] A receptor for a molecule of this size may also be accessible to IgG which has a molecular mass of 200 kDa.

3. Secreted Pathogenic Molecules

The major pathologic changes observed in animal trypanosomiasis are anemia, cachexia, and immunosuppression (see Section III.B). The molecular mechanisms of production of these changes by the parasite infection are not clearly understood. It has been suggested that at least some of the effects of the disease could be caused by parasite-secreted or -released toxins.[182] There is evidence for trypanosome-derived hemolysins, lymphocyte mitogens, inflammatory factors, and hepatotoxins.[182] If such molecules could be purified and used as immunogens, then the host antibody response might remove them from the circulation before they could function. This could moderate the disease significantly. A less generally debilitated animal should be better able to mount an effective immune response to the antigenically varying waves of parasitemia than a more debilitated animal; mortality from the disease could be thus decreased.

4. Procyclic Surface Antigens

Another possible approach to the development of a vaccine for African trypanosomiasis is the "altruistic vaccine" approach. This has been promoted by investigators studying malaria. It has been suggested that antibodies against mosquito gut stages of that parasite might block transmission of the disease by interfering with infection of the vector. It has been found that antibodies directed at antigens on the malaria gametes which fuse in the mosquito gut can prevent the development of animal-infective vector forms of the parasite (sporozoites).[235] While this sort of vaccine would not reduce morbidity or mortality in the vaccinated individual, it would reduce the level of disease in the community by reducing parasite transmission.

Encouraged by the success of the malaria experiment, other investigators have examined the possibility of using a similar approach with African trypanosomes. Feeding trypanosome-infected tsetse flies on animals immunized against the procyclic stage of the parasite significantly reduces the production of animal-infective metacyclic trypanosomes.[236,237] This suppression was observed to be species specific,[237] but the effect was not limited by parasite strain specificity.[236,237] Thus, immunization with antigen from procyclics of one line would potentially reduce transmission of all parasites of a species. Individual antigens on the surface of the uncoated procyclic-form parasites which can induce this type of immunity are now being identified with monoclonal antibodies in several laboratories.

This approach, however, may not be generally useful for African trypanosomiasis. In many areas tsetse flies feed on a variety of different hosts which can be reservoirs of African trypanosomiasis. Many of these are wild animals which cannot be immunized. It is not clear what percentage of fly feeds must be on an immunized host in order to affect parasite transmission. It is possible that the altruistic vaccine approach may work only in areas where most people and domestic animals can be immunized and wild animals can be controlled.

5. Vector Antigens

Finally, instead of vaccinating against the parasite, it might be more practical to vaccinate against the tsetse fly vector. Tsetse flies in the laboratory are fed once daily.[238] In the field they may feed more frequently, as the rate of digestion by flies in the field is much higher.[239] With each blood meal, the tsetse fly takes up its host's antibodies. It has been suggested that it might be possible to affect many hematophagous arthropod vectors of disease adversely by allowing them to imbibe blood containing antibodies directed against themselves. Antibodies that could neutralize digestive enzymes or anticoagulant molecules, or damage the vector gut wall could reduce vector viability by interfering with feeding.[240-242] Also, since the arthropod gut appears permeable to antibodies,[243-245] it might be possible to direct antibody attack at other essential molecules in the vector such as hormones, hormone receptors, or other cell surface molecules in important tissues.[246-248] Another category of potential targets in the vector is the molecules that the African trypanosomes bind to in order to localize themselves to a particular region within the vector for parasite development; such molecules have been observed to be necessary for the cyclical development of *T. cruzi*.[249] Antibodies to these molecules could reduce disease transmission in the same way as an antiprocyclic surface antigen vaccine by interfering with the parasites' cyclical development.

Investigation of the possibility of vaccination against tsetse flies has already begun.[244,246,250] In one study, Nogge[250] showed that the mortality rate of flies could be increased by feeding them on animals immunized with a fly homogenate, and the fecundity rate could be decreased by feeding flies on animals immunized with symbiotic fly gut microbes. Schlein and Lewis[246] found increased mortality of tsetse flies fed on rabbits immunized against tissues of the stable fly, *Stomoxys calcitrans*. Nogge and Giannetti[244] showed that tsetse flies could be killed if fed with antibodies directed at previously ingested host albumin. To decide how practical this approach is, mathematical modeling, using data on behavior of tsetse flies in the field, is required to discover what percentage of potential hosts must be immunized in order to reduce the vector population enough to affect disease transmission.

V. CONCLUSIONS

In these days of the application of the new biotechnologies to the study of parasitic diseases, the possibility of vaccines against many of these diseases seems likely. African trypanosomes have probably been studied with the new technologies more than most other parasites. Indeed, the parasite's surface antigen genes comprise one of the largest bodies of DNA sequence data. However, what we now know about the process of variation of these surface antigens makes it clear that they cannot serve as the basis for a recombinant vaccine; there are simply too many of them. It must be admitted that it is not possible at this time to define a strategy for the development of a vaccine against African trypanosomiasis that is sure to be successful. Nevertheless, it would clearly be very useful to have a means for immunological control of this very important disease. This author has tried to present several different approaches that could be taken in this endeavor, and hopes that this chapter will stimulate others to further thought and work. Instead of focusing on the frustration of our efforts by the remarkable immune evasion tactic of antigenic variation, we should see what a major and exciting challenge African trypanosomiasis presents to biomedical and veterinary research.

REFERENCES

1. **Hoare, C. A.**, *The Trypanosomes of Mammals*, Blackwell Scientific, Oxford, 1972, chap. 1.
2. **Soulsby, E. J. L.**, *Helminths, Arthropods, and Protozoa of Domesticated Animals*, Lea & Febiger, Philadelphia, 1982, 535.
3. **Baker, J. R.**, Systematics of parasitic protozoa, in *Parasitic Protozoa*, Vol. 1, Kreier, J. P., Ed., Academic Press, New York, 1977, chap. 2.
4. **Vickerman, K.**, The diversity of the kinetoplastid flagellates, in *Biology of the Kinetoplastida*, Vol. 1, Lumsden, W. H. R. and Evans, D. A., Eds., Academic Press, London, 1976, chap. 1.
5. **Woo, P. T. K.**, Salivarian trypanosomes producing disease in livestock outside of sub-Saharan Africa, in *Parasitic Protozoa*, Vol. 1, Kreier, J. P., Ed., Academic Press, New York, 1977, chap. 7.
6. **McKelvey, J. J., Jr.**, *Man Against Tsetse: Struggle for Africa*, Cornell University Press, Ithaca, NY, 1973, chap. 1.
7. **Hoeppli, I. R.**, *Parasites and Parasitic infections in Early Medicine and Science*, University of Malaya Press, Singapore, 1959, 490.
8. Anon., African trypanosomiases: 1978—1982, in *6th Programme Rep. Special Programme for Research and Training in Tropical Diseases*, UNDP/World Bank/WHO, Geneva, 1983, chap. 5.
9. Anon., African trypanosomiases, in *7th Programme Rep. Special Programme for Research and Training in Tropical Diseases*, UNDP/World Bank/WHO, Geneva, 1985, chap. 5.
10. **Burke, J.**, Historique de la lutte contre la maladie du sommeil au Congo, *Ann. Soc. Belge Med. Trop.*, 51, 465, 1971.

11. **Steiger, R. F.,** African trypanosomiases, in *The Membrane Pathobiology of Tropical Diseases,* Wallach, D. F. H., Ed., Schwabe, Basel, 1979, chap. 6.

12. **Gashumba, J. K.,** Sleeping sickness in Uganda, *New Sci.,* 89, 164, 1981.

13. **Ford, J.,** *The Role of the Trypanosomiases in African Ecology,* Clarendon Press, Oxford, 1971, chap. 25.

14. **Nantulya, V. M.,** Parasitological and Immunological Aspects of *Trypanosoma (Nannomonas) congolense* Infections of the Mouse (*Mus musculus*) and the Tsetse Fly (*Glossina morsitans morsitans*), Ph.D. thesis, University of Nairobi, Nairobi, 1978.

15. **De Raadt, P. and Seed, J. R.,** Trypanosomes causing disease in man in Africa, in *Parasitic Protozoa,* Vol. 1, Kreier, J. P., Ed., Academic Press, New York, 1977, chap. 5.

16. Anon., Vector Control, in *The African Trypanosomiases,* World Health Organization, Geneva, 1979, chap. 8.

17. **Cuisance, D., Politzar, H., Clair, M., Taze, Y., Bourdoiseau, G., and Fevrier, J.,** La lutte contre *Glossina palpalis gambiensis* Vanderplank par lâchers de mâles irradiés en Houte Volta: étude de paramètres operationnels, in *Isotope and Radiation Research on Animal Diseases and Their Vectors,* Freeman, S. M., Ed., IAEA, Vienna, 1979, 249.

18. **Dame, D. A., Williamson, D. L., Cobb, P. E., Gates, D. B., Warner, P. V., Mtuya, A. G., and Baumgartner, H.,** Integration of sterile insects and pesticides for the control of the tsetse fly, *Glossina morsitans morsitans,* in *Isotope and Radiation Research on Animal Diseases and Their Vectors,* Freeman, S. M., Ed., IAEA, Vienna, 1979, 267.

19. **Williamson, J.,** Review of chemotherapeutic and chemoprophylactic agents, in *The African Trypanosomiases,* Mulligan, H. W., Ed., George Allen and Unwin, London, 1970, chap. 7.

20. **Murray, M. and Urquhart, G. M.,** Immunoprophylaxis against African trypanosomiasis, in *Immunity to Blood Parasites of Animals and Man,* Miller, L. H., Pino, J. A., and McKelvey, J. J., Jr., Eds., Plenum Press, New York, 1977, chap. 11.

21. **Babiker, E. A. and Le Ray, D.,** Adaptation of low virulence stocks of *Trypanosoma brucei gambiense* to rat and mouse, *Ann. Soc. Belge Med. Trop.,* 61, 15, 1981.

22. **Black, S. J., Jack, R. M., and Morrison, W. I.,** Host-parasite interactions which influence the virulence of *Trypanosoma (Trypanozoon) brucei brucei* organisms, *Acta Trop.,* 40, 11, 1983.

23. **Terry, R. J.,** Immunity to African trypanosomiasis, in *Immunology of Parasitic Infections,* Cohen, S. and Sadun, E. H., Eds., Blackwell Scientic, Oxford, 1976, chap. 16.

24. **Johnson, P., Neal, R. A., and Gall, D.,** Protective effect of killed trypanosome vaccines with incorporated adjuvants, *Nature (London),* 200, 83, 1963.

25. **Duxbury, R. E. and Sadun, E. H.,** Resistance produced in mice and rats by inoculation with irradiated *Trypanosoma rhodesiense, J. Parasitol.,* 55, 859, 1969.

26. **Duxbury, R. E., Sadun, E. H., and Anderson, J. S.,** Experimental infections with African trypanosomes. II. Immunization of mice and monkeys with a gamma-irradiated, recently isolated human strain of *Trypanosoma rhodesiense, Am. J. Trop. Med. Hyg.,* 21, 885, 1972.

27. **Wellde, B. T., Schoenbechler, M. J., Diggs, C. L., Langbehn, H. R., and Sadun, E. H.,** *Trypanosoma rhodesiense*: variant specificity of immunity induced by irradiated parasites, *Exp. Parasitol.,* 37, 125, 1975.

28. **Lapièrre, J. and Rousset, J. J.,** Charactères biologiques d'une souche virulente de *Trypanosoma gambiense.* Immunization par vaccins tués, *Bull. Soc. Pathol. Exot.,* 54, 336, 1961.

29. **Soltys, M. A.,** Immunity in trypanosomiasis. V. Immunization of animals with dead trypanosomes, *Parasitology,* 54, 585, 1964.

30. **Seed, J. R.,** The characterization of antigens isolated from *Trypanosoma rhodesiense, J. Protozool.,* 10, 380, 1963.

31. **Seed, J. R. and Gam, A. A.,** The properties of antigens from *Trypanosoma gambiense, J. Parasitol.,* 52, 395, 1966.

32. **Herbert, W. J. and Lumsden, W. H. R.,** Single-dose vaccination of mice against experimental infection with *Trypanosoma (Trypanozoon) brucei, J. Med. Microbiol.,* 1, 23, 1968.

33. **Seed, J. R.,** *Trypanosoma gambiense* and *T. equiperdum.* Characterization of variant specific antigens, *Exp. Parasitol.,* 31, 98, 1972.

34. **Taylor, A. E. R. and Lanham, S. M.,** Partial purification of immunogenic (protective) antigens of *Trypanosoma brucei brucei, Trans. R. Soc. Trop. Med. Hyg.,* 66, 345, 1972.

35. **Franke, E.,** Therapeutische versuche bei trypanosomenerkrankung, *Muench. Med. Wochenschr.,* 52, 2059, 1905.

36. **Ritz, H.,** Ueber rezidive bei experimenteller trypanosomiasis. II. Mitteilung, *Arch. Schiffs Trop. Hyg.,* 20, 397, 1916.

37. **Leupold, F.,** Untersuchungen über rezidivstämme bei trypanosomen mit hilfe des Rieckenberg-phänomens, *Z. Hyg. Infektions-kr.,* 109, 144, 1928.

38. **Russell, H.,** Observations on immunity in relapsing fever and trypanosomiasis, *Trans. R. Soc. Trop. Med. Hyg.,* 30, 179, 1936.

39. **Lourie, E. M. and O'Connor, R. J.,** A study of *Trypanosoma rhodesiense* relapse strains *in vitro, Ann. Trop. Med. Parasitol.,* 31, 319, 1937.

40. **Soltys, M. A.,** Immunity in trypanosomiasis. I. Neutralization reaction, *Parasitology,* 47, 375, 1957.

41. **Shapiro, S. Z. and Pearson, T. W.,** African trypanosomiasis: antigens and host-parasite interactions, in *Parasite Antigens: Toward New Strategies for Vaccines,* Pearson, T. W., Ed., Marcel Dekker, New York, 1986, chap. 7.

42. **Lumsden, W. H. R.,** Principles of viable preservation of parasitic protozoa, *Int. J. Parasitol.,* 2, 327, 1972.

43. **Capbern, A., Giroud, C., Baltz, T., and Mattern, P.,** *Trypanosoma equiperdum:* Étude des variations antigeniques au cours de la trypanosomose experimentale du lapin, *Exp. Parasitol.,* 42, 6, 1977.

44. **Van Der Ploeg, L. H. T., Valerio, D., De Lange, T., Bernards, A., Borst, P., and Grosveld, F. G.,** An analysis of cosmid clones of nuclear DNA from *Trypanosoma brucei* shows that the genes for variant surface glycoproteins are clustered in the genome, *Nucl. Acids Res.,* 10, 5905, 1982.

45. **De Raadt, P.,** Immunity and antigenic variation: clinical observations suggestive of immune phenomena in African trypanosomiasis, in *Trypanosomiasis and Leishmaniasis, with Special Reference to Chagas' Disease,* CIBA Found. Symp. 20 (New Ser.), Elliott, K., O'Connor, M., and Wolstenholme, G. E. W., Eds., Associated Scientific, Amsterdam, 1974, 199.

46. **Vickerman, K.,** Antigenic variation in African trypanosomes, in *Parasites in the Immunized Host: Mechanisms of Survival,* CIBA Found. Symp. 25 (New Ser.), Porter, R. and Knight, J., Eds., Associated Scientific, Amsterdam, 1974, 53.

47. **Doyle, J. J.,** Antigenic variation in the salivarian trypanosomes, in *Immunity to Blood Parasites of Animals and Man,* Miller, L. H., Pino, J. A., and McKelvey, J. J., Jr., Eds., Plenum Press, New York, 1977, chap. 3.

48. **Doyle, J. J., Hirumi, H., Hirumi, K., Lupton, E. N., and Cross, G. A. M.,** Antigenic variation in clones of animals-infective *Trypanosoma brucei* derived and maintained *in vitro, Parasitology,* 80, 359, 1980.

49. **Weitz, B.,** The properties of some antigens of *Trypanosoma brucei, J. Gen. Microbiol.,* 23, 589, 1960.

50. **Miller, J. K.,** Variation of the soluble antigens of *Trypanosoma brucei, Immunology,* 9, 521, 1965.

51. **Vickerman, K.,** On the surface coat and flagellar adhesion in trypanosomes, *J. Cell Sci.,* 5, 163, 1969.

52. **Vickerman, K. and Luckins, A. G.,** Localization of variable antigens in the surface coat of *Trypanosoma brucei* using ferritin-conjugated antibody, *Nature (London),* 224, 1125, 1969.

53. **Allsopp, B. A., Njogu, A. R., and Humphreys, K. C.,** Nature and location of *Trypanosoma brucei* subgroup exoantigen and its relationship to 4S antigen, *Exp. Parasitol.,* 29, 271, 1971.

54. **Cross, G. A. M.,** Identification, purification, and properties of clone-specific glycoprotein antigens constituting the surface coat of *Trypanosoma brucei, Parasitology,* 71, 393, 1975.

55. **Baltz, T., Baltz, D., Pautrizel, R., Richet, E., Lamblin, G., and Degand, P.,** Chemical and immunological characterisation of specific glycoproteins from *Trypanosoma equiperdum* variants, *FEBS Lett.,* 82, 93, 1977.

56. **Olenick, J. G., Travis, R. W., and Garson, S.,** *Trypanosoma rhodesiense:* chemical and immunological characterisation of variant specific surface coat glycoproteins, *Mol. Biochem. Parasitol.,* 3, 227, 1981.

57. **Le Page, R. W. F.,** Further studies on the variable antigens of *T. brucei, Trans. R. Soc. Trop. Med. Hyg.,* 62, 131, 1968.

58. **Cross, G. A. M. and Johnson, J. G.,** Structure and organisation of the variant specific surface antigens of *Trypanosoma brucei,* in *Biochemistry of Parasites and Host-Parasite Relationships,* Van den Bossche, H., Ed., North-Holland, Amsterdam, 1976, 413.

59. **McConnell, J., Cordingley, J. S., and Turner, M. J.,** The extent of variability of variant-specific antigens of *T. b. brucei, Parasitology,* 79, vi, 1979.

60. **Bridgen, P. J., Cross, G. A. M., and Bridgen, J.,** N-Terminal amino acid sequences of variant specific surface antigens from *Trypanosoma brucei, Nature (London),* 263, 613, 1976.

61. **Holder, A. A., and Cross, G. A. M.,** Glycopeptides from variant surface glycoproteins of *Trypanosoma brucei.* C-terminal location of antigenically cross-reacting carbohydrate moieties, *Mol. Biochem. Parasitol.,* 2, 135, 1981.

62. **Borst, P.,** Discontinuous transcription and antigenic variation in trypanosomes, *Annu. Rev. Biochem.,* 55, 701, 1986.

63. **Steinert, M. and Pays, E.,** Selective expression of surface antigen genes in African trypanosomes, *Parasitol. Today,* 2, 15, 1986.

64. **Tonegawa, S.,** Somatic generation of antibody diversity, *Nature (London),* 302, 575, 1983.

65. **Williams, R. O., Young, J. R., and Majiwa, P. A. O.,** Genomic rearrangements correlated with antigenic variation in *Trypanosoma brucei, Nature (London),* 282, 847, 1979.

66. **Hoeijmakers, J. H. J., Frasch, A. C. C., Bernards, A., Borst, P., and Cross, G. A. M.,** Novel expression-linked copies of the genes for variant surface antigens in trypanosomes, *Nature (London),* 284, 78, 1980.

67. **Pays, E., Van Meirvenne, N., Le Ray, D., and Steinert, M.,** Gene duplication and transposition linked to antigenic variation in *Trypanosoma brucei, Proc. Natl. Acad. Sci. U.S.A.,* 78, 2673, 1981.

68. **Longacre, S., Hibner, U., Raibaud, A., Eisen, H., Baltz, T., Giroud, C., and Baltz, D.,** DNA rearrangements and antigenic variation in *Trypanosoma equiperdum:* multiple expression-linked sites in independent isolates of trypanosomes expressing the same antigen, *Mol. Cell. Biol.,* 3, 399, 1983.

69. **Parsons, M., Nelson, R. G., Newport, G., Milhausen, M., Stuart, K., and Agabian, N.,** Genomic organization of *Trypanosoma brucei* variant antigen gene families in sequential parasitemias, *Mol. Biochem. Parasitol.,* 9, 255, 1983.

70. **Pays, E., Lheureux, M., and Steinert, M.,** Analysis of the DNA and RNA changes associated with the expression of isotypic variant-specific antigens of trypanosomes, *Nucl. Acids Res.,* 9, 4225, 1981.

71. **Majiwa, P. A. O., Young, J. R., Englund, P. T., Shapiro, S. Z., and Williams, R. O.,** Two distinct forms of surface antigen gene rearrangement in *Trypanosoma brucei, Nature (London),* 297, 514, 1982.

72. **Milhausen, M., Nelson, R. G., Parsons, M., Newport, G., Stuart, K., and Agabian, N.,** Molecular characterization of initial variants from the IsTat 1 serodeme of *Trypanosoma brucei, Mol. Biochem. Parasitol.,* 9, 241, 1983.

73. **Raibaud, A., Gaillard, C., Longacre, S., Hibner, U., Buck, G., Bernard, G., and Eisen, H.,** Genomic environment of variant surface antigen genes of *Trypanosoma equiperdum, Proc. Natl. Acad. Sci. U.S.A.,* 80, 4306, 1983.

74. **De Lange, T. and Borst, P.,** Genomic environment of the expression-linked extra copies of genes for surface antigens of *Trypanosoma brucei* resembles the end of a chromosome, *Nature (London),* 299, 451, 1982.

75. **Pays, E., Lheureux, M., and Steinert, M.,** The expression-linked copy of surface antigen gene in *Trypanosoma* is probably the one transcribed, *Nature (London),* 292, 265, 1981.

76. **Bernards, A., Van Der Ploeg, L. H. T., Frasch, A. C. C., Borst, P., Boothroyd, J. C., Coleman, S., and Cross, G. A. M.,** Activation of trypanosome surface glycoprotein genes involves a duplication-transposition leading to an altered 3' end, *Cell,* 27, 497, 1981.

77. **Hicks, J., Strathern, J. N., and Klar, A. J. S.,** Transposable mating type genes in *Saccharomyces cerevisiae, Nature (London),* 282, 478, 1979.

78. **Borst, P., Frasch, A. C. C., Bernards, A., Van Der Ploeg, L. H. T., Hoeijmakers, J. H. J., Arnberg, A. C., and Cross, G. A. M.,** DNA rearrangements involving the genes for variant antigens in *Trypanosoma brucei, Cold Spring Harbor Symp. Quant. Biol.,* 45, 935, 1980.

79. **Donelson, J. E., Young, J. R., Dorfman, D., Majiwa, P. A. O., and Williams, R. O.,** The ILTat 1.4 surface antigen gene family of *Trypanosoma brucei, Nucl. Acids Res.,* 10, 6581, 1982.

80. **Young, J. R., Shah, J. S., Matthyssens, G., and Williams, R. O.,** Relationship between multiple copies of a *T. brucei* variable surface glycoprotein gene whose expression is not controlled by duplication. *Cell,* 32, 1149, 1983.

81. **Myler, P. J., Allison, J., Agabian, N., and Stuart, K.,** Antigenic variation in African trypanosomes by gene replacement or activation of alternate telomeres, *Cell,* 39, 203, 1984.

82. **Shea, C., Glass, D. J., Parangi, S., and Van Der Ploeg, L. H. T.,** Variant surface glycoprotein gene expression site switches in *Trypanosoma brucei, J. Biol. Chem.,* 261, 6056, 1986.

83. **Johnson, J. G. and Cross, G. A. M.,** Carbohydrate composition of variant-specific surface antigen glycoproteins from *Trypanosoma brucei, J. Protozool.,* 24, 587, 1977.

84. **Wright, K. A. and Hales, H.,** Cytochemistry of the pellicle of bloodstream forms of *Trypanosoma (Trypanozoon) brucei, J. Parasitol.,* 56, 671, 1970.

85. **Jackson, P. R.,** Lectin binding by *Trypanosoma equiperdum, J. Parasitol.,* 63, 8, 1977.

86. **Johnson, J. G. and Cross, G. A. M.,** Selective cleavage of variant surface glycoproteins from *Trypanosoma brucei, Biochem. J.,* 178, 689, 1979.

87. **Labastie, M. C., Baltz, T., Richet, C., Giroud, C., Duvillier, G., Pautrizel, R., and Degand, P.,** Variant surface glycoproteins of *Trypanosoma equiperdum:* cross-reacting determinants and chemical studies, *Biochem. Biophys. Res. Commun.,* 99, 729, 1981.

88. **De Almeida, M. L. C. and Turner, M. J.,** The membrane form of variant surface glycoproteins of *Trypanosoma brucei, Nature (London),* 302, 349, 1983.

89. **Ferguson, M. A. J. and Cross, G. A. M.,** Myristylation of the membrane form of a *Trypanosoma brucei* variant surface glycoprotein, *J. Biol. Chem.,* 259, 3011, 1984.

90. **Duvillier, G., Nouvelot, A., Richet, C., Baltz, T., and Degand, P.,** Presence of glycerol and fatty acids in the C-terminal end of a variant surface glycoprotein from *Trypanosoma equiperdum, Biochem. Biophys. Res. Commun.,* 114, 119, 1983.

91. **Holder, A. A.,** Carbohydrate is linked through ethanolamine to the C-terminal amino acid of *Trypanosoma brucei* variant surface glycoprotein, *Biochem. J.,* 209, 261, 1983.

92. **Freymann, D. M., Metcalf, P., Turner, M., and Wiley, D. C.,** 6Å-resolution X-ray structure of a variable surface glycoprotein from *Trypanosoma brucei, Nature (London),* 311, 167, 1984.

93. **Cohen, C., Reinhardt, B., Parry, D. A. D., Roelants, G. E., Hirsch, W., and Kanwe, B.,** Alphahelical coiled-coil structures of *Trypanosoma brucei* variable surface glycoproteins, *Nature (London),* 311, 169, 1984.

94. **Auffret, C. A. and Turner, M. J.,** variant specific antigens of *Trypanosoma brucei* exist in solution as glycoprotein dimers, *Biochem. J.,* 193, 647, 1981.

95. **Strickler, J. E. and Patton, C. L.,** *Trypanosoma brucei:* nearest neighbor analysis on the major variable surface coat glycoprotein. Cross-linking patterns with intact cells, *Exp. Parasitol.,* 53, 117, 1982.

96. **Barry, J. D.,** Capping of variable antigens on *Trypanosoma brucei,* and its immunological and biological significance, *J. Cell Sci.,* 37, 287, 1979.

97. **Hall, T. and Esser, K.,** Topologic mapping of protective and nonprotective epitopes on the variant surface glycoprotein of the WRATat 1 clone of *Trypanosoma brucei rhodesiense, J. Immunol.,* 132, 2059, 1984.

98. **Miller, E. N., Allan, L. M., and Turner, M. J.,** Topological analysis of antigenic determinants on a variant surface glycoprotein of *Trypanosoma brucei, Mol. Biochem. Parastiol.,* 13, 67, 1984.

99. **Miller, E. N., Allan, L. M., and Turner, M. J.,** Mapping of antigenic determinants within peptides of a variant surface glycoprotein of *Trypanosoma brucei, Mol. Biochem. Parasitol.,* 13, 309, 1984.

100. **Rice-Ficht, A. C., Chen, K. K., and Donelson, J. E.,** Sequence homologies near the C-termini of the variable surface glycoproteins of *Trypanosoma brucei, Nature (London),* 294, 53, 1981.

101. **Barbet, A. F. and McGuire, T. C.,** Cross-reacting determinants in variant-specific antigens of African trypanosomes, *Proc. Natl. Acad. Sci. U.S.A.,* 75, 1989, 1978.

102. **Barbet, A. F., Musoke, A. J., Shapiro, S. Z., Mpimbaza, G., and McGuire, T. C.,** Identification of the fragment containing cross-reacting antigenic determinants in the variable surface glycoprotein of *Trypanosoma brucei, Parasitology,* 83, 623, 1981.

103. **Holder, A. A.,** Characterisation of the cross-reacting carbohydrate groups on two variant surface glycoproteins of *Trypanosoma brucei, Mol. Biochem. Parasitol.,* 7, 331, 1983.

104. **Campbell, G. H., Esser, K. M., and Weinbaum, F. I.,** *Trypanosoma rhodesiense* infection in B-cell deficient mice, *Infect. Immun.,* 18, 434, 1977.

105. **Hudson, K. M., and Terry, R. J.,** Immunodepression and the course of infection of a chronic *Trypanosoma brucei* infection in mice, *Parasite Immunol.,* 1, 317, 1979.

106. **Seed, J. R. and Gam, A. A.,** Passive immunity to experimental trypanosomiasis, *J. Parasitol.,* 52, 1134, 1966.

107. **Campbell, G. H. and Phillips, S. M.,** Adoptive transfer of variant-specific resistance to *Trypanosoma rhodesiense* with B lymphocytes and serum, *Infect. Immun.,* 14, 1144, 1976.

108. **Nantulya, V. M., Musoke, A. J., Rurangirwa, F. R., and Moloo, S. K.,** Resistance of cattle to tsetse-transmitted challenge with *Trypanosoma brucei* or *Trypanosoma congolense* after spontaneous recovery from syringe-passaged infections, *Infect. Immun.,* 43, 735, 1984.

109. **Nantulya, V. M., Musoke, A. J., and Moloo, S. K.,** Apparent exhaustion of the variable antigen repetoires of *Trypanosoma vivax* in infected cattle, *Infect. Immun.,* 54, 444, 1986.

110. **Wellde, B. T., Hockmeyer, W. T., Kovatch, R. M., Bhogal, M. S., and Diggs, C. L.,** *Trypanosoma congolense:* natural and acquired resistance in the bovine, *Exp. Parasitol.,* 52, 219, 1981.

111. **Gray, A. R.,** Antigenic variation in a strain of *Trypanosoma brucei* transmitted by *Glossina morsitans* and *G. palpalis, J. Gen. Microbiol.,* 41, 195, 1965.

112. **Wilson, A. J., Dar, F. K., and Paris, J.,** Serological studies on trypanosomiasis in East Africa. III. Comparison of antigenic types of *Trypanosoma congolense* organisms isolated from wild flies, *Ann. Trop. Med. Parasitol.,* 67, 313, 1973.

113. **Taussig, M. J.,** Antigenic competition, in *The Antigens,* Vol. 4, Sela, M., Ed., Academic Press, New York, 1977, chap. 5.

114. **Shapiro, S. Z. and Murray, M.,** African trypanosome antigens recognized during the course of infection in N'dama and Zebu cattle, *Infect. Immun.,* 35, 410, 1982.

115. **Shapiro, S. Z.,** Unpublished data, 1982.

116. **Gardiner, P. R., Finerty, J. F., and Dwyer, D. M.,** Iodination and identification of surface membrane antigens in procyclic *Trypanosoma rhodesiense, J. Immunol.,* 131, 454, 1983.

117. **Hunter, S. W., Fugiwara, T., and Brennan, P. J.,** Structure and antigenicity of the major specific glycolipid antigen of *Mycobacterium leprae, J. Biol. Chem.,* 257, 15072, 1982.

118. **Kanfer, J. N.,** Preparation of gangliosides, *Methods Enzymol.,* 14, 660, 1969.

119. **Tetley, L., Vickerman, K., and Moloo, S. K.,** Absence of a surface coat from metacyclic *Trypanosoma vivax:* possible implications for vaccination, *Trans. R. Soc. Trop. Med. Hyg.,* 75, 409, 1981.

120. **Rovis, L., Musoke, A. J., and Moloo, S. K.,** Failure of trypanosomal membrane antigens to induce protection against tsetse-transmitted *Trypanosoma vivax* or *T. brucei* in goats and rabbits, *Acta Trop.,* 41, 227, 1984.

121. **Gardiner, P. R., Webster, P., Jenni, L., and Moloo, S. K.,** Metacyclic *Trypanosoma vivax* possess a surface coat, *Parasitology,* 92, 75, 1986.

122. **Risby, E. and Seed, J. R.,** Purification and properties of purified hexokinase from the African trypanosomes and *Trypanosoma equiperdum, J. Protozool.,* 16, 193, 1969.

123. **Risby, E., and Seed, J. R.,** Purification and properties of purified aldolase from the African trypanosomes and *Trypanosoma equiperdum, Int. J. Biochem.,* 1, 209, 1970.

124. **Risby, E., Seed, T. M., and Seed, J. R.,** *Trypanosoma gambiense, T. rhodesiense, T. brucei, T. equiperdum* and *T. lewisi:* purification and properties of phosphohexose isomerase, *Exp. Parasitol.,* 25, 101, 1969.

125. **Seed, J. R.,** Antigens and antigenic variability of the African trypanosomes, *J. Protozool.,* 21, 639, 1974.

126. **Vickerman, K.,** Morphological and physiological considerations of extracellular blood protozoa, in *Ecology and Physiology of Parasites,* Fallis, A. M., Ed., University Press, Toronto, 1971, 58.

127. **Herbert, W. J. and Parratt, D.,** Virulence of trypanosomes in the vertebrate host, in *Biology of the Kinetoplastida,* Vol. 2, Lumsden, W. H. R. and Evans, D. A., Eds., Academic Press, London, 1979, chap. 10.

128. **MacAskill, J. A. and Holmes, P. H.,** The use of 75-Seleno-methionine labelled *Trypanosoma brucei* to measure parasite replication *in vivo, Tropenmed. Parasitol.,* 34, 197, 1983.

129. **Bowman, I. B. R. and Flynn, I. W.,** Oxidative metabolism of trypanosomes, in *Biology of the Kinetoplastida,* Vol. 1, Lumsden, W. H. R. and Evans, D. A., Eds., Academic Press, London, 1976, chap. 10.

130. **Vickerman, K.,** The mechanism of cyclical development in trypanosomes of the *Trypanosoma brucei* subgroup: an hypothesis based on ultrastructural observations, *Trans. R. Soc. Trop. Med. Hyg.,* 56, 487, 1962.

131. **Tanner, M., Jenni, L., Hecker, H., and Brun, R.,** Characterization of *Trypanosoma brucei* isolated from lymph nodes of rats, *Parasitology,* 80, 383, 1980.

132. **Brown, R. C., Evans, D. A., and Vickerman, K.,** Changes in oxidative metabolism and ultrastructure accompanying differentiation of the mitochondrion in *Trypanosoma brucei, Int. J. Parasitol.,* 3, 691, 1973.

133. **Vickerman, K.,** Polymorphism and mitochondrial activity in sleeping sickness trypanosomes, *Nature (London),* 208, 762, 1965.

134. **Shapiro, S. Z., Naessens, J., Liesegang, B., Moloo, S. K., and Magondu, J.,** Analysis by flow cytometry of DNA synthesis during the life cycle of African trypanosomes, *Acta Trop.,* 41, 313, 1984.

135. **Wijers, D. J. B. and Willett, K. C.,** Factors that may influence the infection rate of *Glossina palpalis* with *Trypanosoma gambiense.* II. The number and the morphology of the trypanosomes present in the blood of the host at the time of the infected feed, *Ann. Trop. Med. Parasitol.,* 54, 341, 1960.

136. **Nantulya, V. M., Doyle, J. J., and Jenni, L.,** Studies on *Trypanosoma (Nannomonas) congolense.* I. On the morphological appearance of the parasite in the mouse, *Acta Trop.,* 35, 329, 1978.

137. **Tsanev, R.,** Cell cycle and liver function, in *Cell Cycle and Cell Differentiation,* Reinert, J. and Holtzer, H., Eds., Springer-Verlag, New York, 1975, 197.

138. **Sendashonga, C. N. and Black, S. J.,** Humoral responses against *Trypanosoma brucei* variable surface antigen are induced by degenerating parasites, *Parasite Immunol.,* 4, 245, 1982.

139. **Black, S. J., Sendashonga, C. N., Lalor, P. A., Whitelaw, D. D., Jack, R. M., Morrison, W. I., and Murray, M.,** Regulation of the growth and differentiation of *Trypanosoma (Trypanozoon) brucei brucei* in resistant (C57B1/6) and susceptible (C3H/He) mice, *Parasite Immunol.,* 5, 465, 1983.

140. **Hoare, C. A.,** Systematic description of the mammalian trypanosomes of Africa, in *The African Trypanosomiases,* Mulligan, H. W., Ed., George Allen and Unwin, London, 1970, chap. 2.

141. **Jenni, L. and Brun, R.,** *In vitro* cultivation of pleomorphic *Trypanosoma brucei* stocks: a possible source of variable antigens for immunization studies, *Trans. R. Soc. Trop. Med. Hyg.,* 75, 150, 1981.

142. **Hoare, C. A.,** The mammalian trypanosomes of Africa, in *The African Trypanosomiases,* Mulligan, H. W. Ed., George Allen and Unwin, London, 1970, chap. 1.

143. **Gibson, W. C., Marshall, F. De C., and Godfrey, D. G.,** Numerical analysis of enzyme polymorphism: a new approach to the epidemiology and taxonomy of trypanosomes of the subgenus *Trypanozoon, Adv. Parasitol.,* 18, 175, 1980.

144. **Tait, A.,** Evidence for diploidy and mating in trypanosomes, *Nature (London),* 287, 536, 1980.

145. **Jenni, L., Marti, S., Schweizer, J., Betschart, B., Le Page, R. W. F., Wells, J. M., Tait, A., Paindavoine, P., Pays, E., and Steinert, M.,** Hybrid formation between African trypanosomes during cyclical transmission, *Nature (London),* 322, 173, 1986.

146. **Luckins, A. G. and Gray, A. R.,** An extravascular site of development of *Trypanosoma congolense, Nature (London),* 272, 613, 1978.

147. **Emery, D. L. and Moloo, S. K.,** The sequential cellular changes in the local skin reaction produced in goats by *Glossina morsitans morsitans* infected with *Trypanosoma (Trypanozoon) brucei, Acta Trop.,* 37, 137, 1980.

148. **Akol, G. W. O. and Murray, M.,** Early events following challenge of cattle with tsetse infected with *Trypanosoma congolense:* development of the local skin reaction, *Vet. Rec.,* 110, 295, 1982.

149. **Emery, D. L. and Moloo, S. K.,** The dynamics of the cellular reactions elicited in the skin of goats by *Glossina morsitans morsitans* infected with *Trypanosoma (Trypanozoon) brucei, Acta Trop.,* 38, 15, 1981.

150. **Cook, R. M.,** The chemotactic response of murine peritoneal exudate cells to *Trypanosoma brucei, Vet. Parasitol.,* 7, 3, 1980.

151. **Tizard, I. R. and Holmes, W. L.,** The release of soluble vasoactive material from *Trypanosoma congolense* in intraperitoneal diffusion chambers, *Trans. R. Soc. Trop. Med. Hyg.,* 71, 32, 1977.

152. **Anon.,** Trypanosomiasis: early events in the skin and lymphatic system, in *ILRAD: Annu. Rep. of the Int. Laboratory for Research on Animal Diseases,* ILRAD, Nairobi, 1984, 33.

153. **Murray, M., Grootenhuis, J. G., Akol, G. W. O., Emery, D. L., Shapiro, S. Z., Moloo, S. K., Dar, F., Bovell, D. L., and Paris, J.,** Potential application of research on African trypanosomiases in wildlife and preliminary studies on animals exposed to tsetse infected with *Trypanosoma congolense,* in *Wildlife Disease Research and Economic Development,* Karstad, L., Nestel, B., and Graham, M., Eds., IDRC, Ottawa, 1981, 40.

154. **Morrison, W. I., Murray, M., and McIntyre, W. I. M.,** Bovine trypanosomiasis, in *Diseases of Cattle in the Tropics: Economic and Zoonotic Relevance,* Ristic, M. and McIntyre, I., Eds., Martinus Nijhoff, Dordrecht, Netherlands, 1981, chap. 36.

155. **Murray, M., Morrison, W. I., Murray, P. K., Clifford, D. J., and Trail, J. C. M.,** Trypanotolerance — a review, *World Anim. Rev.,* 31(Suppl.), 2, 1979.

156. **Balber, A. E.,** *Trypanosoma brucei:* fluxes of the morphological variants in intact and X-irradiated mice, *Exp. Parasitol.,* 31, 307, 1972.

157. **Luckins, A. G.,** Effects of X-irradiation and cortisone treatment of albino rats on infections with *brucei*-complex trypanosomes, *Trans. R. Soc. Trop. Med. Hyg.,* 66, 130, 1972.

158. **Greenblatt, H. C., Diggs, C. L., and Aikawa, M.,** Antibody-dependent phagocytosis of *Trypanosoma rhodesiense* by murine macrophages, *Am. J. Trop. Med. Hyg.,* 32, 34, 1983.

159. **Ngaira, J. M., Nantulya, V. M., Musoke, A. J., and Hirumi, K.,** Phagocytosis of antibody-sensitized *Trypanosoma brucei in vitro* by bovine peripheral blood monocytes, *Immunology,* 49, 393, 1983.

160. **Mayor-Withey, K. S., Clayton, C. E., Roelants, G. E., and Askonas, B. A.,** Trypanosomiasis leads to extensive proliferation of B, T and null cells in spleen and bone marrow, *Clin. Exp. Immunol.,* 34, 359, 1978.

161. **Corsini, A. C., Clayton, C., Askonas, B. A., and Ogilvie, B. M.,** Suppressor cells and loss of B-cell potential in mice infected with *Trypanosoma brucei, Clin. Exp. Immunol.,* 29, 122, 1977.

162. **Murray, P. K., Jennings, F. W., Murray, M., and Urquhart, G. M.,** The nature of immunosuppression in *Trypanosoma brucei* infections in mice. I. The role of the macrophage, *Immunology,* 27, 815, 1974.

163. **Morrison, W. I., Murray, M., Sayer, P. D., and Preston, J. M.,** The pathogenesis of experimentally induced *Trypanosoma brucei* infection in the dog. II. Changes in the lymphoid organs, *Am. J. Pathol.,* 102, 182, 1981.

164. **Morrison, W. I., Murray, M., and Bovell, D. L.,** Response of the murine lymphoid system to a chronic infection with *Trypanosoma congolense.* I. The spleen, *Lab. Invest.,* 45, 547, 1981.

165. **Morrison, W. I., Murray, M., and Hinson, C. A.,** The response of the murine lymphoid system to a chronic infection with *Trypanosoma congolense.* II. The lymph nodes, thymus and liver, *J. Pathol.,* 138, 273, 1982.

166. **Greenwood, B. M. and Whittle, H. C.,** The pathogenesis of sleeping sickness, *Trans. R. Soc. Trop. Med. Hyg.,* 74, 716, 1980.

167. **Luckins, A. G. and Mehiltz, D.,** Observations on serum immunoglobulin levels in cattle infected with *Trypanosoma brucei, T. vivax,* and *T. congolense, Ann. Trop. Med. Parasitol.,* 70, 479, 1976.

168. **Whittle, H. C., Greenwood, B. M., Bidwell, D. E., Bartlett, A., and Voller, A.,** IgM and antibody measurement in the diagnosis and management of gambian trypanosomiasis, *Am. J. Trop. Med. Hyg.,* 26, 1129, 1977.

169. **Hudson, K. M., Byner, C., Freeman, J., and Terry, R. J.,** Immunodepression, high IgM levels, and evasion of the immune response in murine trypanosomiasis, *Nature (London),* 264, 256, 1976.

170. **Urquhart, G. M., Murray, M., Murray, P. K., Jennings, F. W., and Bate, E.,** Immunosuppression in *Trypanosoma brucei* infections in rats and mice, *Trans. R. Soc. Trop. Med. Hyg.,* 67, 528, 1973.

171. **Greenwood, B. M.,** Possible role of a B-cell mitogen in hypergammaglobulinaemia in malaria and trypanosomiasis, *Lancet,* 1, 435, 1974.

172. **Grosskinsky, C. M. and Askonas, B. A.,** Macrophages as primary target cells and mediators of immune dysfunction in African trypanosomiasis, *Infect. Immun.,* 33, 149, 1981.

173. **Ikede, B. O., Lule, M., and Terry, R. J.,** Anaemia in trypanosomiasis: mechanisms of erythrocyte destruction in mice infected with *Trypanosoma congolense* or *T. brucei, Acta Trop.,* 34, 53, 1977.

174. **Dargie, J. D., Murray, P. K., Murray, M., Grimshaw, W. R. T., and McIntyre, W. I. M.,** Bovine trypanosomiasis: the red cell kinetics of Ndama and Zebu cattle infected with *Trypanosoma congolense, Parasitol.,* 78, 271, 1979.

175. **Preston, J. M., Wellde, B. T., and Kovatch, R. M.,** *Trypanosoma congolense:* calf erythrocyte survival, *Exp. Parasitol.,* 48, 118, 1979.

176. **Kobayashi, A., Tizard, I. R., and Woo, P. T. K.,** Studies on the anemia in experimental African trypanosomiasis. II. The pathogenesis of the anemia in calves infected with *Trypanosoma congolense, Am. J. Trop. Med. Hyg.,* 25, 401, 1976.

177. **Amole, B. O., Clarkson, A. B., Jr., and Shear, H. L.,** Pathogenesis of anemia in *Trypanosoma brucei*-infected mice, *Infect. Immun.,* 36, 1060, 1982.

178. **Kobayakawa, T., Louis, J., Izui, S., and Lambert, P. H.,** Autoimmune response to DNA, red blood cells, and thymocyte antigens in association with polyclonal antibody synthesis during experimental African trypanosomiasis, *J. Immunol.,* 122, 296, 1979.

179. **Houba, V. and Allison, A. C.,** M-antiglobulins (rheumatoid-factor-like globulins) and other gammaglobulins in relation to tropical parasitic infections, *Lancet,* 1, 848, 1966.

180. **Houba, V., Brown, K. N., and Allison, A. C.,** Heterophile antibodies, M-antiglobulins and immunoglobulins in experimental trypanosomiasis, *Clin. Exp. Immunol.,* 4, 113, 1969.

181. **Rickman, W. J. and Cox, H. W.,** Trypanosome antigen-antibody complexes and immunoconglutinin interactions in African trypanosomiasis, *Int. J. Parasitol.,* 13, 389, 1983.

182. **Tizard, I., Nielsen, K. H., Seed, J. R., and Hall, J. E.,** Biologically active products from African trypanosomes, *Microbiol. Rev.,* 42, 661, 1978.

183. **Beutler, B. and Cerami, A.,** Cachectin and tumor necrosis factor as two sides of the same biological coin, *Nature (London),* 320, 584, 1986.

184. **Hotez, P. J., Le Trang, N., Fairlamb, A. H., and Cerami, A.,** Lipoprotein lipase suppression in 3T3-L1 cells by a haematoprotozoan-induced mediator from peritoneal exudate cells, *Parasite Immunol.,* 6, 203, 1984.

185. **Fiennes, R. N. T.-W.,** Pathogenesis and pathology of animal trypanosomiases, in *The African Trypanosomiases,* Mulligan, H. W., Ed., George Allen and Unwin, London, 1970, chap. 38.

186. **Ackerman, S. B. and Seed, J. R.,** Immunodepression during *Trypanosoma brucei gambiense* infections in the field vole, *Microtus montanus, Clin. Exp. Immunol.,* 25, 152, 1976.

187. **Jayawardena, A. N. and Waksman, B. H.,** Suppressor cells in experimental trypanosomiasis, *Nature (London),* 265, 539, 1977.

188. **Pearson, T. W., Roelants, G. E., Pinder, M., Lundin, L. B., and Mayor-Withey, K. S.,** Immune depression in trypanosome-infected mice. III. Suppressor cells, *Eur. J. Immunol.,* 9, 200, 1979.

189. **Albright, J. W., Albright, J. F., and Dusanic, D. G.,** Mechanisms of trypanosome-mediated suppression of humoral immunity in mice, *Proc. Natl. Acad. Sci. U.S.A.,* 75, 3923, 1978.

190. **Black, S. J., Sendashonga, C. N., Webster, P., Koch, G. L. E., and Shapiro, S. Z.,** Regulation of parasite-specific antibody responses in resistant (C57B1/6) and susceptible (C3H/He) mice infected with *Trypanosoma (Trypanozoon) brucei brucei, Parasite Immunol.,* 8, 425, 1986.

191. **Askonas, B. A., Corsini, A. C., Clayton, C. E., and Ogilvie, B. M.,** Functional depletion of T- and B-memory cells and other lymphoid cell subpopulations during trypanosomiasis, *Immunology,* 36, 313, 1979.

192. **Soltys, M. A. and Woo, P. T. K.,** Trypanosomes producing disease in livestock in Africa, in *Parasitic Protozoa,* Vol. 1, Kreier, J. P., Ed., Academic Press, New York, 1977, chap. 6.

193. **Ashcroft, M. T., Burtt, E., and Fairburn, H.,** The experimental infection of some African wild animals with *Trypanosoma rhodesiense, T. brucei,* and *T. congolense, Ann. Trop. Med. Parasitol.,* 53, 147, 1959.

194. **Griffin, L. and Allonby, E. W.,** Trypanotolerance in breeds of sheep and goats with an experimental infection of *Trypanosoma congolense, Vet. Parasitol.,* 5, 97, 1979.

195. **Terry, R. J.,** Antibody against *Trypanosoma vivax* present in normal cotton rat serum, *Exp. Parasitol.,* 6, 404, 1957.

196. **Hawking, F.,** The action of human serum upon *Trypanosoma brucei, Protozool. Abstr.,* 3, 199, 1979.

197. **Roelants, G. E.,** Natural resistance to African trypanosomiasis, *Parasite Immunol.,* 8, 1, 1986.

198. **Desowitz, R. S.,** Studies on *Trypanosoma vivax.* X. The activity of some blood fractions in facilitating infection in white rats, *Ann. Trop. Med. Parasitol.,* 48, 142, 1954.

199. **Murray, M. and Morrison, W. I.,** Non-specific induction of increased resistance in mice to *Trypanosoma congolense* and *Trypanosoma brucei* by immunostimulants, *Parasitology,* 79, 349, 1979.

200. **Kissling, K., Karbe, E., and Freitas, E. K.,** *In vitro* phagocytic activity of neutrophils of various cattle breeds with and without *Trypanosoma congolense* infection, *Tropenmed. Parasitol.,* 33, 158, 1982.

201. **Anon.,** Pathogenesis of trypanosome infection, in *ILRAD: Annu. Rep. of the Int. Laboratory for Research on Animal Diseases,* ILRAD, Nairobi, 1984, 41.

202. **Shapiro, S. Z. and Black, S. J.,** Manuscript in preparation.

203. **Koj, A.,** Acute phase reactants: their synthesis, turnover and biological significance, in *Structure and Function of Plasma Proteins,* Allison, A. C., Ed., Plenum Press, New York, 1974, chap. 4.

204. **Mitchell, L. A. and Pearson, T. W.,** Antibody responses induced by immunization of inbred mice susceptible and resistance to African trypanosomes, *Infect. Immun.,* 40, 894, 1983.

205. **Clayton, C. E., Sacks, D. L., Ogilvie, B. M., and Askonas, B. A.,** Membrane fractions of trypanosomes mimic the immunosuppressive and mitogenic effects of living parasites on the host, *Parasite Immunol.,* 1, 241, 1979.

206. **Kaaya, G. P., Tizard, I. R., Maxie, M. G., and Valli, V. E. O.,** Inhibition of leukopoiesis by sera from *Trypanosoma congolense* infected calves: Partial characterization of the inhibitory factor, *Tropenmed. Parasitol.,* 31, 232, 1980.

207. **Brown, K. N.,** Protective immunity to malaria provides a model for the survival of cells in an immunologically hostile environment, *Nature (London),* 230, 163, 1971.

208. **Beat, D. A., Stanley, H. A., Choromanski, L., MacDonald, A. B., and Honigberg, B. M.,** Nonvariant antigens limited to bloodstream forms of *Trypanosoma brucei brucei* and *Trypanosoma brucei rhodesiense, J. Protozool.,* 31, 541, 1984.

209. **Burgess, D. E. and Jerrells, T.,** Molecular identity and location of invariant antigens on *Trypanosoma brucei rhodesiense* defined with monoclonal antibodies reactive with sera from trypanosomiasis patients, *Infect. Immun.,* 50, 893, 1985.

210. **Lonsdale-Eccles, J. D. and Grab, D. J.,** Purification of African trypanosomes can cause biochemical changes in the parasites, *J. Protozool.,* 34, 405, 1987.

211. **Powell, C. N.,** Immunoprotective effects of bound particulate subcellular fractions of *Trypanosoma brucei* and *T. rhodesiense, Med. J. Zambia,* 10, 27, 1976.

212. **Powell, C. N.,** Experimental immunity against trypanosomiasis, *Experientia,* 34, 1450, 1978.

213. **Murray, M., Barry, J. D., Morrison, W. I., Williams, R. O., Hirumi, H., and Rovis, L.,** A review of the prospects for vaccination in African trypanosomiasis. II, *World Anim. Rev.,* 33(Suppl.), 14, 1980.

214. **Shapiro, S. Z.,** Unpublished data, 1980.

215. **Powell, C. N.,** Personal communication, 1986.

216. **Singer, I., Kimble, E. T., III, and Ritts, R. E., Jr.,** Alterations of the host-parasite relationship by administration of endotoxin to mice with infections of trypanosomes, *J. Infect. Dis.,* 114, 243, 1964.

217. **Black, S. J., Shapiro, S. Z., Murray, M., and Borowy, N. K.,** Mechanism of heat-killed *Corynebacterium parvum*-induced non-specific immunity to *Trypanosoma (Trypanozoon) brucei,* in preparation.

218. **Hajduk, S. L. and Vickerman, K.,** Antigenic variation in cyclically transmitted *Trypanosoma brucei.* Variable antigen type composition of the first parasitaemia in mice bitten by trypanosome-infected *Glossina morsitans, Parasitology,* 83, 609, 1981.

219. **Esser, K. M., Schoenbechler, M. J., and Gingrich, J. B.,** *Trypanosoma rhodesiense* blood forms express all antigen specificities relevant to protection against metacyclic (insect form) challenge, *J. Immunol.,* 129, 1715, 1982.

220. **Jenni, L.,** Antigenic variants in cyclically transmitted strains of the *T. brucei* complex, *Ann. Soc. Belge Med. Trop.,* 57, 383, 1977.

221. **Nantulya, V. M., Musoke, A. J., Moloo, S. K., and Ngaira, J. M.,** Analysis of the variable antigen composition of *Trypanosoma brucei brucei* metacyclic trypanosomes using monoclonal antibodies, *Acta Trop.,* 40, 19, 1983.

222. **Crowe, J. S., Barry, J. D., Luckins, A. G., Ross, C. A., and Vickerman, K.,** All metacyclic variable antigen types of *Trypanosoma congolense* identified using monoclonal antibodies, *Nature (London),* 306, 389, 1983.

223. **Nantulya, V. M., Doyle, J. J., and Jenni, L.,** Studies on *Trypanosoma (nannomonas) congolense.* IV. Experimental immunization of mice against tsetse fly challenge, *Parsitology,* 80, 133, 1980.

224. **Akol, G. W. O. and Murray, M.,** *Trypanosoma congolense:* susceptibility of cattle to cyclical challenge, *Exp. Parasitol.,* 55, 386, 1983.

225. **Masake, R. A., Nantulya, V. M., Musoke, A. J., Moloo, S. K., and Nguli, K.,** Characterization of *Trypanosoma congolense* serodemes in stocks isolated from cattle introduced onto a ranch in Kilifi, Kenya, *Parasitology,* 94, 349, 1987.

226. **Silverstein, S. C., Steinman, R. M., and Cohn, Z. A.,** Endocytosis, *Annu. Rev. Biochem.,* 46, 669, 1977.

227. **Goldstein, J. L., Anderson, R. G. W., and Brown, M. S.,** Coated pits, coated vesicles, and receptor mediated endocytosis, *Nature (London),* 279, 679, 1979.

228. **Pearse, B. M. F.,** Coated vesicles from pig brain: purification and biochemical characterization, *J. Mol. Biol.,* 97, 93, 1975.

229. **Langreth, S. G. and Balber, A. E.,** Protein uptake and digestion in bloodstream and culture forms of *Trypanosoma brucei, J. Protozool.,* 22, 40, 1975.

230. **Webster, P. and Grab, D. J.,** Intracellular colocalization of variant surface glycoprotein and transferrin-gold in *Trypanosoma brucei, J. Cell Biol.,* 106, 279, 1988.

231. **Fairlamb, A. H. and Bowman, I. B. R.,** *Trypanosoma brucei:* maintenance of concentrated suspensions of blood stream trypomastigotes *in vitro* using continuous dialysis for measurement of endocytosis, *Exp. Parasitol.,* 49, 366, 1980.

232. **Pastan, I. H. and Willingham, M. C.,** Journey to the center of the cell: role of the receptosome, *Science,* 214, 504, 1981.

233. **Black, S. J., Sendashonga, C. N., O'Brien, C., Borowy, N. K., Naessens, M., Webster, P., and Murray, M.,** Regulation of parasitaemia in mice infected with *Trypanosoma brucei, Curr. Top. Microbiol. Immunol.,* 117, 93, 1985.

234. **Black, S. J. and Shapiro, S. Z.,** Manuscript in preparation.

235. **Rener, J., Carter, R., Rosenberg, Y., and Miller, L. H.,** Anti-gamete monoclonal antibodies synergistically block transmission of malaria by preventing fertilization in the mosquito, *Proc. Natl. Acad. Sci. U.S.A.,* 77, 6797, 1980.

236. **Maudlin, L., Turner, M. J., Dukes, P., and Miller, N.,** Maintenance of *Glossina morsitans morsitans* on antiserum to procyclic trypanosomes reduces infection rates with homologous and heterologous *Trypanosoma congolense* stocks, *Acta Trop.,* 41, 253, 1984.

237. **Murray, M., Hirumi, H., and Moloo, S. K.,** Suppression of *Trypanosoma congolense, T. vivax* and *T. brucei* infection rates in tsetse flies maintained on goats immunized with uncoated forms of trypanosomes grown *in vitro, Parasitology,* 91, 53, 1985.

238. **Nash, T. A. M., and Jordan, A. M.,** Methods for rearing and maintaining *Glossina* in the laboratory, in *The African Trypanosomiases,* Mulligan, H. W., Ed., George Allen and Unwin, London, 1970, chap. 20.

239. **Bursell, E.,** Feeding, digestion and excretion, in *The African Trypanosomiases,* Mulligan, H. W., Ed., George Allen and Unwin, London, 1970, chap. 12.

240. **Alger, N. E. and Cabrera, E. J.,** An increase in death rate of *Anopheles stephensi* fed on rabbits immunized with mosquito antigen, *J. Econ. Entomol.,* 65, 165, 1972.

241. **McGowan, M. J., Homer, J. T., O'Dell, G. V., McNew, R. W., and Barker, R. W.,** Performance of ticks fed on rabbits inoculated with extracts derived from homogenized ticks *Amblyomma maculatum* Koch (Acarina: Ixodidae), *J. Parasitol.,* 66, 42, 1980.

242. **Kemp, D. H., Agbede, R. I. S., Johnston, L. A. Y., and Gough, J. M.,** Immunization of cattle against *Boophilus microplus* using extracts derived from adult female ticks: feeding and survival of the parasite on vaccinated cattle, *Int. J. Parasitol.,* 16, 115, 1986.

243. **Schlein, Y., Spira, D. T., and Jacobson, R. L.,** The passage of serum immunoglobulins through the gut of *Sarcophaga falculata,* Pand., *Ann. Trop. Med. Parasitol.,* 70, 227, 1976.

244. **Nogge, G. and Giannetti, M.,** Specific antibodies: a potential insecticide, *Science,* 209, 1028, 1980.

245. **Ackerman, S., Clare, F. B., McGill, T. W., and Sonenshine, D. E.,** Passage of host serum components, including antibody, across the digestive tract of *Dermacentor variabilis* (Say), *J. Parasitol.,* 67, 737, 1981.

246. **Schlein, Y. and Lewis, C. T.,** Lesions in haematophagous flies after feeding on rabbits immunized with fly tissues, *Physiol. Entomol.,* 1, 55, 1976.

247. **Ackerman, S., Floyd, M., and Sonenshine, D. E.,** Artificial immunity to *Dermacentor variabilis* (ACARI: IXODIDAE): vaccination using tick antigens, *J. Med. Entomol.,* 17, 391, 1980.

248. **Mongi, A. O., Shapiro, S. Z., Doyle, J. J., and Cunningham, M. P.,** Immunization of rabbits with *Rhipicephalus appendiculatus* antigen-antibody complexes, *Insect Sci. Appl.,* 7, 471, 1986.

249. **Sher, A. and Snary, D.,** Specific inhibition of the morphogenesis of *Trypanosoma cruzi* by a monoclonal antibody, *Nature (London),* 300, 639, 1982.

250. **Nogge, G.,** Aposymbiotic tsetse flies, *Glossina morsitans morsitans* obtained by feeding on rabbits immunized specifically with symbionts, *J. Insect Physiol.,* 24, 299, 1978.

Chapter 8

COCCIDIA VACCINES

Harry D. Danforth and Patricia C. Augustine

TABLE OF CONTENTS

I. INTRODUCTION

Coccidiosis is a complex intestinal disease induced by the *Eimeria* species that is of major economic importance in domestic animals. The areas to be covered in this chapter will concern the methods and technologies in use to develop vaccines for the eventual immunological control of this disease. The beginning section will describe the basic biology and immunology of the coccidia. This will be followed by a discussion of current and projected vaccine approaches in which the parasite itself or its extracted and expressed proteins are used as antigenic sources. Many of these areas are still in their infancy, so some supposition of their use in vaccine development is necessary. In the concluding remarks some problems involved with the development of such vaccines including absence of cross-protection due to species and strain variability will be discussed.

II. BIOLOGY AND IMMUNOLOGY

A. Impact

The coccidia have their greatest impact on the domestic poultry and cattle industries of the world because of the mortality, morbidity, and weight loss that the infection produces. It is difficult to put a figure on the present day cost of these parasites, but a conservative estimate based on 6- to-16-year-old data is at least $1 billion annually.[1,2] Control of these parasites is primarily through the use of anticoccidial compounds that are usually administered in the feed. The importance of these compounds is evidenced by the direct relationship seen between the growth of the poultry industry and the development of these anticoccidials.[3] However, there are some problems associated with this type of control. These include the increased cost of clearing a compound for use in food animals, which has resulted in fewer companies screening for and developing anticoccidials, and the resistance of the coccidia to a number of compounds already on the market. The combination of these factors may eventually have a catastrophic effect on the domestic livestock industry.

B. Life Cycle

The life cycles of the *Eimeria* are complex (Figure 1). Sporulated oocysts are ingested by the animal, and sporozoites are subsequently released into the intestinal lumen where they invade the epithelium to begin the infection. There are a number of asexual reproductive cycles or generations that produce large numbers of merozoites. It is during this stage of parasite development that intestinal lesions are seen and weight loss or death of the animal may occur. The merozoites eventually develop into sexual stages that form the oocysts. The oocysts are shed in the fecal material, sporulate, and can initiate the entire cycle again upon reingestion. There are further complications involved with these life cycles in that the *Eimeria* are species specific with regard to the animals they infect, the different species of coccidia not only infect different areas of the intestine, but also migrate within the gut after infection. This migration apparently can even include an extraintestinal phase, since some species have been found in a variety of tissues after oocyst inoculation.[4]

C. Immunity

This type of infection induces a large variety of immune responses in the host, many of which are nonprotective. Despite a great amount of work done on characterizing the immune response to *Eimeria*, no clear picture has emerged to indicate how resistance to the parasite is acquired.[4] What is known is that immunity against the coccidia is species specific such that an animal which is protected against one species of *Eimeria* after infection is completely susceptible to other infective species. The two major manifestations of this protective immune response are cell-mediated immunity (CMI) and antibody response. Studies with different

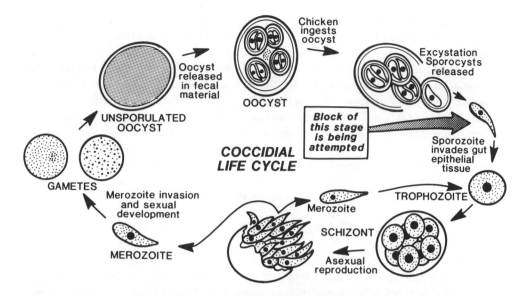

FIGURE 1. The life cycle of *Eimeria* begins with the ingestion of oocysts by the animal. Excystation then occurs and the sporozoites are released into the intestinal lumen, invade host cells, and round up to form trophozoites. Asexual generations or schizonts then develop intracellularly and produce merozoites which reinvade host cells. These merozoites either develop into other schizont generations or the sexual stages (micro- or macrogametes). Following fertilization of the macrogamete by the microgametes, oocysts form and are shed in the fecal material. These become sporulated and infective within 2 to 3 d after shedding. The entire life cycle takes from 5 to 7 d, depending on the species. It is against the early stages (sporozoites and trophozoites) that the protective immune response of the animal is directed, and it is at these stages that attempts at vaccine development is centered.

Eimeria species *in vivo* have demonstrated that resistance to reinfection is T cell dependent and predominately CMI. A recent investigation,[5] utilizing for the first time an *in vitro* T cell proliferation assay, has suggested that species-specific protective immunity in chickens may be due to the development of a species-specific T cell response. The T cell response to the parasite is apparently dependent both on the presence of an optimal number of antigen-presenting cells and the number and developmental stage of the infecting parasite. Soluble products obtained from Concanavalin A-stimulated T cells, identified as lymphokines, have also been implicated in providing protection against the mammalian *Eimeria*[6,7] and are apparently involved in the immune response to avian *Eimeria*.[8] Antibodies may play an important role in protective immune response, probably more in the control of primary infections. This is evidenced by the ability of serum from immunized chickens to passively protect birds not previously exposed to the parasite against coccidiosis, the ability of IgA from cecal extracts of immunized chickens to inhibit invasion and development by sporozoites *in vitro*, and the greater susceptibility to primary infection of bursectomized chickens and antibody-deficient mice.[4,9] Specific antiserum and monoclonal IgG antibodies raised against avian coccidia have also been found to inhibit sporozoite penetration and/or development in cultured cells.[4,10,11]

It is generally accepted that immunity to *Eimeria* affects the sporozoite stage of the parasite, but that the protection-eliciting antigens are found in the asexual stages.[4] The sporozoites, however, are not totally inhibited from penetrating host cells. In fact, there is a variation of the effect of immunity on invasion depending on the species studied and the particular area of the gut that the species invades.[12] Invasion by lower intestinal species (*E. tenella* and *E. adenoeides*) are inhibited up to 50%, while invasion by the upper intestinal species (*E. acervulina* and *E. meleagrimitis*) is not decreased in immunized animals. The parasites that invade do not usually develop further. Resistance to coccidial infection is not limited to the sporozoite stage. Merozoites introduced into immune hosts have also been shown to be affected in a similar manner through inhibition of penetration and blocking of development.[4] This is not surprising since the asexual stages which give rise to the merozoites contain the protective antigens. Some of the immune effects on the coccidia are even reversible, as has been demonstrated by the development of parasites transferred from immune hosts to naive animals.[4] The kinetics of the appearance and disappearance of different subclasses of immunoglobin in the serum and bile of animals after primary, secondary, and tertiary inoculations show that the animals undergo different stages of development of the immune response.[13] Thus, it seems likely that the mechanisms which limit primary infections may differ from those which limit challenge infections. What all this is showing is that the host animal has evolved a variety of different responses to the parasite, many of which are interacting, that allows it to combat the various development stages at any stage of immunity.

III. VACCINE APPROACHES

A. Criteria

There are certain basic criteria which must be included if a vaccine is going to be successful in the animal industry. The animal is a commodity, and as such it is raised for profit. Therefore, it is most important that a vaccine be reasonable in cost and be as effective as other current methods used for control of the disease. Poultry and livestock producers will not be interested in something that costs more and does less. Vaccines should have a relatively long shelf life and be easily integrated into the routine used by the producers for handling and raising the animals. The immunization must give solid protection within a short period following vaccination, be of long duration, and not cause any adverse reaction to the animal. The greatest progress in the development of vaccines against coccidia has been with the poultry industry, so it is from these studies that the following section on vaccine approaches is centered.

B. Use of Parasite

The first attempts at vaccination against coccidia was to simply expose the animals to a controlled number of viable oocysts and thereby induce a protective immune response through natural infection. This approach has led to the development of the only vaccines currently on the market, CocciVac® and CocciVac T® (Sterwin Laboratories), which are used exclusively in the poultry industry.[14,15] These vaccines usually contain all the pathogenic species of coccidia that infect chickens and turkeys, although they can be formulated to include only certain species as dictated by the needs of the poultry producers. Their use has been limited mainly to the immunization of replacement birds in the breeder and layer or egg-producing flocks, but CocciVac T® is now being utilized in some commercial turkey flocks raised for meat consumption.[16] The birds are either inoculated individually or a prescribed number of oocysts are mixed in the feed or water to expose the animals to the parasite at 1 to 14 d of age. These animals usually develop a significant degree of immunity after the initial dose and have the added benefit of reexposing themselves 3 to 4 weeks postimmunization through ingestion of oocysts they have shed. Use of CocciVac® requires careful flock management. There must be enough cycling of the coccidia to ensure that the infection is heavy enough to produce a good immune response, yet not so heavily that clinical effects and the associated weight loss or drop in egg production occur. A high degree of resistance to the immunizing species does not usually develop without this low-dose reinfection.

A refinement of the use of a live vaccine has recently occurred with the development and use of attenuated strains of poultry coccidia for immunization. This has been accomplished by either embryo adaptation of the coccidia or selection for precociousness.[12] Embryo adaptation requires that the *Eimeria* be serially propagated in the chorioallantoic membrane of embryonated chicken eggs. With continuous passage of these strains, there is a gradual loss of pathogenicity due to a change in the asexual development of the coccidia without a loss in immunogenicity. Usually the size of these stages are reduced, resulting in the production of fewer merozoites. However, not all pathogenic species of coccidia infecting chickens, significantly *E. acervulina* and *E. maxima*, can be grown in the embryonated chicken eggs, indicating that other means of attenuating individual species are needed. Selection for precociousness requires that only the earliest-produced oocysts collected from infected animals be used as the inoculum for each succeeding generation or passage of the strain in birds. As few as ten generations of selection for precociousness has resulted in a decrease in the prepatent period of these strains, which was accompanied by a reduction in pathogenicity when compared with the original parent strain. The shortened prepatent times of the precocious lines are either the result of the loss of one or two entire asexual generations of the life cycle or the accelerated maturation of the asexual generations.[18]

Emphasis has been placed on use of the precocious lines as components of a coccidial vaccine consisting of viable oocysts.[17,19] These lines are extremely stable, retain excellent immunogenicity, and do not revert to the pathogenicity of the original parent strain. At the present time, precocious lines have been produced for seven of the economically important species that infect chickens (*E. acervulina*, *E. brunetti*, *E. maxima*, *E. mitis*, *E. necatrix*, *E. praecox*, and *E. tenella*).[19] Since virulence of these strains has been significantly reduced, the major problem in using these lines as immunizing agents is to insure that birds raised under commercial conditions are exposed to sufficient number of oocysts to elicit a protective immune response. The problem has been solved at least at the laboratory trial level by use of the "trickle dose" vaccination technique. In this procedure, oocysts encapsulated in calcium alginate beads are mixed in the feed, and the birds are continuously dosed with low levels of oocysts during the first few weeks after hatching.[17,20,21] This type of exposure has been shown to produce solid or complete immunity with virulent parasites.[17] Incorporating the oocysts into alginate maintains their viability throughout the immunization period and allows for even distribution in the feed. Any number of oocysts from each species can be encapsulated in each bead so the exact dosage per kilogram of feed is known. Since the birds consume more feed as they grow, there is a linear increase in the trickle vaccination dose that they ingest; this essentially imitates the natural pattern of exposure. In trials with birds trickle-immunized with the precocious strains, significant protection to coccidial challenge by the original parent strains was seen.[20,21] These immunized birds even performed as well as unimmunized, challenged birds fed anticoccidial compounds. Problems associated with this type of immunization will be discussed in a later section of this chapter, but trickle infection with precocious strains currently appears to be the most promising development in the use of protective immune response for control of avian coccidia at the commercial level.

C. Subunit Vaccines

There have been numerous attempts to immunize animals against coccidial infections using killed parasites or parasite-derived material, but these had been largely unsuccessful.[9] Recently, vaccination trials utilizing either sporulated oocyst extracts of *E. acervulina* and *E. tenella*,[22,23] an upper and lower intestinal species, respectively, or an isolated surface protein from sporozoites of *E. tenella* have now demonstrated protection against challenge, as determined by a significant reduction in intestinal lesions in the birds. An analysis of the immunoreactivity of the polypeptides in the extracts of *E. tenella* with antisporozoite sera showed nine prominent bands of 235, 105, 94, 82, 71, 68, 45, and 26 kDa.[22,23] No report was made on the number or type of polypeptides present in the *E. acervulina* extract. Titration experiments determined that the immune response was antigen dose dependent, with maximal protection seen at a dose of 10 μg of protein per bird for each species extract.

However, protection was seen for as low as 0.1 µg of protein with the *E. tenella* extract and 1 µg with the *E. acervulina*. Birds vaccinated by either a single intramuscular injection of 10 µg of protein at 2 d of age or by three intramuscular injections of the same amount of material at 2, 9, and 16 d of age without adjuvant were protected from at least 3 weeks of age till the end of grow-out at 7 weeks for both extracts. The level of protection for the *E. tenella* extract was also shown to be highly dependent on the number of oocysts given in the challenge dose, with numbers of 100,000 or greater overwhelming the protective response. The protection produced by the *E. tenella* extract was specific for only that species. However, immunization with the *E. acervulina* antigens provided immunity not only against a homologous challenge, but also against two other species (*E. tenella* and *E. maxima*). This was the first indication that cross-protection to other species could be produced by exposure to antigens from a single species, an observation that differed markedly from the species-specific immunity induced by oocyst infections.

There is not much information concerning the *E. tenella* sporozoite surface protein antigen described by Paul et al.[24] It was isolated by use of a monoclonal antibody that inhibited penetration of sporozoites *in vitro*.[25] In nonreduced SDS-polyacrylamide gels, the protein migrates as a single band to an apparent molecular weight of 21 to 23 Kda, while in reduced gels two bands are seen at 17 and 8 Kda.[25] This antigen elicited serum antibodies in immunized chicken that neutralized the *in vitro* penetration of *E. tenella* sporozoites and reportedly protected the chickens against *E. tenella* challenge.[24,25] How much protection was seen in the birds has not been reported, and only a few animals were used in these experiments because only minute amounts of the antigen could be isolated from the sporozoite stage.

These findings suggest that subunit vaccine against avian coccidiosis is feasible, if effective coccidial antigens could be obtained inexpensively in the unlimited quantities needed for large scale immunization. The use of molecular cloning techniques to produce coccidial antigens has recently shown some promise toward solving this problem.[26] Messenger RNA has been isolated from sporulated oocysts of *E. tenella* and used as a template to synthesize cDNA which was cloned into a λgtll bacteriophage. Chicken antiparasite sera and monoclonal antibodies raised against the parent line of *E. tenella* from which the mRNA was isolated have identified a number of clones that are producing different coccidial proteins. Following subcloning and protein synthesis induction, a relatively high production of four of these proteins fused to β-galactosidase has been achieved.

One of these, designated 5401-fusion protein, when injected subcutaneously in 0.5 ml of Freund's complete adjuvant into 4-week-old chickens, was found to stimulate a detectable serum antibody response against *E. tenella* sporozoites (immunofluorescent antibody [IFA] titer greater than 1:500) by 2 weeks postimmunization. This titer peaked at 4 weeks postimmunization (1:1000), and then rapidly dropped off by the 5th week. The effectiveness of the immunity appeared to be dependent, in part, on the size of the oocyst challenge. Birds immunized with one injection of 2400 to 4800 ng of the fusion protein and challenged 4 weeks postimmunization with 75,000 oocysts of *E. tenella* per bird demonstrated partial protection against the infection.[26] This was evidenced by a decrease in intestinal lesions and increased weight gain when compared to unimmunized, challenged controls. Although substantial, the protection was recognized as incomplete because the immunized birds did not gain as much weight as unchallenged controls and some intestinal lesions were still present. With a lighter challenge of 25,000 oocysts per bird, the weight gains of the 5401-immunized birds were similar to the unchallenged controls, and the intestinal lesions were reduced to the level seen with mild, nonpathogenic infections. In all these studies, protection generated by the 5401-fusion protein could be overwhelmed with a challenge dose of 300,000 or more oocysts of *E. tenella*. No cross-protection was seen against challenge with three other species of coccidia (*E. acervulina*, *E. necatrix*, and *E. maxima*).

Analysis of the 5401-fusion protein showed that it is approximately 150 Kda in size, of which about 35 Kda is coccidial protein. There is a region of highly repetitive amino acid sequences where nine amino acids (Ala-Glu-Glu-Leu-Pro-Gly-Glu-Glu-Gly) are repeated five times. In Western blot analysis of solubilized *E. tenella* sporozoite proteins, serum from birds immunized with the protein reacted with a number of antigens greater than 150 Kda. Since the coccidial portion of the 5401-fusion protein has a size of 35 Kda, it is probable that only part of the full-length gene is present in the isolated clone. Whether increased protection would be achieved with the complete protein is not known, but at least one epitope involved with the development of protective immunity against *E. tenella* is present in the portion of the gene so far isolated. It has not yet been determined if the repetitive sequences are the epitope sites for the 5401-fusion protein.

D. Receptor Molecules

Recent studies suggest the feasibility of another area for development of anticoccidial vaccines: blockage of receptor or recognition sites for invasion. These studies have shown that receptor or recognition molecules of the host cell and parasite function during invasion and that removal or neutralization of these molecules partially inhibits invasion of the cells by the parasites.[10,27,28] Receptor molecules could consititute prime candidates as immunogens against the coccidia. A single molecule might even serve as a universal immunogen for all the coccidial species infecting chickens and turkeys for the following reasons. The coccidia exhibit quite rigid site specificity for invasion and development in the natural host, but all species will also invade foreign hosts and cultured kidney epithelial cells. This suggests that there may be common recognition molecules for all of the *Eimeria*. These common recognition molecules might constitute the basis for or be incorporated into a vaccine, the efficacy of which is based at least partly on inhibition of parasite invasion.

E. Practical Applications

The results from all of these approaches show that a commercial vaccine for use against avian coccidiosis is not only feasible, but could be practically applied within the constraints of the poultry industry. Birds could be immunized at an early age either at the hatchery by injection of a subunit vaccine or, as has already been documented, at the poultry farms by use of CocciVac® and the attenuated strains given in the water or feed. Evidence of protection against the coccidia was seen in birds immunized with all types of antigens by 2 to 4 weeks postimmunization, and the protection apparently lasted throughout the grow-out period. Both CocciVac® and the attenuated strains can be stored at 4°C for periods of up to 6 months without any apparent loss in antigenicity.[15,19] The molecular cloned antigens have been held at −20°C for periods of 1 year without loss of effectiveness. Provided that the dosage of coccidial antigen or oocysts numbers are carefully monitored, little pathology is seen with the birds following immunization. The fact that CocciVac® is already on the market and that the attenuated strains have been patented and are now being developed for use in the industry demonstrates that they are cost-effective for the poultry industry. It can only be assumed that a future subunit vaccine could also be produced at a cost-effective level.

IV. PROBLEMS

There are numerous problems associated with the use of any anticoccidial vaccination program when it is expanded to include the billions of commercial birds raised on poultry farms. Most of the current problems are concerned with the control of pathology during the induction of immunity; these have been extensively reviewed by other authors. There are a few which represent major hurdles in the development of a vaccine.[4,8,17,21,22,29,31] One of the most important of these is associated with the use of CocciVac® vaccines and the attenuated or precocious strains. Their use requires that no anticoccidial compounds can be fed concomitantly to the birds. An anticoccidial will interrupt or block the life cycle of the introduced strains as effectively as it does the "natural" strains found in the poultry houses and thereby limit the development of immunity. The question then comes as to how to prevent the birds from acquiring coccidiosis from oocysts already present in the houses during the 2 to 4 weeks that is required to develop protective immunity. There are also selection pressures on the coccidia in the poultry houses, which, at least with the CocciVac® vaccines, may not allow certain species to become established and produce the oocysts needed for reinfection and the production of solid immunity. There are no easy solutions to these problems. The poultry producer may have to, and sometimes does, accept moderate losses for the development of immunity.

These problems that are associated with the live vaccines could be solved with the use of a subunit vaccine, since no oocysts would be needed to produce immunity, and an anticoccidial could be fed during this early period before the effective immune response is developed. However, an additional problem, one that cannot be solved as readily and represents a complication for both live and subunit vaccines, is the presence of immunological diversity among the *Eimeria* species. It has been documented with at least two chicken species (*E. acervulina* and *E. maxima*) that isolates from different geographical areas do not always show cross-protection following oral immunization.[9] In the use of CocciVac®, this problem has been offset somewhat by the addition of more strains of particular species to increase the chances that sufficient protection would then be developed. Presumably the same approach could be taken with the attenuated strains, if the attenuation technique would work for all the different isolated species. One attempt at producing a precocious line of *E. maxima* derived from a mixture of field strains in order to produce cross-protection was not successful.[32] Both live and subunit vaccines may have to be formulated for the particular area in which they are to be used. However, even if the strain diversity with regard to protective immunity is solved, the coccidia may still possess other means for altering their antigenicity. It is not known how the coccidial antigens may change if subjected to long-term selective immunization pressure.

Studies done with the 5401-fusion protein, the only molecular-engineered coccidial protein reported to elicit some protection to challenge, have indicated still other problems with vaccine candidates. This fusion protein was not able to completely block the intestinal lesions after challenge, and the immunity that it elicited could be overwhelmed by a large challenge dose of *E. tenella*. More importantly, the protection afforded the birds was effective only against *E. tenella*. It is possible that the level of protection could be enhanced through improved antigen presentation and delivery or use of immunopotentiators,[33,34] but it may be necessary to identify, isolate, and produce a number of antigens from each species, and even the different developmental stages, to produce a successful, multivalent subunit vaccine. Although little is known about the antigenic makeup of the various species of coccidia or the strains within these species, production of such a vaccine is not beyond the realm of possibility. A number of groups have reported the construction of both genomic and cDNA libraries for at least eight pathogenic *Eimeria* species.[25,35-37] In some instances, coccidial protein has been elicited, and at least 23 different clone isolates have been identified.[25,36,37] The fusion proteins produced have ranged from 9 to 77 Kda, but no information about their potential as immunogens is available. A 23-Kda protein that is reportedly analogous to a protein isolated from the *E. tenella* sporozoite surface that produced a protective response against coccidial infection has been produced.[25] The entire gene for this protein is now cloned, but it has not been reported whether this protein has elicited the same type of protective immunity in birds.

There have been several studies which have shown that host genetic factors have a role in controlling disease susceptibility in chickens to coccidial infections, but it is not known how this genetic control is exerted within the immune system.[5,38-43] The most recent investigations have examined the influence of the major histocompatibility complex (MHC)-linked β-genes on the host response to coccidia through of use of inbred and partially congenic lines of birds.[41-43] The results of these studies have been conflicting. The β-complex is apparently not linked with susceptibility to certain species of coccidia and it does not appear to be involved with the innate response of the bird to initial infection. However, certain β-haplotypes of birds do show varying levels of immunocompetency with regard to coccidial challenge. Congenic lines are now available to evaluate how the MHC is directly involved with bird response to parasite challenge and subunit vaccine candidates. Preliminary evidence has shown that there is a difference in both the degree and onset of immune response to the 5401-fusion protein with different haplotype birds. Until these investigations are finished, it is difficult to do much more than speculate as to what effect the genetic differences within bird populations will have on the development of a subunit vaccine. Since the resistance to coccidial reinfection appears to be predominately dependent on the cellular mechanisms of immunity, then the role of the MHC and other factors involved with immunological responsiveness becomes more important and equally intriguing. There will undoubtedly be variation in the immunological recognition of molecular-engineered antigens, which may block the effectiveness of vaccination. It is simply not known if immunization will help the birds to overcome genetically linked inadequacies in effector mechanisms. In order to determine which subset of individuals will respond to a subunit vaccine, further research such as mapping of the Class I and II regions of the MHC using defined populations must be undertaken. This knowledge can then be applied to the heterogenous populations found in the poultry industry to hopefully establish a successful vaccination program.

ACKNOWLEDGMENTS

The authors wish to acknowledge the cooperation of the Genex Corporation, Gaithersburg, MD, in providing the genetically engineered 5401-fusion protein. The excellent technical assistance of Diane Adger-Johnson, Lourdes F. Carson, Lawrence Spriggs, Gary Wilkins, Keith A. Gold, and Paul Anakis in the 5401-fusion protein studies is also noted.

REFERENCES

1. **Fitzgerald, P. R.,** The economics of bovine coccidiosis, *Feedstuffs,* 44, 28, 1972.
2. **Biggs, P. M.,** The world of poultry diseases, *Avian Pathol.,* 11, 281, 1982.
3. **Reid, W. M., Long, P. L., and McDougald, L. R.,** Coccidiosis, in *Diseases of Poultry,* 8th ed., Hofstad, M. S., Barnes, H. J., Reid, W. M., and Yoder, H. W., Jr., Eds., Iowa State University Press, Ames, 1984, 692.
4. **Rose, M. E.,** Immune responses to *Eimeria* infection, in *Research in Avian Coccidiosis,* McDougald, L. R., Joyner, L. P., and Long, P. L., Eds., University of Georgia, Department of Poultry Science, Athens, 1986, 449.
5. **Lillehoj, H. S.,** Immune response during coccidiosis in SC and FP chickens. In vitro assessment of T cell proliferation response to stage-specific parasite antigens, *Vet. Immunol. Immunopathol.,* 13, 321, 1986.
6. **Speer, C. A., Reduker, D. W., Burgess, D. E., Whitmire, W. M., and Splitter, G. A.,** Lymphokine-induced inhibition of growth of *Eimeria bovis* and *Eimeria papillata* (Apicomplexa) in cultured bovine monocytes, *Infect. Immun.,* 50, 566, 1985.
7. **Hughes, H. P. A., Speer, C. A., Kyle, J. E., and Dubey, J. P.,** Activation of murine macrophages and a bovine monocyte cell line by bovine lymphokine to kill intracellular pathogens *Eimeria bovis* and *Toxoplasma gondii, Infect. Immun.,* 53, in press.
8. **Kogut, M. H.,** Effect of avian lymphokines on the in vitro development of *Eimeria tenella,* Paper 1735/00, 71st Annu. Meet. Federation Proc. Federation of American Societies for Experimental Biology, Washington, D.C., 1987.

9. **Rose, M. E.,** Host immune response in *The Biology of the Coccidia,* Long, P. L., Ed., University Park Press, Baltimore, 1982, 330.

10. **Augustine, P. C. and Danforth, H. D.,** Effects of hybridoma antibodies on invasion of cultured cells by sporozoites of *Eimeria, Avian Dis.,* 29, 1212, 1985.

11. **Danforth, H. D.,** Use of hybridoma antibodies combined with genetic engineering in the study of protozoan parasites: a review, in *Research in Avian Coccidiosis,* McDougald, L. R., Joyner, L. P., and Long, P. L., Eds., University of Georgia, Department of Poultry Science, Athens, 1986, 574.

12. **Augustine, P. C., and Danforth, H. D.,** A study of the dynamics of the invasion of immunized birds by *Eimeria* sporozoites, *Avian Dis.,* 30, 347, 1985.

13. **Mockett, A. P. A. and Rose, M. E.,** Immune responses to *Eimeria:* quantification of antibody isotypes to *Eimeria tenella* in chicken serum and bile by means of the ELISA, *Parasite Immunol.,* 8, 481, 1986.

14. **Edgar, S. A.,** Coccidiosis of chickens and turkeys and control by immunization, presented at Avicultura Moderna del XI Congreso Mundial, Mexico City, La Prensa Medica Mexicana, 1958, 415.

15. **Reid, W. M.,** Progress in the control of coccidiosis with anticoccidials and planned immunization, *Am. J. Vet. Res.,* 36, 593, 1975.

16. **Edgar, S. A.,** Practical immunization of chickens and turkeys against coccidia, in *Research in Avian Coccidiosis,* McDougald, L. R., Joyner, L. P., and Long, P. L., Eds., University of Georgia, Department of Poultry Science, Athens, 1986, 617.

17. **Jeffers, T. K.,** Attenuation of coccidia — a review, in *Research in Avian Coccidiosis,* McDougald, L. R., Joyner, L. P., and Long, P. L., Eds., University of Georgia, Department of Poultry Science, Athens, 1986, 482.

18. **McDonald, V. and Shirley, M. W.,** The asexual development of precocious lines of *Eimeria* spp. in the chicken, in *Research in Avian Coccidiosis,* McDougald, L. R., Joyner, L. P., and Long, P. L., Eds., University of Georgia, Department of Poultry Science, Athens, 1986, 502.

19. **Shirley, M. W. and Millard, B. J.,** Studies on the immunogenicity of seven attenuated lines of *Eimeria* given as a mixture to chickens, *Avian Pathol.,* 15, 629, 1986.

20. **Davis, P. J., Barratt, M. E. J., Morgan, M., and Parry, S. H.,** Immune response of chickens to oral immunization by "trickle" infections with coccidia, in *Research in Avian Coccidiosis,* McDougald, L. R., Joyner, L. P., and Long, P. L., Eds., University of Georgia, Department of Poultry Science, Athens, 1986, 634.

21. **Johnson, J. K., McKenzie, M. E., Perry, E., and Long, P. L.,** The immune response of young chickens given "graded" or "trickle" infections with coccidia, in *Research in Avian Coccidiosis,* McDougald, L. R., Joyner, L. P., and Long, P. L., Eds., University of Georgia, Department of Poultry Science, Athens, 1986, 642.

22. **Murrey, P. K., Bhogal, B. S., Crane, M. St. J., and McDonald, T. T.,** *Eimeria tenella* — in vivo immunization studies with sporozoite antigen, in *Research in Avian Coccidiosis,* McDougald, L. R., Joyner, L. P., and Long, P. L., Eds., University of Georgia, Department of Poultry Science, Athens, 1986, 564.

23. **Murrey, P. K. and Galuska, S.,** European patent application 85401246.5, 1985.

24. **Paul, L. S., Brothers, V. M., Files, J. G., Tedesco, J. L., Newman, K. Z., and Gore, T. C.,** Purification and characterization of a major surface antigen of *Eimeria tenella, J. Cell Biochem.,* 10A, 169, 1986.

25. **Files, J. G., Paul, L. S., and Gabe, J. D.,** Identification and characterization of the gene for a major surface antigen of *Eimeria tenella* in *Molecular Strategies of Parasitic Invasion,* Vol. 42, Agabian, N., Goodman, H., and Nogueira, N., Eds., Alan R., Liss, New York, 1987, 713.

26. **Danforth, H. D. and Augustine, P. C.,** Use of hybridoma antibodies and recombinant DNA technology in protozoan vaccine development, *Avian Dis.,* 30, 37, 1985.

27. **Augustine, P. C.,** Effect of polyions, Ca^{++}, and enzymes on penetration of cultured cells by *Eimeria meleagrimitis* sporozoites, *J. Parasitol.,* 66, 498, 1980.

28. **Augustine, P. C. and Danforth, H. D.,** Use of monoclonal antibodies to study surface antigens of *Eimeria* sporozoites, *Proc. Helminthol. Soc. Wash.,* 54, in press.

29. **Davis, P. J.,** Immunity to coccidia, in *Avian Immunology, British Poultry Science Symp. No. 16,* Rose, M. E., Payne, L. N., and Freeman, B. M., Eds., British Poultry Science, Edinburgh, 1981, 361.

30. **Mielke, D.,** Immunbiologische aspekte bei kokzidien-infektionen, *Monatsh. Veterinaermed.,* 37, 108, 1982.

31. **Fayer, R. and Reid, W. M.,** Control of coccidiosis, in *The Biology of the Coccidia,* Long, P. L., Ed., University Park Press, Baltimore, 1982, chap. 11.

32. **McDonald, V., Shirley, M. W., and Bellatti, M. A.,** *Eimeria maxima:* characteristics of attenuated lines obtained by selection for precocious development in the chicken, *Exp. Parasitol.,* 61, 192, 1986.

33. **Klesius, P. H.,** Immunopotentiation against internal parasites, *Vet. Parasitol.,* 10, 239, 1982.

34. **Klesius, P. H. and Giambrone, J. J.,** Modulation of the immune responses to coccidia, in *Research in Avian Coccidiosis,* McDougald, L. R., Joyner, L. P., and Long, P. L., Eds., University of Georgia, Department of Poultry Science, Athens, 1986, 555.

35. **Binger, M.-H., McAndrew, S. J., and Schildknecht, G.,** Cloning and expression of *E. tenella* merozoite cDNA, *J. Cell Biochem.,* 10A, 144, 1986.

36. **Clarke, L. E., Messer, L. I., and Wisher, M. H.,** Antigens of *Eimeria* cloned and expressed in *E. coli, J. Cell Biochem.,* 10A, 145, 1986.
37. **Rose, M. E.,** Parasitology, Report of the Houghton Poultry Research Station 1985—86, Houghton Poultry Research Station, Houghton, Huntingdon, Cambridgeshire, 1986.
38. **Long, P. L.,** The effect of breed of chickens on resistance to *Eimeria* infections, *Br. Poult. Sci.,* 9, 71, 1968.
39. **Klesius, P. H. and Giambrone, J. J.,** Strain-dependent differences in murine susceptibility to coccidia, *Infect. Immun.,* 26, 1111, 1979.
40. **Ruff, M. D. and Bacon, L. D.,** Coccidiosis in 15 B-congenic chicks, *Poult. Sci.,* 63, 172, 1984.
41. **Clare, R. A., Strout, R. G., Taylor, R. L., Collins, W. M., and Briles, W. E.,** Major histocompatibility (B) complex effects on acquired immunity to cecal coccidiosis, *Immunogenetics,* 22, 593, 1985.
42. **Lillehoj, H. S.,** Effects of immunosuppression on avian coccidiosis: cyclosporin A but not hormonal bursectomy abrogates host protective immunity, *Infect. Immun.,* 55, 1616, 1987.
43. **Lillehoj, H. S. and Ruff, M. D.,** Comparison of disease susceptibility and subclass-specific antibody response in SC and FP chickens experimentally inoculated with *Eimeria tenella E. acervulina,* or *E. maxima, Avian Dis.,* 31, 112, 1986.

Chapter 9

TOXOPLASMA VACCINES

Alan M. Johnson

TABLE OF CONTENTS

I. INTRODUCTION

It is now well established that the intracellular protozoon *Toxoplasma gondii* may cause congenital infection in humans and many species of domestic animals. In particular, world-wide losses from congenital toxoplasmosis in sheep probably amount to millions of dollars annually. It has been postulated that toxoplasmosis is a prominent disease in sheep because they are bred synchronously and thereby provide large, susceptible populations.[1] In addition, rates of infection are higher in intensively managed, high-density flocks.[2,3] However, cattle appear to have an innate resistance to the parasite,[4,5] and it is possible that an innate susceptibility, as well as their breeding patterns and management, makes toxoplasmosis such an important infection in sheep. The vast literature on toxoplasmosis in sheep[6] and pigs[7] has recently been extensively reviewed, and although cases of congenital infection in goats[8–10] are not as numerous, even small losses in a flock of pure-bred goats could be quite costly. The development of a vaccine that prevented fetal loss from toxoplasmosis in sheep, goats, and pigs would therefore be quite worthwhile.

The purpose of this chapter is to review progress to date in vaccination against toxoplasmosis, to compare and contrast the various strategies that are being used to develop successful *Toxoplasma* vaccines, and to suggest the directions in which future development should proceed.

The major pathways of the life cycle of the parasite have now been fully elucidated[11,12] and are outlined in the schematic diagram in Figure 1. However, one aspect of the life cycle warrants comment in relation to the possibility of vaccinating herbivores. Apart from congenital infection, it has appeared very unlikely that strict herbivores such as sheep are infected in any way other than via ingestion of the sporozoites in the oocysts excreted in cat feces. In several trials, acutely infected sheep have been mixed with susceptible animals and there has been no evidence of cross-infection (reviewed by Blewett and Watson[1]), even when the groups lambed or aborted together. Although experimental studies have suggested that venereal transmission may occur, this route is unlikely to be important in natural infections.[1] Sheep-to-sheep passage has also been considered unlikely.[1,13] Even if nonimmune animals ate placenta (from which *T. gondii* is often cultured[14]), the proliferating tachyzoite, the form of the parasite most likely to be present in this tissue, is not resistant to digestive juices.[15] However, the recent finding of a digestive juice-resistant form of the parasite in ovine placentas[16] raises the possibility of this route for infection. Whether this is important in natural infections or not, a vaccine that could protect animals against toxoplasmosis resulting from the ingestion of sporozoites, bradyzoites, or tachyzoites would be more efficacious than one which protected against sporozoite challenge only. Therefore, development of a successful vaccine against toxoplasmosis in herbivores should ideally be carried out with this in mind, but it is essential for any vaccine that would be considered for use in humans, where it is well documented that infection can occur via ingestion of either sporozoites or bradyzoites.[17,18] To do this we need to consider the antigenic structure of the parasite.

II. ANTIGENIC STRUCTURE

Because *T. gondii* is obligately intracellular, host cell contamination is almost always present to some degree. This has been the major problem with obtaining accurate information about the antigenic structure of the parasite. Even the best method of purification, filtration of parasites and host cells through a 3-μm Nuclepore® polycarbonate membrane,[19] rarely achieves 100% purity. However, the use of monoclonal antibodies (MAb) to the parasite has greatly expanded knowledge on the antigens of *T. gondii*. Information obtained until mid-1984 has been comprehensively reviewed by Johnson[20] and Hughes.[21] Work carried out since then has confirmed that the tachyzoite stage of the virulent RH strain of *T. gondii*

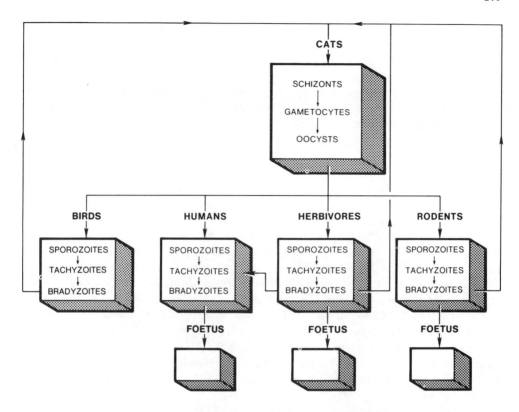

FIGURE 1.   Major pathways of the life cycle of *Toxoplasma gondii*.

has only four major antigens that can be radioiodinated on its cell membrane surface. These have molecular sizes of 43, 35/30, 22, and 14 kDa. Although the second largest of these antigens was originally found to have a molecular size of 35 kDa,[22,23] recent workers have suggested that this dominant antigen has a molecular size of 30 kDa, and it has now been designated P30.[24,25] For the sake of clarity, this designation will be continued here. P30 is about 5% of the total protein of the RH strain tachyzoite,[24] is not significantly glycosylated,[25] and appears to contain an immunodominant region with repetitive epitopes.[26] It is probably present on most strains of parasite found in the environment, because almost all sera tested from naturally infected humans recognize either the purified P30[27] or compete with MAb produced against the antigen.[28,29] It is interesting that sera from mice infected with another protozoon, *Hammondia hammondi*, also immunoprecipitate an antigen of about 30 kDa and one of about 43 kDa is present on the surface of radioiodinated RH strain tachyzoites.[30] *H. hammondi* has been used to vaccinate animals against toxoplasmosis, and these experiments will be discussed in more detail in Section III.B. The significance of the recent finding that sera from some humans who do not show serological reactivity in the usual diagnostic tests indicating past exposure to *T. gondii* react with a range of parasite antigens,[31] including P30, after Western blotting, remains to be determined. When injected intraperitoneally into mice, purified P30 and a frozen-thawed lysate of RH strain tachyzoites do cause a significant increase in natural killer cell activity, whereas *T. gondii* antigen secreted into tissue culture, host cell antigens, and phosphate buffered saline (PBS) do not.[32] This work is probably the first wherein the cell-mediated immune (CMI) response to a specific *T. gondii* antigen has been studied in this way.

The CMI response to intact *T. gondii* or parasite fractions has been reviewed in depth by several authors.[21,33-35] These publications have discussed the relative importance of the CMI response compared with the humoral response to *T. gondii*, and it is clear that although the

CMI response to the parasite is the host's major defense, it cannot be considered to stand alone. The immune system of the intact host is a multifactorial combination of both CMI and humoral responses, and both should be considered when developing vaccines against toxoplasmosis.

Although early work suggested antigenic similarity among all five stages (schizont, gametocyte, sporozoite, tachyzoite, and bradyzoite) of the life cycle of the parasite,[36] it now appears that there are distinct antigenic differences among some stages. Fluorescein-labeled antiserum raised against bradyzoites of the ME49 strain of *T. gondii* has been found not to react against RH strain tachyzoites, whereas antiserum raised against the tachyzoite does show partial reaction against the bradyzoite.[37] Strain variation did not appear to influence these results, as no difference was found in the reaction of tachyzoites of both the RH and ME49 strains with fluorescein-labeled antisera raised against RH strain tachyzoites or ME49 strain bradyzoites.

Sporozoites of the C strain of *T. gondii* have been found to possess two membrane proteins of 65 and 25 kDa that are not present on the tachyzoite stage of either the C strain or RH strain.[38] These proteins can be radioiodinated and are antigenic, as murine antiserum raised against sporozoites reacted with them on Western blotting. In addition, the tachyzoite membrane proteins P30 and P22 were not detectable, and P43 was markedly reduced on the surface of radioiodinated sporozoites. This was later confirmed using antisera from humans naturally infected via an oocyst-transmitted infection. The antisera, but not normal control sera, also recognized prominent sporozoite proteins of 190, 67, and 25 kDa. Although the antisera did react with proteins of >100, 62, and 30 kDa on tachyzoites, they did not identify these antigens on sporozoites. Major antigens of 97, 66, 44, and 22 kDa were common to both sporozoites and tachyzoites.[39]

At least five studies have been unable to find qualitative antigenic differences among RH strain, C37 strain, and C56 strain tachyzoites,[22,40] among RH strain, C strain, and P strain tachyzoites,[39] between RH strain and ME49 strain tachyzoites,[37] or between RH strain and C strain tachyzoites,[38] although minor qualitative differences were noted. However, contrary to their previous findings,[38] Kasper et al.[41] have now reported qualitative antigenic differences among the RH strain, C strain, and P strain tachyzoites. Specifically, C strain and P strain tachyzoites possessed antigens of 62, 37, 23, 8, and 7 kDa, while RH strain tachyzoites did not. In addition, each strain had unique antigens of 39 and 26 kDa (RH strain), 54 kDa (P strain), and 88 and 25 kDa (C strain). The results appeared to vary depending on which type of test was used to detect the differences. The significance of these results for vaccination against toxoplasmosis remains to be determined, but strain-specific variation does not appear to have been an obvious problem with research to date.

III. ACTIVE IMMUNIZATION

A. Nonviable Vaccines
1. Whole Parasites

Attempts to protect animals against toxoplasmosis by vaccination with nonviable, whole tachyzoites have produced conflicting results. Vaccination of guinea pigs with either heat-, formalin-, or phenol-killed RH strain tachyzoites conferred resistance to challenge with an otherwise lethal dose of RH strain tachyzoites.[41,42] However, Foster and McCulloch[43] were unable to protect guinea pigs against a virulent challenge unless the phenol-killed tachyzoites were given in an adjuvant. The resistance to challenge in guinea pigs induced by phenol-killed parasites appeared to be dependent on the antibody titer produced. Guinea pigs with dye test titers >1:1024 survived, whereas others succumbed to infection. However, the authors did recognize the necessity for an adjuvant rather than large numbers of organsims in the vaccine.

Several groups have been unable to protect mice by immunization with heat-killed[44-45] and heat- or formalin-killed[46] homologous tachyzoites, in or without adjuvant, against a highly virulent parasite challenge. Krahenbuhl et al[47] were able to induce resistance to a parasite challenge in mice immunized with formalin-killed RH strain tachyzoites given in or without adjuvant. However, their challenge was only the medium-virulence C56 strain, and as many as 24% of control mice survived.

Rabbits could not be protected by immunization with heat-killed parasites,[48] and the histopathology of rabbits which had been immunized with formalin-killed RH strain tachyzoites was identical to that of controls which showed no signs of protective immunity to the RH strain challenge infection.[49] On the contrary, Wildfuhr[50] had previously reported being able to confer resistance to a virulent challenge in rabbits, but not hamsters, by injection of heat-killed tachyzoites.

As there were large differences among some of the parameters used in the experiments discussed here, comparisons must be general. However, the trend appears to be that a certain degree of protection against medium virulence *T. gondii* strains is obtainable by injection of killed tachyzoites, but protection against highly virulent strains is not as certain. This may not be a large problem, however, as the RH strain used in most vaccination experiments was isolated in 1940[51] and has been passaged in the peritoneal cavity of mice or in tissue culture almost every week for the last 47 years. It can therefore be considered to be a laboratory strain with extreme virulence. The intraperitoneal LD_{100} dose of this strain for the mouse is less than 10 tachyzoites.[52] It is widely used to prepare antigen[53] because it is easy to obtain large numbers of tachyzoites, but it is much more virulent than other strains of *T. gondii* such as pork 1 and ovine 2[54] and ME49, C37, C56, C, and P strains, all isolated from the environment more recently. We should now be attempting to use oral challenges with the oocysts or cysts of such strains to test vaccines, as this is much closer to the natural infection. Previous studies done during the last 30 years do provide a framework upon which to base our future research, and most are therefore worthy of consideration. In particular, more recent workers have gone deeper than the whole, nonviable tachyzoite vaccines considered here, and have tried defined parasite fractions to vaccinate against the disease.

2. Defined Parasite Fractions

The more common methods used to obtain defined parasite fractions have been osmotic lysis and/or sonication of RH strain tachyzoites. These preparations have then been used as crude lysates or centrifuged to give a soluble supernatant fluid and a particulate fraction.

Injection of a lysed parasite fraction or a soluble parasite fraction, both given in adjuvant, have been found to give a longer time to death and lower percentage mortality in mice challenged with the medium-virulence C56 strain.[47] The particulate parasite fraction also gave significant protection against the challenge, but to a lesser degree than that obtained with the other two fractions. These results were confirmed and extended several years later when Araujo and Remington[55] found that these parasite fractions protected mice against a C56 strain challenge, irrespective of whether they were given in or without adjuvant. In addition, mice immunized with at least 200 μg of purified *T. gondii* ribonucleic acid (RNA) had longer times to death and lower total mortality compared with nonimmunized controls. Previous to this, Preston and Dumonde[56] had reported that *Leishmania* ribosomes could be used to protect guinea pigs against homologous challenge, and Leon et al.[57] had shown that *Trypanosoma* polyribosomes could protect against homologous challenge in mice. However, the effect of the *T. gondii* RNA appeared to be nonspecific, as 200 μg of RNA extracted from uninfected murine peritoneal macrophages and 100 μg of synthetic polyribonucleotide polycytidylic acid, both given without adjuvant, conferred significant protection against the C56 strain challenge. This nonspecific nature of some *T. gondii* vaccines had also been shown several years earlier when it was found that formalin-killed, whole RH strain tachyzoites given alone or in adjuvant conferred resistance to mice against an otherwise fatal challenge with the intracellular bacterium *Listeria monocytogenes*.[52] However, the protective effect of the parasite vaccine against the heterologous challenge was considerably greater when given in the adjuvant.

Three groups have carried out extensive studies on the effects of different adjuvants on the efficacy of defined parasite fractions as vaccines. Masihi et al.[58] found a significant prolongation in mean survival time in mice injected intravenously with a *T. gondii* proto-plasm/oil/*Mycobacterium tuberculosis* trehalose dimycolate adjuvant mix and challenged intraperitoneally with virulent tachyzoites of the same BK strain. However, *T. gondii* antigens in saline had no effect, while Freund's complete adjuvant and a mycobacterial glycolipid/oil-in-water emulsion both gave a significant prolongation in mean survival time in the absence of parasite antigen. The other groups used a more natural vaccination and challenge route. McLeod et al.[59] immunized mice parenterally and orally with an RH strain tachyzoite lysate fraction in a variety of combinations of adjuvants. The animals were then challenged orally with cysts of the nonvirulent ME49 strain of *T. gondii*. The effect of the antigen and/or adjuvant administration on peritoneal macrophage antiparasite activity before and after the ME49 challenge was determined by calculating the percentage of cells infected and the number of *T. gondii* per host cell vacuole. No differences were found among vaccinated and nonvaccinated controls. Waldeland and Frenkel[60] immunized mice with lysates of ta-chyzoites of the RH or M-7741 strains of *T. gondii* and challenged them orally with oocysts of the M-7741 strain. The tachyzoite lysates alone, or when given in one of a range of adjuvants, gave slight protection against low numbers of oocysts, but this effect disappeared as the number of oocysts in the challenge increased.

An RH strain lysate fraction has proved successful in the prevention of ovine abortion due to toxoplasmosis.[61] Ewes were immunized subcutaneously, mated 7 d after a second dose of the vaccine, and challenged subcutaneously when the animals were 59 to 90 d pregnant, with cysts of a parasite strain originally isolated from an aborted lamb; 45% percent of the immunized ewes that had developed elevated (>1:128) antibody titers, but only 15% of the nonimmunized controls, delivered live lambs. Only 20% of a group of ewes which were injected but did not develop elevated antibody titers delivered live lambs. Of the nonvaccinated, unchallenged group, 86% delivered live lambs.

As stated previously, *T. gondii* is obligately intracellular, and so when we consider the results of experiments using defined tachyzoite fractions, we must also consider the possibility that host cell components may be at least partly responsible for any protection found. This problem is illustrated by the results in Table 1. The unfiltered peritoneal exudates of *T. gondii*-infected male LACA mice were pooled and contained 3×10^{10} RH strain tachyzoites, 4×10^9 red blood cells, and 2×10^8 white blood cells. The peritoneal exudates of uninfected male LACA mice were pooled and contained 5.25×10^9 red blood cells and 7.9×10^8 white blood cells. These fractions were spun at $40,000 \times g$ for 60 min, and the pellets were resuspended in 1% formalin saline and incubated at 22°C for 60 min. The suspensions were again spun at $40,000 g$ for 60 min, and the pellets were resuspended in 10 ml PBS. These fractions were ultrasonicated (50 W, 20 KHz for 5 min) and contained 13.3 mg ml^{-1} protein (*T. gondii* fraction) and 1.35 mg ml^{-1} protein (control fraction). The pellets were further diluted 1:10, 1:100, and 1:1000 in PBS, and 0.2 ml of each dilution was injected intraper-itoneally into groups of male LACA mice. Mice in another group were injected with 0.2 ml of PBS. After 1 month, all mice were challenged with an intraperitoneal inoculation of 5×10^2 RH strain tachyzoites and survival times were recorded. Although a significant prolongation in the mean survival time was obtained in the mice immunized with the *T. gondii* fractions containing 2.7×10^3 and 2.7×10^2 μg of protein, a similar prolongation was also obtained in mice immunized with 2.7×10^2 and 2.7×10^1 μg of control protein. In fact, immunization with 2.7×10^1 μg of control protein conferred significant protection against challenge, whereas 2.7×10^1 μg of *T. gondii* protein did not.

Table 1
MEAN SURVIVAL TIME OF GROUPS OF VACCINATED
MICE CHALLENGED WITH *TOXOPLASMA* TACHYZOITES

Group	Fraction injected	Amount of protein injected (μg)	Mean survival time (h $\pm$ 1 SD)	Significance with respect to PBS group (Mann-Whitney U test)
1	PBS	—	175.3 $\pm$ 9.2	—
2	*Toxoplasma*	2.7×10^3	207.4 $\pm$ 20.8	$p <0.0001$
3	*Toxoplasma*	2.7×10^2	190.6 $\pm$ 22.0	$p = 0.003$
4	*Toxoplasma*	2.7×10^1	188.3 $\pm$ 30.2	N.S.[a]
5	*Toxoplasma*	2.7	176.1 $\pm$ 25.7	N.S.
6	Control cells	2.7×10^2	195.7 $\pm$ 20.3	$p <0.0001$
7	Control cells	2.7×10^1	187.4 $\pm$ 15.4	$p = 0.002$
8	Control cells	2.7	187.5 $\pm$ 25.0	N.S.

[a] N.S. = not significant $p >0.05$.

There are several ways to overcome the problem of host cell contamination when trying to identify parasite antigens that may be suitable as a vaccine against toxoplasmosis. One way is to use an alternative vaccination schedule such as those to be discussed in the following sections, and another is to attempt to purify the parasite antigens out of the *T. gondii*/host cell mixture. This latter approach has been used by several groups.

Araujo and Remington[62] used polyacrylamide gel electrophoresis to size separate RH strain tachyzoite proteins, cut the gel into small strips, and then eluted the proteins in each strip. Groups of mice were injected intraperitoneally with one of the eluates and challenged intraperitoneally with C56 strain tachyzoites. Only 35% of mice injected with proteins in the range 45 to 25 kDa, and only 20% of mice injected with protein of about 14 kDa died, whereas 100% of control mice died. While very encouraging, the success of this experiment could theoretically have been due to host cell proteins between 45 to 25 and 14 kDa. An even more rigorous test is to immunopurify the parasite antigen and use this specific defined fraction as a vaccine. Sharma et al.[63] used a MAb to an internal cytoplasmic antigen of RH strain tachyzoites to immunopurify the antigen, against which the MAb was directed. Mice were then immunized with the affinity-purified antigen alone or in adjuvant, and 1 week later were challenged intraperitoneally with C56 strain tachyzoites. Mice immunized with antigen in adjuvant were completely protected, and only 30% of mice injected with antigen alone died, whereas 60% of mice injected with PBS died. In a similar manner, P30 has been immunopurified and injected intraperitoneally with adjuvant into BALB/c mice.[64] The strain of mice used is specifically mentioned here because, as will be discussed in Section IV.B.2, they are not ideal for research on toxoplasmosis. The animals were then challenged intraperitoneally with C strain tachyzoites. Of the mice injected with antigen, 90% died, whereas only 45% of mice injected with PBS and adjuvant died. When immunizations were given subcutaneously, 80% of the mice injected with antigen died, whereas 60% of the mice injected with PBS died. In addition, the mice injected with P30 that survived had more cysts in their brains than did mice that had survived the challenge and had been injected with PBS only. It is interesting that bradyzoites of the C strain did not react with the MAb to P30, and the authors inferred that circulating antibody to P30 may have selected out parasites lacking P30, allowing them to proliferate unchecked. Unfortunately, it is not clear whether the bradyzoites in only the mice injected with P30 lacked this antigen, which is what we would expect if a selection process had taken place in only those mice and not the control mice injected with PBS.

Comparison of the results described in this section on nonviable vaccines and comment on their success or otherwise is made difficult by the large differences in host animals, routes of inoculation and challenge, methods used to kill the tachyzoites or prepare the defined parasite fractions, and the virulence of the challenge. However, the biggest problem with this schedule of vaccine research, particularly with regard to defined parasite fractions, is obtaining enough pure antigen to test. A way around this problem is to test live vaccines against toxoplasmosis.

B. Live Vaccines

It is now well established that mice, guinea pigs, and rabbits which survive a challenge with a low-virulence *T. gondii* strain are protected against challenges with more virulent strains that would otherwise be fatal.[44,46,65-70] The degree of protection afforded does vary, as immunization with more virulent strains is less effective, while strains with similar initial pathogenicity for mice can give differing levels of protection.[65] In addition, mice, rabbits, and hamsters immunized with virulent parasites and treated with anti-*T. gondii* drugs also develop resistance to further, otherwise fatal parasite challenges.[44,71-73] The protection afforded by injection of a viable parasite strain is of the premunition type. It is possible to isolate organisms with the virulence of the RH strain from the brains of mice immunized with the nonvirulent Beverley strain for up to 1 year after the challenge with the RH strain.[66,69,74] Such injection of viable *T. gondii* strains has been used to try and vaccinate against ovine abortion.

As early as 1964, Jacobs and Hartley[75] showed that inoculation of sheep with the Beverley strain of *T. gondii* before mating conferred some protection against congenital transmission following an experimental challenge with the 56-48 strain of *T. gondii*. A variety of parasite stages and routes of inoculation and challenge were tried. Unfortunately, because toxoplasmosis is endemic in New Zealand, about half of the animals had been naturally infected before the start of the experiment. However, none of the "in-contact" control ewes, which were neither experimentally infected nor challenged, but just run with the test animals, showed signs of congenital toxoplasmosis. Natural infection was also a problem in a field trial conducted in England.[76] Ewes given cysts subcutaneously 7 weeks before mating were challenged subcutaneously when 60 to 90 d pregnant, with cysts of the same parasite strain; 7% of the inoculated ewes, 53% of the noninoculated ewes, and 30% of the noninoculated, unchallenged group, aborted. Although all animals were seronegative at the start of the experiment, two lambs born to one ewe in the noninoculated, unchallenged group suffered congenital toxoplasmosis, and 64% of all ewes in this group developed antibodies to a presumed natural challenge during the experiment. The protective effect of inoculation of viable *T. gondii* found in these two experiments was confirmed in another trial conducted in Scotland.[77] Seronegative or seropositive (inoculated subcutaneously the previous year with cysts of the M1 strain of *T. gondii*) ewes in the 4th month of pregnancy were challenged subcutaneously with one of an increasing range of doses of M1 cysts; 4 of the 15 seronegative ewes aborted, and the lambs that survived showed rising antibody levels to *T. gondii* in the first 3 months. The lambs born to the seropositive ewes showed no signs of congenital toxoplasmosis.

While the injection of viable *T. gondii* prior to mating would appear to be an ideal method of preventing congenital toxoplasmosis in sheep, there are several disadvantages with this schedule. Although the parasite strains used to inoculate animals have been of low virulence, it is theoretically possible that an increase in virulence may occur and that the animals may die as a result of the attempted vaccination. This could be extremely expensive in a flock of purebred goats. In addition, chronic toxoplasmosis can depress the immune response of experimental mice to a clostridial and a louping-ill virus vaccine,[78] both of which are commonly used in the management of sheep. In fact, it has been found that concomitant *T.*

gondii/louping-ill virus infection causes a greater mortality than that found in sheep just infected with the virus alone.[79] Consequently, vaccine research has been directed at developing or finding strains of *T. gondii* that are characterized and well documented as being nonvirulent.

One such strategy has been to inoculate animals with *H. hammondi*. This protozoon is remarkably similar to *T. gondii*, except that the tissue stages of *T. gondii* in the intermediate host are infectious for other intermediate hosts, whereas those of *H. hammondi* are not.[80] Although a suggestion to classify *H. hammondi* as a separate species of *T. gondii*[81] was not widely accepted,[82] recent work at the molecular level has supported the hypothesis that the two organisms belong to the same genus.[83] Whatever proves to be the case, *H. hammondi* (continued here for clarity) infection does protect animals against a subsequent, otherwise fatal *T. gondii* challenge. Initial laboratory studies showed that almost all mice and hamsters inoculated orally with *H. hammondi* oocysts survived an oral challenge with one of a range of otherwise fatal doses of oocysts of the M-7741 strain of *T. gondii*.[84] The significance of these results is heightened by the fact that protection was afforded against an oral challenge of oocysts. Therefore, this is one of the earlier works to show vaccination against the natural route of infection. In a later study, goats were orally inoculated with the CR-4 strain of *H. hammondi* 17 to 73 d before mating, and then challenged orally between the 51st and 119th d of pregnancy with oocysts of the GT-1 strain of *T. gondii*.[85] Four of the five vaccinated, challenged does gave birth to eight healthy kids, whereas the two nonvaccinated, challenged does either retained dead fetuses or aborted. However, *T. gondii* was isolated from a range of tissues from all kids born to vaccinated does, suggesting that they survived the worst effects of congenital toxoplasmosis, but were still infected nonetheless.

Jones et al.[86] attenuated the virulence of the Pe strain of *T. gondii* by continuous chronic carriage in mice, and also increased its virulence by rapid passage in mice. The PeC and PeV strains, respectively, that resulted were used to orally inoculate mice and then to challenge them intraperitoneally. Depending on how long after inoculation of PeC the PeV challenge was given, all mice survived. While indicative of the fact that the virulence of *T. gondii* strains can be reduced and used to vaccinate, this study also highlights the fact that the virulence of the parasite can also be increased, thereby possibly causing death of inoculated animals. Another manipulation of the virulence of a *T. gondii* strain resulted in the apparent loss of the ability to form cysts.[87] The avirulent 119 strain of *T. gondii* was made virulent by rapid passage in mice, and the virulence was then attenuated by incubation of the tachyzoites in a "high-molecular-weight fraction" of an extract of *T. gondii*. Cysts could not be found in mice or rabbits inoculated with tachyzoites treated in this way, although it was possible to isolate viable parasites from the animals up to 12 weeks after inoculation. Rabbits inoculated in this way were challenged with the virulent 119 strain 1 year later, and all animals survived. Another "incomplete" strain of *T. gondii* without the capacity to sexually reproduce in cats has been reported to increase lambing percentages in a flock of sheep.[88] This preliminary report needs substantiation because it was affected by the problem mentioned previously, in that as many as 50% of the flock had been naturally exposed prior to the experiment.

Several groups have used irradiation of RH strain tachyzoites to obtain attenuated parasites for use as a vaccine.[89-91] In general, intraperitoneal inoculation of such attenuated strains has provided total protection to mice against an otherwise lethal intraperitoneal challenge of RH strain tachyzoites. However, the degree of protection obtainable appears to be dependent on the amount of irradiation, number of inoculations, and size of challenge.[92] Perhaps the most significant finding of these experiments was that viable parasites could not be isolated from animals inoculated with the attenuated strains. If the parasites had only a limited life in the host after attenuation it would circumvent the possibility of animals eventually succumbing to the inoculation. This effect has also been sought by chemical mutation as well as irradiation of tachyzoites.

Pfefferkorn and Pfefferkorn[93] used *N*-methyl-*N'*-nitro-*N* nitrosoguanidine to mutate RH strain tachyzoites. The parasites were still able to grow well at 33°C; however, growth was slowed down as the temperature was raised, and at 40°C the ability of the strain to form plaques in tissue culture was lost. One particular mutant termed *ts*-4 was found to be at least 10^4 times less virulent than the original RH strain. The authors suggested that a febrile response caused by the intraperitoneal injection of *ts*-4 into the mice may raise their body temperature above 36.5°C (their normal body temperature) and restrict the growth of the mutant. However, this was later found to be unlikely, and it was then considered that the host's immune response was responsible for the limited persistence of *ts*-4.[94] This mutant of the RH strain was found to produce high circulating antibody titers, but not be recoverable from host mice more than 3 months after inoculation. A subcutaneous injection of *ts*-4 also had the ability to protect at least 80% of mice against oral challenges of M-7741 oocysts that proved fatal to all nonvaccinated mice.[60] In addition, 60% of hamsters immunized subcutaneously with *ts*-4 survived challenge with 2×10^3 RH strain tachyzoites, whereas all hamsters immunized with a lysed RH strain tachyzoite preparation, and all negative control animals, succumbed to the challenge.

Temperature-sensitive and irradiated RH strain mutants that do not persist in the host appear to possess several advantages that would be required in an ideal vaccine against toxoplasmosis. These advantages and also the disadvantages will be discussed in Section V. If the *ts*-4 strain does prove to be temperature sensitive, it may be of use in attempting to vaccinate rarer or endangered Australian marsupials, which are extremely susceptible to toxoplasmosis and have a low normal body temperature (usually in the range 35 to 38°C, depending on the species).

IV. PASSIVE IMMUNIZATION

Because of the difficulties associated with production of sufficient purified tachyzoites or defined tachyzoite fractions to use for vaccination studies, passive immunization has been tried as an alternative schedule for vaccination against toxoplasmosis.

A. Polyclonal Antibody
1. In Vitro Studies

RH strain tachyzoites exposed *in vitro* to heat-inactivated sera from naturally infected humans or mice chronically infected with the PeC strain, could no longer enter murine fibroblasts or HeLa cells, but were rapidly ingested by murine macrophages.[87] More than 90% of these ingested parasites were rapidly killed, but without specific antibody incubation, only 50% of the tachyzoites entering the macrophages were killed. The remaining 50% of the tachyzoites multiplied in phagocytic vacuoles that did not acquire lysosomal factors. The specific antibody affected the tachyzoites in such a way that most were no longer capable of preventing fusion of lysosomes with the phagocytic vacuoles in which they resided. Several groups reported similar results with murine experimental systems and virulent tachyzoites,[96-98] and Anderson and Remington[99] looked at a natural human infection system. When RH strain tachyzoites were preincubated with either *T. gondii* antibody positive or negative human serum and added to human macrophage monolayers, the tachyzoites treated with specific antibody were inactivated or killed upon entering the macrophages. Parasites not treated with specific antibody multiplied and destroyed the cells. It has recently been found that the ability of specific antibody-coated or heat-killed tachyzoites to survive inside phagocytic vacuoles is correlated with acidification of the phagosome,[100] whereas vacuoles containing live parasites fail to acidify.

Viable *T. gondii* enter a wide range of both phagocytic and nonphagocytic cells by an active, energy-requiring process,[101,102] while nonviable organisms are probably only phagocytosed (discussed in Reference 103). The results of the experiments described here are therefore consistent with the hypothesis that antisera to *T. gondii* contain factors, which are probably specific antibody, but other possibilities such as lymphokines will be discussed in Section IV.D, that have the ability to kill the parasite or render it inactive in the absence of functional complement. The targets of these factors would appear to be ideal candidates for a vaccine against toxoplasmosis. Passive immunization of whole animals to identify these targets has an advantage over *in vitro* studies in that it involves the host's immune responses that are not found in tissue culture.

2. In Vivo Studies

Results of studies to test the ability of passive transfer of antiserum to *T. gondii* to protect mice against a challenge with either a highly or moderately virulent *T. gondii* strain have been conflicting. Nakayama[45] could not vaccinate mice against an intraperitoneal challenge with 3×10^3 RH strain tachyzoites by intraperitoneal injection of 0.3 ml of rabbit antiserum (dye test titer 1:16,384), although some protection was noted against a challenge with only 10^2 tachyzoites. Others have also reported being unable to vaccinate mice against a subcutaneous challenge with 2×10^4 RH strain tachyzoites by an intraperitoneal injection of 1 ml of rabbit antiserum (dye test titer 1:65,336).[71] On the other hand, Foster and McCulloch[43] immunized mice with an injection of 0.2 ml of guinea pig antiserum (dye test titer 1:16,384) and found that these mice had a prolonged mean time to death when challenged intraperitoneally with 2×10^4 RH strain tachyzoites. Although Strannegard and Lycke[104] could not protect mice against an intraperitoneal challenge with 10^3 tachyzoites of the moderately virulent GBG-1 strain of *T. gondii* by intraperitoneal injection of 0.2 ml of rabbit antiserum (dye test titer 1:8000), Krahenbuhl et al.[47] were able to confer some protection to mice against an intraperitoneal challenge with 5×10^4 tachyzoites of the medium-virulence C56 strain by 3 intraperitoneal injections of 0.5 ml of murine antiserum (dye test titer 1:16,000). Perhaps the most significant protection of mice against a moderately virulent parasite challenge was that obtained by immunization with 5 0.3-ml intraperitoneal injections of rabbit antiserum (dye test titer 1:256,000 to 1:10^6) and challenge with 10 cysts of the Alt strain of *T. gondii*; 25% of control mice died, whereas 90% of immunized mice survived.[65]

Hafizi and Modabber[105] found that when cyclophosphamide-treated mice were passively immunized by intraperitoneal and intravenous injection of murine antiserum, 70% survived an intraperitoneal challenge with five to ten cysts of the T strain of *T. gondii*, whereas only 20% of mice immunized with normal mouse serum survived. Survival of mice only occurred when the level of circulating anti-*T. gondii* antibody was >1:512.

We have found that the mean survival time of mice passively immunized with murine antiserum was significantly prolonged compared with that of mice given PBS or normal mouse serum. Groups of 20 LACA mice were immunized with intraperitoneal injections of 0.5 ml of either murine antiserum, PBS, or normal mouse serum. After 18 h, five mice from each group were picked at random, lightly anesthetized, bled from the retroorbital venous plexus, and their sera were tested for *T. gondii* antibody by indirect immunofluorescence. At 24 h after passive immunization, all mice were challenged with 5×10^2 viable RH strain tachyzoites, and survival times were recorded. The results of this experiment, which are similar to those found by Foster and McCulloch,[43] are contained in Table 2. A generalization from all of these experiments might be that although passive transfer of antiserum confers little or no real protection against challenge with highly virulent *T. gondii* strains (even though statisically very significant prolongations in the mean survival times have been recorded), it does confer signficant protection to mice against parasite strains of moderate to low virulence. It is not surprising that this situation appears to be analogous to that found *in vitro*. Specific antibody renders a percentage of the parasites inactive, but some survive and eventually kill the host, although the reduction in the numbers of viable

Table 2
MEAN SURVIVAL TIME OF GROUPS OF PASSIVELY
IMMUNIZED MICE CHALLENGED WITH *TOXOPLASMA*
TACHYZOITES

Group	Immunized with	Serum antibody titer[a]	Mean survival time (h ± 1 SD)	Significance with respect to normal mouse serum (Mann-Whitney U test)
1	PBS	<1:16	162.6 ± 8.5	N.S.[b]
2	Normal mouse serum	<1:16	167.0 ± 12.4	—
3	Anti-*Toxoplasma* immune serum	1:9,410	185.1 ± 3.7	$p < 0.0001$

[a] Geometric mean titer of five mice.
[b] N.S. = not significant $p > 0.05$.

parasites results in a prolongation of the survival time. Pavia[106] has recently looked at this question by examining individual organs for the number of viable parasites present compared with controls. The liver, lungs, spleen, brain, and skin in guinea pigs injected intravenously with guinea pig antiserum contained significantly less parasites than did the organs of animals injected with normal serum and challenged intradermally with RH strain tachyzoites.

Unfortunately, however, there are also theoretical disadvantages with injecting polyclonal antiserum to *T. gondii* to try and passively immunize animals against toxoplasmosis. The anti-*T. gondii* humoral immune response of an animal may be suppressed by the passive transfer of specific antiserum,[107] and the volumes of antiserum required would make it extremely unlikely that it could be used as a practical veterinary vaccination schedule. Even from an experimental point of view, different animals of the same species may produce different antibodies to the same antigen,[108] and guinea pigs or rabbits may recognize different determinants in the tachyzoite to those recognized by mice. Therefore, if there are antigens of *T. gondii* that are important in protection against infection, the amount of antibody to those specific antigens in a polyclonal antiserum may be very low even though the antiserum may have a very high antibody titer. The use of MAb to *T. gondii* to try and identify specific antigens that may be important in protection against infection overcomes this problem.

B. Monoclonal Antibody
1. In Vitro Studies

To date, two groups have investigated the ability of MAb to *T. gondii* to inhibit infection of tissue culture monolayers. Hauser and Remington[109] incubated RH strain tachyzoites with one of three MAbs to the parasite, a polyclonal-specific antiserum, or normal mouse serum, and added them to monolayers of normal murine peritoneal macrophages. At 2 and 18 h after infection, the percentage of macrophages containing intracellular organisms was calculated by microscopy. The percentage of cells containing viable parasites in the monolayers infected with tachyzoites preincubated with any of the MAb or the polyclonal-specific antiserum was significantly less than the cells in the monolayer infected with tachyzoites pretreated with normal mouse serum. Unfortunately, not all parasites were killed, and some were able to persist. Two of the MAbs (called 2G11 and 3E6) were directed against antigens of 35 (P30) and 14 kDa, and the third (1E11) was directed against an antigen of 27 kDa, all of which were present on the tachyzoite surface membrane. Sethi et al.[110] conducted a similar experiment with tachyzoites of the BK strain, but used supernatant fluid from a murine myeloma cell line as a negative control. All five MAbs against *T. gondii* caused increased intracellular killing of the tachyzoites preincubated with them. The specificity of the reactions was tested by absorbing the MAb with tachyzoites and repeating the experiment. The absorbed MAbs were no longer capable of rendering parasites susceptible to the anti-*T. gondii* effect of macrophages. Analysis of the antigens against which the MAbs were directed showed that five membrane surface antigens, one of which was a 35- to 45-kDa component functional in the dye test, could be immunoprecipitated from radiolabeled tachyzoites.

In a detailed study using a slightly different rationale, Schwartzman[111] used partially purified, penetration-enhancing factor (PEF)[112] to develop an assay for enhancement of the penetration of RH strain tachyzoites into human fibroblast monolayers. Although none of the four MAbs against *T. gondii* tested were able to block parasite penetration into the monolayers in the normal manner, all of them significantly reduced the effect PEF had on the monolayer parasite cultures. They caused a decrease in the number of plaques seen when PEF and a standard inoculum of tachyzoites were added in the absence of the MAb. However, plaque numbers were not reduced to those seen in monolayer/parasite cultures without added PEF. All of the four MAbs were found to be directed against bodies, that appeared to be rhoptries, in the cytoplasm of the anterior of tachyzoites.

Although this line of research has the advantage of being able to look at very specific reactions, it has the disadvantage in that it is distant from the natural course of the disease even in laboratory animals and cannot involve the large range of host responses not found in tissue cultures. Therefore, Johnson et al.[23] used a different rationale to investigate the effect of MAb binding to the parasite. RH strain tachyzoites were incubated with an equal volume of PBS, MOPC 21, or a MAb called FMC 22.[113] MOPC 21 is the MAb produced by the murine myeloma parent cell line, and FMC 22 is a MAb of the IgG1 isotype that reacts with an antigen of 14 kDa found on the tachyzoite membrane surface.[23] The tachyzoite suspensions were then diluted to give 5×10^4 viable parasites ml^{-1}, and 0.2 ml of each fraction was injected intraperitoneally into groups of LACA mice. Preincubation of the tachyzoites with FMC 22 conferred a significantly greater mean survival time to mice than preincubation with either PBS or MOPC 21. Although the prolongation was only about 7 h, the mean generation time of the RH strain in the murine peritoneal cavity is about half this,[53] and so the prolongation obtained was consistent with the inactivation of greater than 50% of the tachyzoites injected.

A further refinement to this technique is the passive immunization of mice with specific MAb.

2. *In Vivo Studies*

Johnson et al.[23] passively immunized mice by intraperitoneal injection of each of six MAbs to *T. gondii* or MOPC 21 as a negative control. Groups of immunized mice were then challenged intraperitoneally with either RH strain tachyzoites or ten cysts of the Mt. Pleasant 73/2930 strain of *T. gondii*. Two MAbs (FMC 19 and FMC 22) conferred total protection against the moderately virulent challenge, with all mice surviving, whereas 90% of control mice died. FMC 19 and FMC 22 also conferred significant protection against the highly virulent RH strain challenge, as measured by a prolonged mean survival time of immunized compared with control groups of mice. One MAb (FMC 23) gave significant protection (80% survival) against the moderately virulent challenge only. Passive immunization with dilutions of FMC 22 indicated that the lowest serum titer needed to confer significant protection (>55% survival) to mice against the moderately virulent challenge was 1:640. Immunoprecipitation and autoradiography of radiolabeled, intact tachyzoites confirmed that FMC 19 was directed against an antigen of 35 kDA (P30) and that FMC 22 and FMC 23 were directed against an antigen of 14 kDA.

Table 3

**RESPONSE OF MICE PASSIVELY IMMUNIZED WITH
A MONOCLONAL ANTIBODY AND CHALLENGED
WITH *TOXOPLASMA* OOCYSTS**

Group	Immunized with	Serum antibody titer[a]	Mortality (dead/total)	Significance with respect to control antibody group (Fisher exact test)
1	FMC 3	<1:10	10/10	—
2	FMC 19	1:2,030	6/10	$p = 0.04$
3	MOPC 21	<1:10	10/10	—
4	FMC 22	1:16,255	10/10	N.S.[c]

[a] Geometric mean titer of three mice.
[b] 6 weeks after challenge.
[c] N.S. = not significant $p > 0.05$.

Table 4

**RESPONSE OF MICE PASSIVELY IMMUNIZED WITH
A MONOCLONAL ANTIBODY AND CHALLENGED
WITH *TOXOPLASMA* CYSTS**

Group	Immunized with	Serum antibody titer[a]	Mortality[b] (dead/total)	Significance with respect to control antibody group (Fisher exact test)
1	FMC 3	<1:10	10/10	—
2	FMC 19	1:2,030	10/10	N.S.[c]
3	MOPC 21	<1:10	10/10	—
4	FMC 22	1:16,255	10/10	N.S.

[a] Geometric mean titer of three mice.
[b] 6 weeks after challenge.
[c] N.S. = not significant $p > 0.05$.

Although these results were encouraging, the criticism leveled at other works which did not use a natural route of challenge also applies to this study. Therefore, we passively immunized mice and challenged them orally with either cysts or oocysts of the ovine 2 strain of *T. gondii*.[54] Mice were immunized with an intraperitoneal injection of either FMC 19 or FMC 22, or control MAb of irrelevant specificity, but identical isotype. MOPC 21 is an IgG1 myeloma protein[114,115] that is unrelated to *T. gondii*, and FMC 3 is an IgG3 (as is FMC 19) MAb that reacts against a subpopulation of human T, B, and null cells.[116] At 18 h after passive immunization, three mice from each group were chosen at random, lightly anesthetized, and bled from the retroorbital venous plexus in order to determine their serum *T. gondii* antibody level by indirect immunofluorescence. At 24 h after immunization, all mice were challenged orally with either 35 cysts or 200 oocysts. As can be seen in Tables 3 and 4, neither FMC 19 nor FMC 22 protected mice against a challenge with *T. gondii* cysts, but mice with a serum anti-*T. gondii* titer of FMC 19 of at least 1:2000 were significantly protected against a lethal oocyst challenge. Because of the several significant changes, particularly route and strain of challenge, it is difficult to accurately identify what the reasons for the lack of protection found in several groups in this experiment may be. However, further studies using different strains of *T. gondii* will be necessary in order to confirm the usefulness of the antigens, against which FMC 19 and FMC 22 are directed, as vaccines.

Another study also found that passive immunization with a MAb to P30 failed to protect mice against *T. gondii*.[64] In fact, an increased mortality in the mice immunized with the MAb to P30 and increased numbers of cysts in the brains of mice surviving was reported. However, there are several factors that lessen the impact of these findings. The negative control animals were not immunized with an irrelevant MAb or treated in any way. The intraperitoneal challenge of C strain tachyzoites was given only 1 h after either intraperitoneal or intravenous immunization with the MAb ascitic fluid, and even the authors suggested that ascitic fluid could enhance rapid growth of the tachyzoite. In addition, the BALB/c strain of mice used in this study has been shown to be not ideal for research on toxoplasmosis. Of the common strains available, this strain of mice appears to be the most resistant to toxoplasmosis.[117-120] Therefore, a small but significant protective host response against the parasite detectable in other, more susceptible strains, such as C57BL/6, may not be found using BALB/c mice.

A MAb (F3G3) to cytoplasmic components of *T. gondii* has been found to confer protection against toxoplasmosis.[63] Groups of test mice were injected intraperitoneally with F3G3, and control mice received crude preparations of a murine myeloma cell culture supernatant fluid 1 d before, and 3, 7, and 10 d after intraperitoneal challenge with either 10^3, 10^4, or 10^5 C56 strain tachyzoites. Of mice immunized with F3G3 and challenged with either 10^3 or 10^4 tachyzoites, 90% survived, compared with the survival of only 30% of control mice challenged with 10^3, and 50% of control mice challenged with 10^4 tachyzoites. Only 50% of control mice and 70% of mice immunized with F3G3 survived the challenge with 10^5 tachyzoites. The results of this study highlight another important factor to be considered with studies on vaccination against toxoplasmosis, particularly those that have given no protection. Although the obvious conclusion, i.e., the fraction is not protective, may be correct; the challenge dose needs to be carefully quantitated so that a possible significant but small protective effect is not negated by an overwhelming parasite challenge. While it can be argued that such "fine-tuning" will not be found in natural infections, the actual doses received by animals in the field is unknown, and laboratory studies should be designed to show the possible protective effect of a particular antigen or antibody. Having identified the putative vaccine component in the laboratory, further studies can then be designed to try and protect animals with it in field studies.

C. Anti-Idiotypic Vaccines

Although anti-idiotypic antibodies have not been used to vaccinate against toxoplasmosis, they have had marked protective effects against infection with *Trypanosoma*[121] and *Leishmania*[122] and should at least be considered in relation to *T. gondii*. The potential for using anti-idiotypic antibodies as vaccines has been reviewed by Bona and Moran[123] and Monroe and Greene.[124]

An anti-idiotypic antibody to *T. gondii* has been made by extensive absorption of a rabbit antiserum to human anti-*T. gondii* antibody.[125] In inhibition experiments on 24 human sera containing different amounts of antibody to *T. gondii*, 60% of them cross-reacted with the anti-idiotypic antibody.

We have prepared monoclonal anti-idiotypic antibodies to FMC 19 and FMC 22,[126] with the aim of being able to use these to mimic the antigens that FMC 19 and FMC 22 are directed against and, hence, achieve protection without the need for the *T. gondii* antigens. The difficulty in obtaining enough parasite material to undertake large vaccination trials has already been mentioned, and the ability to use anti-idiotypic antibodies to mimic parasite antigens will be a major advantage with vaccination schedules using such antibodies.

D. Lymphokines

Although by definition lymphokines are not vaccines, some have been found to have anti-*T. gondii* effects and should therefore be considered here.

Toxoplasmosis in mice leads to the generation of serum interferon (INF) titers[177] which have the capacity to protect mice against an otherwise fatal viral infection.[128] INF was also found to have an anti-*T. gondii* effect.[129] L cell monolayers treated with murine INF before the addition of RH strain tachyzoites were significantly protected from *T. gondii* lytic destruction as compared with infected monolayers not treated with INF. As with other tissue culture studies, some tachyzoites persisted in even the INF-treated cells. The protective effect was host cell species specific. It was not found with murine INF and chicken embryo fibroblast monolayers, nor chicken INF and L cells.

This early research on INF and toxoplasmosis appears to have slowed down, probably because of the negative reports of several authors,[130,131] and also the difficulties associated with obtaining sufficient purified INF at that time. The recent availability of recombinant INF has led to a resurgence of this research. McCabe et al.[132] found that mice given murine gamma INF (INF-γ) intraperitoneally were significantly protected, as measured by survival, when challenged with an intravenous injection of C56 strain tachyzoites. In addition, C57BL/6 mice, a good producer of IFN-γ, showed significant prolongation of mean survival time following primary infection with tachyzoites of the S-273 strain of *T. gondii* or secondary infection with RH strain tachyzoites, compared with that of BALB/c mice, which were consistently poor producers of IFN-γ.[72] However, there was no direct correlation between susceptibility to *T. gondii* and serum IFN-γ levels.

Human recombinant IFN-γ has been found to suppress the growth of RH strain tachyzoites in treated human fibroblasts, although the effect was dependent on the tryptophan concentration in the medium.[133] This may have been the reason for the negative results of others.[131] Further studies suggested that the anti-*T. gondii* effect of IFN-γ was the result of starvation of the tachyzoites for tryptophan.[134] Although tryptophan depletion is unlikely to be achievable in whole animals and, therefore, INF-γ administration would not appear to be a practical schedule to prevent toxoplasmosis, this line of research is providing valuable information on the interaction between *T. gondii* and host cells. In addition, the previous *in vivo* studies did show some protective effect, and *in vivo* administration of murine IFN-γ does induce macrophage activation against RH strain tachyzoites in mice.[135] Perhaps IFN-γ has an *in vivo* effect in addition to that, or different from that, found *in vitro*.

Another lymphokine, murine recombinant granulocyte-macrophage colony-stimulating factor has been found to have no inhibitory effect on RH strain tachyzoites in the presence of murine macrophages,[136] although it can induce killing of another intracellular protozoon, *Leishmania tropica*.[137] On the other hand, a third lymphokine, interleukin 2 (IL 2), has been found to be active against *T. gondii*. It was shown tht 40% of mice injected intraperitoneally with a total of 300 U of recombinant human IL 2 and 30% of mice that received a total of 500 U of IL 2 died after intraperitoneal challenge with C56 strain tachyzoites, while all control mice injected with only PBS succumbed to the challenge.[138] In addition, mice given a total of 300 U of IL 2 after intraperitoneal injection of cysts of the nonvirulent Beverley strain of *T. gondii* developed about half as many cysts in their brains than did untreated mice.

The action of IFN-γ and IL 2 are considered to be nonspecific because they affect organisms other than *T. gondii*. However, there are other less well-defined lymphokines that also have inhibitory effects on *T. gondii*. Chinchilla and Frenkel[139] found that lymphokine-like mediator was released by immune hamster lymphocytes following stimulation with freeze-thaw-lysed T-45 strain tachyzoites of *T. gondii*. It not only protected macrophages, as measured by reduced numbers of tachyzoites per 100 host cells, but also fibroblasts and kidney cells infected with T-45 strain. The mediator had a molecular size of between 4 and 5 kDa and was parasite specific as well as being species specific for the cells in which it was produced. A recent study confirmed these results.[73] Another lymphokine-like mediator called obioactin, also between 3 to 5 kDa, has been found to have anti-*T. gondii* effects.[140] Obioactin was found in the sera of cattle challenged intravenously with RH strain tachyzoites. However, unlike the mediator found by Chinchilla and Frenkel, obioactin inhibited the growth of RH strain tachyzoites in a wide range of heterologous cells such as murine macrophages and kidney cells, canine monocytes, and human heart and brain cells. Therefore, the exact relationship between obioactin and the lymphokine found by Chinchilla and Frenkel remains to be determined.

Shirahata et al.[97] found that when murine peritoneal macrophages were exposed to the culture supernatant fluid from immune murine spleen lymphocytes that had been incubated with RH strain tachyzoite antigens, the intracellular multiplication of tachyzoites of the Beverley-Shimizu strain of *T. gondii* in these cells was greatly inhibited. In further studies, they called the factor responsible for this effect *Toxoplasma* growth inhibitory factor (TOXO-GIF).[141,142] TOXO-GIF was synthesized *de novo* by sensitized T lymphocytes in response to specific antigen stimulation for as short a period as 6 h. However, it was also induced by concanavalin A and phytohemagglutinin. It had a molecular size of between 30 and 40 kDA. Sethi and Brandis[143] found a lymphokine between 50 to 100 kDa which was released upon specific *T. gondii* antigen stimulation of lymphocytes, but not by concanavalin A stimulation. The relationship between this lymphokine and TOXO-GIF is therefore unknown. Although TOXO-GIF and IL 2 have common characteristics, the fact that a microassay can differentiate between the two[144] suggests that they are separate entities. It therefore appears that as well as INF-γ and IL 2, there are at least two lymphokines (one less than 5 kDA, the other greater than 30 kDa), that inhibit the multiplication of *T. gondii*.

V. HOW TO VACCINATE?

We can conclude from the results of previous experiments in which immunization against toxoplasmosis was attempted that there are parasite antigens that are likely to protect against the disease. However, the problems associated with production of sufficient purified specific parasite antigens have already been discussed. Recent developments in eukaryotic gene cloning have enabled large amounts of some recombinant parasite proteins, particularly those of *Plasmodium falciparum*, to be produced[145] and used to immunize against infection.[146,147] Such techniques are now being used to produce *T. gondii* antigens in *Escherichia coli*.[148,149]

We have cloned a 2-kilobase gene of *T. gondii* that codes for an immunodominant antigen in the expression vector bacteriophage λgtll. The β-galactosidase fusion protein produced reacts with immune rabbit serum after western blotting, and this recombinant fusion protein can be purified from *E. coli* lysogens (Figure 2) to provide large amounts of parasite antigen. An immunization trial of this recombinant parasite protein in goats is being planned at present.[150] We have also attempted to get expression of the recombinant *T. gondii* antigen in *Campylobacter fetus* as lysates of *C. fetus* are injected subcutaneously into sheep to prevent campylobacteriosis. If the recombinant bacteriophage could be made to express *T. gondii* in this organism, a dual immunization could be achieved with one injection. However, the *C. fetus* could not be transfected with the recombinant bacteriophage, probably because it did not carry the appropriate receptors. Therefore, we intially intend to inject the recombinant *T. gondii* protein, purified from *E. coli*, subcutaneously into goats. Although this schedule is technically easy and avoids the current controversies[151] surrounding genetically engineered vaccines, the humoral and CMI responses that will be generated by this route of injection are likely to be low. In addition, as discussed in the Introduction, under natural conditions found in the pasture, toxoplasmosis is acquired by ingestion of the parasite. Therefore, the establishment of a first-line immune response in the gut would appear to be a desirable schedule for immunization trials.

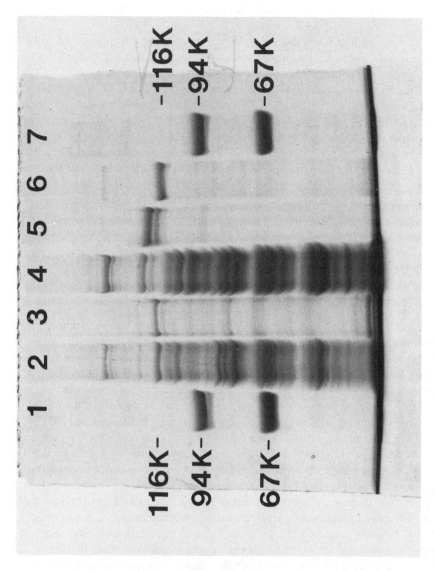

FIGURE 2. A coomassie blue-stained polyacrylamide gel electrophoretic analysis of M_r standards (Lanes 1 and 7); lysate of *E. coli* containing λgt11(Lane 2); purified lysate of *E. coli* containing λgt11 (Lane 3) — note marked purification of β-galactosidase at 116K; lysate of *E. coli* containing recombinant λgt11 with *T. gondii* antigen gene (Lane 4) — note increase in size of β-galactosidase fusion protein to about 125K; purified lysate of *E. coli* containing recombinant λgt11 with *T. gondii* antigen (Lane 5)— note marked purification of recombinant β-galactosidase fusion protein; and commercial preparation of β-galactosidase (Lane 6). (K = kilodalton.)

Secretory IgA specific for *T. gondii* has been found in the intestinal secretion of mice injected orally with ME49 strain cysts.[152] Even though it was detectable only between 4 to 8 weeks after immunization, the parasite probably passes from the intestine very quickly, and so the antigenic stimulation is present for only a short period of time. Schedules that allowed a continuing antigenic stimulation over a longer period would be expected to provoke a significantly greater immune response. The continued presence of cysts in the host's tissues is probably the major reason why premunition immunity works so well against toxoplasmosis.

One way to achieve an extended antigen presence in the gastrointestinal tract is to orally immunize animals with a *galE Salmonella typhimurium* strain.[153] Such strains can transiently colonize the Peyer's patches of mice and deliver specific antigenic stimuli without adversely affecting the host. Although it is not yet known whether such strains have a similar effect in sheep and goats, one recombinant strain containing genes for fimbrial antigens of porcine enterotoxigenic *E. coli* has been orally injected into pigs, and it did result in specific immune responses in the pig sera.[154] In addition, a *galE* mutant of *Salmonella typhi* has been found to be stable and safe when fed to humans.[155]

Although even the specific secretory IgA response evoked by this type of vaccination schedule may not lead to eventual destruction of all parasites ingested, even a small reduction in the numbers of viable parasites crossing out of the intestine would be worthwhile. Of course, a combination of oral vaccination with a recombinant bacterium expressing *T. gondii* antigen, followed at a later date by a subcutaneous injection of the purified recombinant parasite protein, may be better than just a single immunization.

Alternatively, immunization with a recombinant vaccinia virus,[156] such as that used to vaccinate foxes against rabies[157] or that found to provoke antibodies to *Plasmodium knowlesi* in rabbits,[158] may be useful for vaccination against toxoplasmosis.

Other groups are pursuing naturally attenuated or experimentally altered live vaccines, while still others see lymphokines as being worth pursuing as a potential vaccine. As well as these, there will probably be other schedules devised and pursued before a successful vaccine is developed. Of course, it may be that a combination of schedules is required to ensure adequate protection against toxoplasmosis.

No matter how we choose to vaccinate against toxoplasmosis, the ultimate, overriding factor will be cost. A successful vaccine must not only prevent losses from abortions caused by toxoplasmosis, it must cost no more than several cents per dose. Otherwise, it will not be economical, and the desirability of its use will be markedly decreased. Table 5 contains a very approximate rating of several vaccination schedules that appear to be viable at present. They each have their own advantages and disadvantages.

VI. CONCLUSION

This chaper attempts to review the literature pertinent to vaccination against toxoplasmosis published up to mid 1987. It presents one individual's views on the past research and, based on this, briefly attempts to suggest likely schedules that may provide a successful vaccine against toxoplasmosis. For this reason, not all workers in the area, particularly those committed to other vaccination schedules, may agree with some of the statements made here. However, it is clear that *T. gondii* does not possess many of the problems associated with other protozoan parasites, such as *Trypanosoma* and *Plasmodium*. Therefore, it would not be overly optimistic to suggest that a successful vaccine against toxoplasmosis in animals should be commercially available within the next 5 years. Whether it is or not remains to be seen. Toxoplasmosis is a serious cause of livestock loss, and a vaccination schedule against it is certainly worth developing.

Table 5
RATING OF POTENTIAL VACCINE SCHEDULES FOR VARIOUS PARAMETERS

Parameter	Vaccine					
	Recombinant plasmid in bacterium (oral inoculation)	Recombinant protein (s.c. inoculation)	Live attenuated, persisting (oral or s.c. inoculation)	Live attenuated, nonpersisting (oral or s.c. inoculation)	Lympho-kine	Anti-idio typic antibody
Storage/stability	6[a,b]	6[b]	2	1	6[b]	6[b]
Continuity of stimulus from one inoculation	5	3	6	4	3	3
Potential biohazard	1	4	2	3	6	6
Ease of large-scale manufacture	5	4	2	1	3	6
Possible adverse effects in host	3	6	1	2	6	6
Cost	5	4	2	1	3	6

[a] Six-best schedule, one-worst schedule, for that parameter.
[b] Schedules with similar rating are tied.

REFERENCES

1. **Blewett, D. A. and Watson, W. A.,** The epidemiology of ovine toxoplasmosis. II. Possible sources of infection in outbreaks of clinical disease, *Br. Vet. J.,* 139, 546, 1983.
2. **Sharman, G. A. M., Williams, K. A. B., Thorburn, H., and Williams, H.,** Studies of serological reactions in ovine toxoplasmosis encountered in intensively-bred sheep, *Vet. Rec.,* 91, 670, 1972.
3. **Plant, J. W., Freeman, P., and Saunders, E.,** Serological survey of the prevalence of *Toxoplasma gondii* antibodies in rams in sheep flocks in New South Wales, *Aust. Vet. J.,* 59, 87, 1982.
4. **Beverley, J. K. A., Henry, L., Hunter, D., and Brown, M. E.,** Experimental toxoplasmosis in calves, *Res. Vet. Sci.,* 23, 33, 1977.
5. **Munday, B. L. and Corbould, A.,** Serological responses of sheep and cattle exposed to natural *Toxoplasma* infection, *Aust. J. Exp. Biol. Med. Sci.,* 57, 141, 1979.
6. **Dubey, J. P. and Towle, A.,** Toxoplasmosis in Sheep: A Review and Annotated Bibliography, Misc. Publ. No. 10, Commonwealth Institute of Parasitology, C. A. B. International, Farnham Royal, U.K., 1986
7. **Dubey, J. P.,** A review of toxoplasmosis in pigs, *Vet. Parasitol.,* 19, 181, 1986.
8. **Munday, B. L. and Mason, R. W.,** Toxoplasmosis as a cause of perinatal death in goats, *Aust. Vet. J.,* 55, 485, 1979.
9. **Dubey, J. P., Miller, S., Desmonts, G., Thulliez, P., and Anderson, W. R.,** *Toxoplasma gondii* induced abortion in dairy goats, *J. Am. Vet. Med. Assoc.,* 188, 159, 1986.
10. **Nurse, G. H. and Lenghaus, C.,** An outbreak of *Toxoplasma gondii* abortion, mummification and perinatal death in goats, *Aust. Vet. J.,* 63, 27, 1986.
11. **Levine, N. D.,** Whatever became of *Isospora bigemina?*, *Parasitol. Today,* 3, 101, 1987.
12. **Levine, N. D.,** *The Protozoan Phylum Apicomplexa,* CRC Press, Boca Raton, FL, in press.
13. **Frenkel, J. K. and Wallace, G. D.,** Transmission of toxoplasmosis by tachyzoites: possibility and probability of a hypothesis, *Med. Hypoth.,* 5, 529, 1979.
14. **Watson, W. A. and Beverley, J. K. A.,** Epizootics of toxoplasmosis causing ovine abortion, *Vet. Rec.,* 88, 120, 1971.
15. **Jacobs, L., Remington, J. S., and Melton, M. L.,** The resistance of the encysted form of *Toxoplasma gondii, J. Parasitol.,* 46, 11, 1960.
16. **Dubey, J. P.,** *Toxoplasma gondii* cysts in placentas of experimentally infected sheep, *Am. J. Vet. Res.,* 48, 352, 1987.
17. **Benenson, M. W., Takafuji, E. T., Lemon, S. M., Greenup, R. L., and Sulzer, A. J.,** Oocyst-transmitted toxoplasmosis associated with ingestion of contaminated water, *N. Engl. J. Med.,* 307, 666, 1982.
18. **Kean, B. H., Kimball, A. C., and Christenson, W. N.,** An epidemic of acute toxoplasmosis, *JAMA,* 208, 1002, 1969.
19. **Dahl, R. J. and Johnson, A. M.,** Purification of *Toxoplasma gondii* from host cells, *J. Clin. Pathol.,* 36, 602, 1983.
20. **Johnson, A. M.,** The antigenic structure of *Toxoplasma gondii:* a review, *Pathology,* 17, 9, 1985.
21. **Hughes, H. P. A.,** Toxoplasmosis: the need for improved diagnostic techniques and accurate risk assessment, *Curr. Top. Microbiol. Immunol.,* 120, 105, 1985.
22. **Handman, E., Goding, J. W., and Remington, J. S.,** Detection and characterization of membrane antigens of *Toxoplasma gondii, J. Immunol.,* 124, 2578, 1980.
23. **Johnson, A. M., McDonald, P. J., and Neoh, S. H.,** Monoclonal antibodies to *Toxoplasma* cell membrane surface antigens protect mice from toxoplasmosis, *J. Protozool.,* 30, 351, 1983.
24. **Kasper, L. H., Crabb, J. H., and Pfefferkorn, E. R.,** Purification of a major membrane protein of *Toxoplasma gondii* by immunoabsorption with a monoclonal antibody, *J. Immunol.,* 130, 2407, 1983.
25. **Johnson, A. M., Haynes, W. D., Leppard, P. J., McDonald, P. J., and Neoh, S. H.,** Ultrastructural and biochemical studies on the immunohistochemistry of *Toxoplasma gondii* antigens using monoclonal antibodies, *Histochemistry,* 77, 209, 1983.
26. **Rodriguez, C., Afchain, D., Capron, A., Dissous, C., and Santoro, F.,** Major surface protein of *Toxoplasma gondii* (p30) contains an immunodominant region with repetitive epitopes, *Eur. J. Immunol.,* 15, 747, 1985.
27. **Santoro, F., Afchain, D., Pierce. R., Cesbron, J. Y., Ovlaque, G., and Capron, A.,** Serodiagnosis of *Toxoplasma* infection using a purified parasite protein (p30), *Clin Exp. Immunol.,* 62, 262, 1985.
28. **Dahl, R. J., Woods, W. H., and Johnson, A. M.,** Recognition by the human immune system of candidate vaccine epitopes of *Toxoplasma gondii* measured by a competitive ELISA, *Vaccine,* 5, 187, 1987.
29. **Santoro, F., Charif, H., and Capron, A.,** The immunodominant epitope of the major membrane tachyzoite protein (P30) of *Toxoplasma gondii, Parasite Immunol.,* 8, 631, 1986.
30. **Araujo, F. G., Dubey, J. P., and Remington, J. S.,** Antigenic similarity between the Coccidian parasites *Toxoplasma gondii* and *Hammondia hammondi, J. Protozool.,* 31, 145, 1984.

31. **Potasman, I., Araujo, F. G., and Remington, J. S.,** *Toxoplasma* antigens recognized by naturally occurring human antibodies, *J. Clin. Microbiol.,* 24, 1050, 1986.
32. **Hughes, H. P. A.,** Personal communication, 1987.
33. **Sharma, S. D. and Remington, J. S.,** Macrophage activation and resistance to intracellular infection, in *Lymphokines: Lymphokines in Macrophage Activation,* Vol 3, Pick, E., Ed., Academic Press, New York, 1981, 181.
34. **Frenkel, J. K.,** Experimental analysis of tissue immunity in toxoplasmosis, *Lyon Med.,* 248 (Suppl. 17), 67, 1982.
35. **Krahenbuhl, J. L. and Remington, J. S.,** The immunology of *Toxoplasma* and toxoplasmosis, in *Immunology of Parasitic Infections,* 2nd ed., Cohen, S. and Warren, K. S., Eds., Blackwell Scientific, Oxford, 1982.
36. **Dubey, J. P., Miller, N. L., and Frenkel, J. K.,** The *Toxoplasma gondii* oocyst from cat feces, *J. Exp. Med.,* 132, 636, 1970.
37. **Lunde, M. N. and Jacobs, L.,** Antigenic differences between endozoites and cystozoites of *Toxoplasma gondii, J. Parasitol.,* 69, 806, 1983.
38. **Kasper, L. H., Bradley, M. S., and Pfefferkorn, E. R.,** Identification of stage-specific sporozoite antigens of *Toxoplasma gondii* by monoclonal antibodies, *J. Immunol.,* 132, 443, 1984.
39. **Kasper, L. H. and Ware, P. L.,** Recognition and characterization of stage-specific oocyst/sporozoite antigens of *Toxoplasma gondii* by human antisera, *J. Clin. Invest.,* 75, 1570, 1985.
40. **Handman, E. and Remington, J. S.,** Antibody responses to *Toxoplasma* antigens in mice infected with strains of different virulence, *Infect. Immun.,* 29, 215, 1980.
41. **Cutchins, E. C. and Warren, J.,** Immunity patterns in the guinea pig following *Toxoplasma* infection and vaccination with killed *Toxoplasma, Am. J. Trop. Med. Hyg.,* 5, 197, 1956.
42. **Jacobs, L.,** Propagation, morphology, and biology of *Toxoplasma, Ann. N.Y. Acad. Sci.,* 64, 154, 1956.
43. **Foster, B. G. and McCulloch, W. F.,** Studies of active and passive immunity in animals inoculated with *Toxoplasma gondii, Can. J. Microbiol.,* 14, 103, 1968.
44. **Stahl, W. and Akao, S.,** Immunity in experimental toxoplasmosis, *Keio J. Med.,* 13, 1, 1964.
45. **Nakayama, I.,** Effects of immunization procedures in experimental toxoplasmosis, *Keio J. Med.,* 14, 63, 1965.
46. **Stadtsbaeder, S., Nguyen, B. T., and Calvin-Preval, M. C.,** Respective role of antibodies and immune macrophages during acquired immunity against toxoplasmosis in mice, *Ann. Immunol. (Paris),* 126C, 461, 1975.
47. **Krahenbuhl, J. L., Ruskin, J., and Remington, J. S.,** The use of killed vaccines in immunization against an intracellular parasite *Toxoplasma gondii, J. Immunol.,* 108, 425, 1972.
48. **Levaditi, C., Lepine, P., and Schoen, R.,** Anti-*Toxoplasmic* immunity (in French), *C. R. Soc. Biol.,* 2, 1130, 1928.
49. **Huldt, G.,** Experimental toxoplasmosis: effect of inoculation of *Toxoplasma* in seropositive rabbits, *Acta Pathol. Microbiol. Scand.,* 68, 592, 1966.
50. **Wildfuhr, G.,** Experimental immunizations with *Toxoplasma gondii* (in German), *Z. Immunitatsforsch.,* 113, 435, 1956.
51. **Sabin, A. B.,** Toxoplasmic encephalitis in children, *JAMA,* 116, 801, 1941.
52. **Ruskin, J. and Remington, J. S,** Resistance to intracellular infection in mice immunized with *Toxoplasma* vaccine and adjuvant, *J. Reticuloendothel. Soc.,* 9, 465, 1971.
53. **Johnson, A. M., McDonald, P. J., and Neoh, S. H.,** Kinetics of the growth of *Toxoplasma gondii* (RH strain) in mice, *Int. J. Parasitol.,* 9, 55, 1979.
54. **Rothe, J., McDonald, P. J., and Johnson, A. M.,** Detection of *Toxoplasma* cysts and oocysts in an urban environment in a developed country, *Pathology,* 17, 497, 1985.
55. **Araujo, F. G. and Remington, J. S.,** Protection against *Toxoplasma gondii* in mice immunized with *Toxoplasma* cell fractions, RNA and synthetic polyribonucleotides, *Immunology,* 27, 711, 1974.
56. **Preston, P. M. and Dumonde, D. C.,** Immunogenicity of a ribosomal antigen of *Leishmania enriettii, Trans. R. Soc. Trop. Med. Hyg.,* 65, 18, 1971.
57. **Leon, L. L., Leon, W., Chaves, L., Costa, S. C. G., Querioz Cruz, M., Brascher, H. M., and Oliveira-Lima, A.,** Immunization of mice with *Trypanosoma cruzi* polyribosomes, *Infect. Immun.,* 27, 38, 1980.
58. **Masihi, K. N., Brehmer, W., and Werner, H.,** The effect of *Toxoplasma* cell fractions and mycobacterial immunostimulants against virulent *Toxoplasma gondii* in mice, *Zentralbl. Bakteriol. Parasitenkd. Infektionskr. Hyg. Abt. I Orig. Reihe A,* 245, 377, 1979.
59. **McLeod, R., Estes, R. G., and Mack, D. G.,** Effects of adjuvants and *Toxoplasma gondii* antigens on immune response and outcome of peroral *T. gondii* challenge, *Trans. R. Soc. Trop. Med. Hyg.,* 79, 800, 1985.
60. **Waldeland, H. and Frenkel, J. K.,** Live and killed vaccines against toxoplasmosis in mice, *J. Parasitol.,* 69, 60, 1983.

61. **Beverley, J. K. A., Archer, J. F., Watson, W. A., and Fawcett, A. R.,** Trial of a killed vaccine in the prevention of ovine abortion due to toxoplasmosis, *Br. Vet. J.,* 127, 529, 1971.
62. **Araujo, F. G. and Remington, J. S.,** Partially purified antigen preparations of *Toxoplasma gondii* protect against lethal infection in mice, *Infect. Immun.,* 45, 122, 1984.
63. **Sharma, S. D., Araujo, F. G., and Remington, J. S.,** *Toxoplasma* antigen isolated by affinity chromatography with monoclonal antibody protects mice against lethal infection with *Toxoplasma gondii, J. Immunol.,* 133, 2818, 1984.
64. **Kasper, L. H., Currie, K. M., and Bradley, M. S.,** An unexpected response to vaccination with a purified major membrane tachyzoite antigen (P30) of *Toxoplasma gondii, J. Immunol.,* 134, 3426, 1985.
65. **Masihi, K. N. and Werner, H.,** Immunization of NMRI mice against virulent *Toxoplasma gondii.* Differing efficacy of eleven cyst-forming *Toxoplasma* strains, *Z. Parasitenkd.,* 54, 209, 1977.
66. **Reikvam, A. and Lorentzen-Styr, A. M.,** Virulence of different strains of *Toxoplasma gondii* and host response in mice, *Nature (London),* 261, 508, 1976.
67. **Wolf, A., Cowen, D., and Paige, B.,** Human toxoplasmosis: occurrence in infants as an encephalomyelitis verification by transmission to animals, *Science,* 89, 226, 1939.
68. **Frenkel, J. K.,** Effect of vaccination and sulfonamide therapy on experimental toxoplasmosis, *Fed. Proc. Fed. Am. Soc. Exp. Biol.,* 11, 468, 1952.
69. **Nakayama, I.,** Persistence of the virulent RH strain of *Toxoplasma gondii* in the brains of immune mice, *Keio J. Med.,* 13, 7, 1964.
70. **Krahenbuhl, J. L., Blazkovec, A. A., and Lysenko, M. G.,** The use of tissue culture-grown trophozoites of an avirulent strain of *Toxoplasma gondii* for the immunization of mice and guinea pigs, *J. Parasitol.,* 57, 386, 1971.
71. **Gill, H. S. and Prakash, O.,** A study on the active and passive immunity in experimental toxoplasmosis, *Indian J. Med. Res.,* 58, 1157, 1970.
72. **Shirahata, T., Mori, A., Ishikawa, H., and Goto, H.,** Strain differences of interferon-generating capacity and resistance in *Toxoplasma*-infected mice, *Microbiol. Immunol.,* 30, 1307, 1986.
73. **Reyes, L. and Frenkel, J. K.,** Specific and nonspecific mediation of protective immunity to *Toxoplasma gondii, Infect. Immun.,* 55, 856, 1987.
74. **Nakayama, I.,** On the survival of high virulent strain of *Toxoplasma gondii* inoculated intravenously into immune mice, *Keio J. Med.,* 15, 13, 1966.
75. **Jacobs, L. and Hartley, W. J.,** Ovine toxoplasmosis: studies on parasitaemia, tissue infection, and congenital transmission in ewes infected by various routes, *Br. Vet. J.,* 120, 347, 1964.
76. **Beverley, J. K. A. and Watson, W. A.,** Prevention of experimental and of naturally occurring ovine abortion due to toxoplasmosis, *Vet. Rec.,* 88, 39, 1971.
77. **Blewett, D. A., Miller, J. K., and Buxton, D.,** Response of immune and susceptible ewes to infection with *Toxoplasma gondii, Vet. Rec.,* 111, 175, 1982.
78. **Buxton, D., Reid, H. W., and Pow, I.,** Immunosuppression in toxoplasmosis: studies in mice with a clostridial vaccine and louping-ill virus vaccine, *J. Comp. Pathol.,* 89, 375, 1979.
79. **Reid, H. W., Buxton, D., Gardiner, A. C., Pow, I., Finlayson, J., and MacLean, M. J.,** Immunosuppression in toxoplasmosis: studies in lambs and sheep infected with louping-ill virus, *J. Comp. Pathol.,* 92, 181, 1982.
80. **Frenkel, J. K. and Dubey, J. P.,** *Hammondia hammondi* gen. nov., sp. nov., from domestic cats, a new coccidian related to *Toxoplasma* and *Sarcocystis, Z. Parasitenkd.,* 46, 3, 1975.
81. **Levine, N. D.,** Taxonomy of *Toxoplasma, J. Protozool.,* 24, 36, 1977.
82. **Hendricks, L. D., Ernst, J. V., Courtney, C. H., and Speer, C. A.,** *Hammondia pardalis* sp. n. (Sarcocystidae) from the ocelot, *Felis pardalis,* and experimental infection of other felines, *J. Protozool.,* 26, 39, 1979.
83. **Johnson, A. M., Illana, S., Dubey, J. P., and Dame, J. B.,** *Toxoplasma gondii* and *Hammondia hammondi:* DNA comparison using cloned rRNA gene probes, *Exp. Parasitol.,* 63, 272, 1987.
84. **Christie, E. and Dubey, J. P.,** Cross-immunity between *Hammondia* and *Toxoplasma* infections in mice and hamsters, *Infect. Immun.,* 18, 412, 1977.
85. **Dubey, J. P.,** Prevention of abortion and neonatal death due to toxoplasmosis by vaccination of goats with the nonpathogenic coccidium *Hammondia hammondi, Am. J. Vet. Res.,* 42, 2155, 1981.
86. **Jones, T. C., Len, L., and Hirsch, J. G.,** Assessment in vitro of immunity against *Toxoplasma gondii, J. Exp. Med.,* 141, 466, 1975.
87. **Pettersen, E. K.,** Experimental toxoplasmosis in mice and rabbits: virulence and cyst formation of *Toxoplasma gondii, Acta Pathol. Microbiol. Scand. Sect. B,* 85, 95, 1977.
88. **Wilkins, M. F. and O'Connell, E.,** Effect on lambing percentage of vaccinating ewes with *Toxoplasma gondii* (letter), *N. Z. Vet. J.,* 31, 181, 1983.
89. **Seah, S. K. K. and Hucal, G.,** The use of irradiated vaccine in immunization against experimental murine toxoplasmosis, *Can. J. Microbiol.* 21, 1379, 1975.
90. **Chhabra, M. B., Mahajan, R. C., and Ganguly, N. K.,** Effects of ^{60}Co irradiation on virulent *Toxoplasma gondii* and its use in experimental immunization, *Int. J. Radiat. Biol.,* 35, 433, 1979.

91. **Mas Bakal, P. and Veld, N.,** Response of white mice to inoculation of irradiated organisms of the *Toxoplasma* strain RH, *Z. Parasitenkd.,* 59, 211, 1979.
92. **Tran Manh Sung, R.,** Vaccination in toxoplasmosis (radio-vaccine) (in French), *Lyon Med.,* 248, 101, 1982.
93. **Pfefferkorn, E. R. and Pfefferkorn, L. C.,** *Toxoplasma gondii:* isolation and preliminary characterization of temperature-sensitive mutants, *Exp. Parasitol.,* 39, 365, 1976.
94. **Waldeland, H., Pfefferkorn, E. R., and Frenkel, J. K.,** Temperature-sensitive mutants of *Toxoplasma gondii:* pathogenicity and persistence in mice, *J. Parasitol.,* 69, 171, 1983.
95. **Elwell, M. R. and Frenkel, J. K.,** Immunity to toxoplasmosis in hamsters, *Am. J. Vet. Res.,* 45, 2668, 1984.
96. **Sethi, K. K., Pelster, B., Suzuki, N., Piekarski, G., and Brandis, H.,** Immunity to *Toxoplasma gondii* induced in vitro in non-immune mouse macrophages with specifically immune lymphocytes, *J. Immunol.,* 115, 1151, 1975.
97. **Shirahata, T., Shimizu, K., and Suzuki, N.,** Effects of immune lymphocyte products and serum antibody on the multiplication of *Toxoplasma* in murine peritoneal macrophages, *Z. Parasitenkd.,* 49, 11, 1976.
98. **Anderson, S. E., Bautista, S. C., and Remington, J. S.,** Specific antibody-dependent killing of *Toxoplasma gondii* by normal macrophages, *Clin. Exp. Immunol.,* 26, 375, 1976.
99. **Anderson, S. E. and Remington, J. S.,** Effect of normal and activated human macrophages on *Toxoplasma gondii, J. Exp. Med.,* 139, 1154, 1974.
100. **Sibley, L. D., Weidner, E., and Krahenbuhl, J. L.,** Phagosome acidification blocked by intracellular *Toxoplasma gondii, Nature (London),* 315, 416, 1985.
101. **Nichols, B. A. and O'Connor, G. R.,** Penetration of mouse peritoneal macrophages by the protozoon *Toxoplasma gondii, Lab. Invest.,* 44, 324, 1981.
102. **Werk, R. and Bommer, W.,** *Toxoplasma gondii:* membrane properties of active energy-dependent invasion of host cells, *Tropenmed. Parasitol.,* 31, 417, 1980.
103. **Edelson, P. J.,** Intracellular parasites and phagocytic cells: cell biology and pathophysiology, *Rev. Infect. Dis.,* 4, 124, 1982.
104. **Strannegard, O. and Lycke, E.,** Effect of antithymocyte serum on experimental toxoplasmosis in mice, *Infect. Immun.,* 5, 769, 1972.
105. **Hafizi, A. and Modabber, F. Z.,** Effect of cyclophosphamide on *Toxoplasma gondii* infection: reversal of the effect by passive immunization, *Clin. Exp. Immunol.,* 33, 389, 1978.
106. **Pavia, C. S.,** Protection against experimental toxoplasmosis by adoptive immunotherapy, *J. Immunol.,* 137, 2985, 1986.
107. **Araujo, F. G. and Remington, J. S.,** Induction of tolerance to an intracellular protozoan (*Toxoplasma gondii*) by passively administered antibody, *J. Immunol.,* 113, 1424, 1974.
108. **Crowle, A. J.,** *Immunodiffusion,* Academic Press, New York, 1973, chap. 2.
109. **Hauser, W. E. and Remington, J. S.,** Effect of monoclonal antibodies on phagocytosis and killing of *Toxoplasma gondii* by normal macrophages, *Infect. Immun.,* 32, 637, 1981.
110. **Sethi, K. K., Endo, T., and Brandis, H.,** *Toxoplasma gondii* trophozoites precoated with specific monoclonal antibodies cannot survive within normal murine macrophages, *Immunol. Lett.,* 2, 343, 1981.
111. **Schwartzman, J. D.,** Inhibition of a penetration-enhancing factor of *Toxoplasma gondii* by monoclonal antibodies specific for rhoptries, *Infect. Immun.,* 51, 760, 1986.
112. **Norrby, R. and Lycke, E.,** Factors enhancing the host-cell penetration of *Toxoplasma gondii, J. Bacteriol.,* 93, 53, 1967.
113. **Johnson, A. M., McNamara, P. J., Neoh, S. H., McDonald, P. J., and Zola, H.,** Hybridomas secreting monoclonal antibody to *Toxoplasma gondii, Aust. J. Exp. Biol. Med. Sci.,* 59, 303, 1981.
114. **Svasti, J. and Milstein, C.,** The complete amino acid sequence of a mouse light chain, *Biochem. J.,* 128, 427, 1972.
115. **Kohler, G. and Milstein, C.,** Continuous cultures of fused cells secreting antibody of predefined specificity, *Nature (London),* 256, 495, 1975.
116. **Zola, H., Beckman, I. G. R., Bradley, J., Brooks, D. A., Kupa, A., McNamara, P. J., Smart, I. J., and Thomas, M. E.,** Human lymphocyte markers defined by antibodies derived from somatic cell hybrids. III. A marker defining a subpopulation of lymphocytes which cuts across the normal T-B-null classification, *Immunology,* 40, 143, 1980.
117. **Williams, D. M., Grumet, F. C., and Remington, J. S.,** Genetic control of murine resistance to *Toxoplasma gondii, Infect. Immun.,* 19, 416, 1978.
118. **Johnson, A. M.,** Strain-dependent, route of challenge-dependent, murine susceptibility to toxoplasmosis, *Z. Parasitenkd.,* 70, 303, 1984.
119. **McLeod, R., Estes, R. G., Mack, D. G., and Cohen, H.,** Immune response of mice to ingested *Toxoplasma gondii:* a model of *Toxoplasma* infection acquired by ingestion, *J. Infect. Dis.,* 149, 234, 1984.

120. **Jones, T. C. and Erb, P.**, H-2 complex-linked resistance in murine toxoplasmosis, *J. Infect. Dis.*, 151, 739, 1985.

121. **Sacks, D. L., Esser, K. M., and Sher, A.**, Immunization of mice against African trypanosomiasis using anti-idiotypic antibodies, *J. Exp. Med.*, 155, 1108, 1982.

122. **Van Brunt, J.**, More hope for anti-idiotypic vaccines, *Biotechnology*, 5, 421, 1987.

123. **Bona, C. and Moran, T.**, Idiotype vaccines, *Ann. Immunol. (Paris)*, 136C, 21, 1985.

124. **Monroe, J. G. and Greene, M. I.**, Anti-idiotypic antibodies and disease, *Immunol. Invest.*, 15, 263, 1986.

125. **de Saint-Basile, G., Lisowska-Grospierre, B., and Griscelli, C.**, Immune response to *Toxoplasma gondii* in man. I. Common idiotypic determinants of *Toxoplasma* - specific human antibodies, *Ann. Immunol. (Paris)*, 132C, 351, 1981.

126. **Hohmann, A., Comacchio, R., Bradley, J., and Johnson, A. M.**, Unpublished data, 1987.

127. **Rytel, M. W. and Jones, T. C.**, Induction of interferon in mice infected with *Toxoplasma gondii*, *Proc. Soc. Exp. Biol. Med.*, 123, 859, 1966.

128. **Freshman, M. M., Merigan, T. C., Remington, J. S., and Brownlee, I. E.**, *In vitro* and *in vivo* antiviral action of an interferon-like substance induced by *Toxoplasma gondii*, *Proc. Soc. Exp. Biol. Med.*, 123, 862, 1966.

129. **Remington, J. S. and Merigan, T. C.**, Interferon: protection of cells infected with an intracellular protozoan (*Toxoplasma gondii*), *Science*, 161, 804, 1968.

130. **Schmunis, G., Weissenbacher, M., Chowchuvech, E., Sawicki, E., Galin, M. A., and Baron, S.**, Growth of *Toxoplasma gondii* in various tissue cultures treated with In.Cn or interferon, *Proc. Soc. Exp. Biol. Med.*, 143, 1153, 1973.

131. **Ahronheim, G. A.**, *Toxoplasma gondii:* human interferon studies by plaque assay, *Proc. Soc. Exp. Biol. Med.*, 161, 522, 1979.

132. **McCabe, R. E., Luft, B. J., and Remington, J. S.**, Effect of murine interferon gamma on murine toxoplasmosis, *J. Infect. Dis.*, 150, 961, 1984.

133. **Pfefferkorn, E. R.**, Interferon-γ blocks the growth of *Toxoplasma gondii* in human fibroblasts by inducing the host cell to degrade tryptophan, *Proc. Natl. Acad. Sci. U.S.A.*, 81, 908, 1984.

134. **Pfefferkorn, E. R., Eckel, M., and Rebhun, S.**, Interferon-γ suppresses the growth of *Toxoplasma gondii* in human fibroblasts through starvation for trytophan, *Mol. Biochem. Parasitol.*, 20, 215, 1986.

135. **Murray, H. W., Spitalny, G. L., and Nathan, C. F.**, Activation of mouse peritoneal macrophages in vitro and in vivo by interferon-γ, *J. Immunol.*, 134, 1619, 1985.

136. **Hughes, H. P. A., Speer, C. A., Kyle, J. E., and Dubey, J. P.**, Activation of murine macrophages and a bovine monocyte cell line by bovine lymphokines to kill the intracellular pathogens *Eimeria bovis* and *Toxoplasma gondii*, *Infect. Immun.*, 55, 784, 1987.

137. **Handman, E. and Burgess, A. W.**, Stimulation by granulocyte-macrophage colony-stimulating factor of *Leishmania tropica* killing by macrophages, *J. Immunol.*, 122, 1134, 1979.

138. **Sharma, S. D., Hofflin, J. M., and Remington, J. S.**, In vivo recombinant interleukin 2 administration enhances survival against a lethal challenge with *Toxoplasma gondii*, *J. Immunol.*, 135, 4160, 1985.

139. **Chinchilla, M. and Frenkel, J. K.**, Mediation of immunity to intracellular infection (*Toxoplasma* and *Besnoitia*) within somatic cells, *Infect. Immun.*, 19, 999, 1978.

140. **Suzuki, N., Izumo, A., Sakurai, H., Saito, A., Miura, H., and Osaki, H.**, Toxplasmacidal activity of obioactin derived from hydrolyzed *Toxoplasma* immune bovine serum in heterologous cell cultures, *Zentralbl. Bakteriol. Parasitenkd. Infektionsk. Hyg. Abt. I Orig. Reihe A.*, 256, 356, 1984.

141. **Shirahata, T., Shimizu, K., Noda, S., and Suzuki, N.**, Studies on production of biologically active substance which inhibits the intracellular multiplication of *Toxoplasma* within mouse macrophages, *Z. Parasitenkd.*, 53, 31, 1977.

142. **Nagasawa, H., Igarashi, I., Matsumoto, T., Sakurai, H., Marbella, C., and Suzuki, N.**, Mouse spleen cell-derived *Toxoplasma* growth inhibitory factor: its separation from macrophage migration inhibitory factor, *Immunobiology*, 157, 307, 1980.

143. **Sethi, K. K. and Brandis, H.**, Characteristics of soluble T-cell derived factor(s) which can induce non-immune murine macrophages to exert anti-*Toxoplasma* activity, *Z. Immunitatsforsch.*, 154, 226, 1978.

144. **Baker, P. E., Hagemo, A., Knoblock, K., and Dubey, J. P.**, *Toxoplasma gondii:* microassay to differentiate *Toxoplasma* inhibiting factor and interleukin 2, *Exp. Parasitol.*, 55, 320, 1983.

145. **Dame, J. B., Williams, J. L., McCutchan, T. F., Weber, J. L., Wirtz, R. A., Hockmeyer, W. T., Maloy, W. L., Haynes, J. D., Schneider, I., Roberts, D., Sanders, G. S., Reddy, E. P., Diggs, C. L., and Miller, L. H.**, Structure of the gene encoding the immunodominant surface antigen on the sporozoite of the human malaria parasite *Plasmodium falciparum*, *Science*, 225, 593, 1984.

146. **Collins, W. E., Anders, R. F., Pappaioanou, M., Campbell, G. H., Brown, G. V., Kemp, D. J., Coppel, R. L., Skinner, J. C., Andrysiak, P. M., Favaloro, J. M., Corcoran, L. M., Broderson, J. R., Mitchell, G. F., and Campbell, C. C.**, Immunization of *Aotus* monkeys with recombinant proteins of an erythrocyte surface antigen of *Plasmodium falciparum*, *Nature (London)*, 323, 259, 1986.

147. **Herrington, D. A., Clyde, D. F., Losonsky, G., Cortesia, M., Murphy, J. R., Davis, J., Baqar, S., Felix, A. M., Heimer, E. P., Gillessen, D., Nardin, E., Nussenzweig, R. S., Nussenzweig, V., Hollingdale, M. R., and Levine, M. M.,** Safety and immunogenicity in man of a synthetic peptide malaria vaccine against *Plasmodium falciparum* sporozoites, *Nature (London)*, 328, 257, 1987.

148. **Prince, J. B., Remington, J. S., and Sharma, S.,** Characterization and cDNA cloning of a protective antigen from *Toxoplasma gondii*, presented at Molecular Strategies of Parasitic Invasion Conf., Park City, UT, January 26 to 31, 1986; as abstracted in *J. Cellular Biochem.*, 10A, 152, 1986.

149. **Johnson, A. M., Illana, S., Andersons, S., and McDonald, P. J.,** Expression of an immunodominant *Toxoplasma gondii* antigen in *Escherichia coli*, submitted.

150. **Munday, B. L., Obendorf, D. L., Dubey, J. P., and Johnson, A. M.,** Unpublished data, 1987.

151. **Marx, J. L.,** Assessing the risks of microbial release, *Science*, 237, 1413, 1987.

152. **McLeod, R. and Mack, D. G.,** Secretory IgA specific for *Toxoplasma gondii*, *J. Immunol.*, 136, 2640, 1986.

153. **Hone, D., Morona, R., Attridge, S., and Hackett, J.,** Construction of defined *galE* mutants of *Salmonella* for use as vaccines, *J. Infect. Dis.*, 156, 167, 1987.

154. **Hackett, J.,** Personal communication, 1987.

155. **Wahdan, M. H., Serie, C., Cerisier, Y., Sallam, S., and Germanier, R.,** A controlled field trial of live *Salmonella typhi* strain Ty 21a oral vaccine against typhoid: three-year results, *J. Infect. Dis.*, 145, 292, 1982.

156. **Brown, F., Schild, G. C., and Ada, G. L.,** Recombinant vaccinia viruses as vaccines, *Nature (London)*, 319, 549, 1986.

157. **Blancou, J., Kieny, M. P., Lathe, R., Lecocq, J. P., Pastoret, P. P., Soulebot, J. P., and Desmettre, P.,** Oral vaccination of the fox against rabies using a live recombinant vaccinia virus, *Nature (London)*, 322, 373, 1986.

158. **Smith, G. L., Godson, G. N., Nussenzweig, V., Nussenzweig, R. S., Barnwell, J., and Moss, B.,** *Plasmodium knowlesi* sporozoite antigen: expression by infectious recombinant vaccinia virus, *Science*, 224, 397, 1984.

Chapter 10

VACCINATION AGAINST *THEILERIA ANNULATA* THEILERIOSIS

E. Pipano

TABLE OF CONTENTS

I. INTRODUCTION

Theileria annulata (Dschunkowsky and Luhs 1904)[1] is a tick-transmitted protozoan parasite of cattle. The parasites first invade cells of the lymphatic system where they develop into schizonts. About 8 d later they appear in the peripheral blood as intraerythrocytic forms. The intraerythrocytic stages of *T. annulata* were first reported by Dschunkowsky and Luhs from cattle in Transcaucasia (U.S.S.R.).[1] These parasites resembled *Piroplasma parvum* (= *T. parva*) described previously by Koch[2] in East Africa, but since the Transcaucasian parasites were mostly rounded instead of rod-shaped, like *P. parvum,* they they were named *P. annulatum.*

Schizonts had been observed previously in cattle infected with either the East African or the Transcaucasian species, but the relationship of schizonts to the intraerythrocytic parasites was only later established by Gonder.[3] This discovery led to the reclassification of these organisms from the genus *Piroplasma* to the newly created genus *Theileria.*[4]

In the following decades, diseases associated with *T. annulata*- like parasites were reported from the Mediterranean area and the Near East. Occasionally they were incorrectly diagnosed as East Coast Fever (caused by *T. parva*) or as benign bovine theileriosis (caused by *T. mutans*).[5-7]

During nearly 2 decades of intensive work, a group of French investigators, headed by Edmond Sergent,[8] studied the biological and immunological properties of a *Theileria* species in North Africa. They concluded that the organisms observed in North African cattle were distinct from other *Theileria* spp. and they named it *T. dispar.* The results of their studies have remained of capital importance for the control of theileriosis to this very day. However, they later recognized the similarity between the Algerian and Transcaucasian species and, in accordance with the International Code of Zoological Nomenclature, the name *T. annulata* was assigned to the North African form.[9]

In the latest revised classification of the protozoan parasites of domestic animals,[10] this parasite was classified as follows:

Subkingdom	Protozoa
Phylum	Apicomplexa
Class	Sporozoea
Order	Piroplasmida
Family	Theileridae
Genus	*Theileria*
Species	*Theileria annulata*

Infection with *T. annulata* has been called tropical piroplasmosis (theileriosis),[1] "Mediterranean fever'',[5] and "Egyptian fever''.[6] The disease is widespread in cattle from the Mediterranean basin, through the Near and Middle East, into India and West China.[11]

Vector eradication has not been used as an organized measure for control of *T. annulata,* and specific chemotherapy has not been available until only very recently when some potentially useful drug had been synthesized.[12] On the other hand, vaccination against this infection has been practiced on a small scale in isolated geographical regions for several decades.[13-15] Control by vaccination has tended to become more widespread in recent years thanks to the development of a safe, cell-culture vaccine.[16]

The present chapter, therefore, deals mainly with preparation and application of anti-*T. annulata* vaccines. Characteristics of the parasite and the response of cattle to the infection are also discussed here to the extent that they are relevant to the subject of vaccination.

II. LIFE CYCLE

The various development stages of *T. annulata* appear to differ in their pathogenic and immunogenic attributes, and thus in their roles in causing disease, inducing an immune response, and in elaboration of vaccination procedures.[17]

T. annulata is transmitted by ticks of the genus *Hyalomma.* In the gut of immature ticks that have ingested infected blood, spindle-shaped and spherical bodies are formed. Microgamont-like stages develop from the spindle-shape organisms, while the spherical forms are still present. The zygote-like stage which is later detected in the gut wall[18] transforms into a motile kinete[19] that penetrates the salivary glands usually after the moult of the ticks. The kinete transforms into fission bodies. No further development occurs until the tick is "activated" by feeding or by environmental changes like relatively high temperatures or the proximity of cattle.[20,21] The division of the fission bodies yields sporozoites.[22] Of all the developmental stages of *T. annulata* in the tick, only the sporozoites are infective for cattle.

The primary stage following the inoculation of *T. annulata* sporozoites into cattle is the intralymphocytic schizont. Electron microscopic studies have shown that penetration of sporozoites into lymphoid cells is completed within 10 min.[23] Inside the host cell, the sporozoite becomes enlarged and loses the two inner membranes of its pellicle. Simultaneously, nuclear division starts, resulting in the formation of schizonts. During the early stages, the schizonts possess relatively large and sometimes irregular-shaped chromatin granules (nuclei) up to 1.9 μm in size. This is the macroschizont (= macromeront) stage. This is followed by schizonts with relatively smaller nuclei, up to 0.8 μm, the microschizonts (= micromeronts).[24] The schizonts produce merozoites by a budding-like process of the nuclei.[25] Beginning from the 8th day after infection, the merozoites are released from the lymphocytes and invade red blood cells. *T. annulata* isolates that have been passaged for prolonged periods may lose their capacity to produce erythrocytic stages.[26,27]

In addition to the sporozoites from ticks, the cattle stages (schizonts and erythrocytic merozoites) can also produce infection in bovines when inoculated intravenously, intramuscularly or subcutaneously.

III. PATHOLOGY

Cattle (*Bos taurus* and *Bos indicus*) of all breeds appear to be susceptible to *T. annulata* infection. The parasite can also complete its cycle in the water buffalo (*Bubalus bubalis*), but usually without causing clinical disease or death.

The most frequent clinical symptoms of theileriosis — fever, accelerated pulse, anorexia, anemia, icterus, and bilirubinuria — are not sufficient in themselves to allow a firm diagnosis of *T. annulata* infection. Swollen peripheral lymph nodes accompanied by the above clinical symptoms strengthens the suspicion of theileriosis. Laboratory confirmation of acute theileriosis is obtained by detecting schizonts in smears of lymph nodes or liver material obtained by needle biopsy.[8]

Erythrocytic merozoites can be detected in blood films 2 to 3 d after the appearance of schizonts and may persist for years after recovery. In heavy infections a few schizonts may be detected in peripheral lymphocytes on blood films prepared during the acute stage of the disease.

The main macroscopical lesions are petechia on subcutaneous tissues and seroses, hyperplasia of lymph nodes, splenomegaly, yellowish enlarged liver with distended bladder, petechia and ecchymoses of the myocard, gray-white foci on the cortex of the kidney, and typical ulcers of abomasum.

Mortality rates caused by *T. annulata* vary with the virulence of the parasite strain and with the sensitivity of the cattle. Rates from 35 to 40[8,14] up to 90%[28] have been reported.

The basic laboratory technique for detecting *T. annulata* in cattle was described by Sergent et al.[8] (1945). This included the identification of erythrocytic merozoites in thin blood films and of schizonts in material from lymph node or liver needle biopsy. Later, serological techniques for assessing antibody levels were developed. These are discussed in Section V of the present chapter.

Thin blood films are prepared in the usual manner, with capillary blood obtained from the tip of the ear or the tip of the tail by means of small cut or needle prick.

For lymph node biopsies, a 16- or 18-gauge needle is attached to a 2-ml syringe. Prior to inserting the needle, the skin around the area of the biopsy is clipped, shaved, and disinfected. Mild aspiration with the syringe is exercised as the needle makes puncture movements. For liver tissue biopsy, an aspirating hypodermic needle (1.4 × 110 mm) is inserted through the cleansed skin on the next to the last intercostal space, about 10 to 15 cm below an imaginary horizontal line traced on the level of the tuber coxae. Without withdrawing the needle, eight to ten puncture movements are made about 4 to 6 cm deep in the direction of the sternum. The tissues obtained from both types of biopsy are smeared on microscope slides and dried. All preparations — blood films and biopsy smears — are fixed in methanol and stained with Giemsa or May-Grunwald.

Material obtained from biopsies can be used also for initiating cultures (Section VII). In that case, a 0.5-cm incision of the skin is made before inserting the needle in order to decrease the danger of bacterial contamination from surface skin tissue.

IV. EPIZOOTOLOGICAL CONSIDERATIONS

T. annulata theileriosis has a seasonal distribution that is modulated by the ecology of the vectors.[29,30] In North Africa the majority of outbreaks were reported from June to September.[31] Recent observations in India have associated clinical outbreaks with the rainy summer months (April-October).[32] Sporadic cases have been encountered in various geographical regions all year round.

Transmission of *T. annulata* has been discussed by Robinson[33] in a recently published review. About 15 species of the genus *Hyalomma* have been incriminated as vectors. Species of major importance appear to be *H. detritum, H. anatolicum, H. excavatum,* and *H. dromerdari.* The taxonomy of the *Hyalomma* group in various geographical areas is not entirely crystallized, and there is no certainty that all cited species are authentic.

It seems that most, if not all, *Hyalomma* species that infest cattle are able to transmit *T. annulata.* The role of any particular species appears to be limited by host preference under natural conditions rather than by the intrinsic ability of the tick to serve as a host of the parasite.

T. annulata theileriosis may appear as a field- or barn-transmitted infection.[8] In barns with clay or stone walls, nymphs of *H. detritum* moult in the cracks and crevices to produce adults that infest introduced cattle. In areas where the vector develops in the field, the infection affects grazing cattle primarily. In enzootic areas, cattle on zero grazing may be kept free of theileriosis. However, even in such cases, infected ticks may be introduced accidently by hay from enzootic fields or by stray grazing cattle. Adult ticks issued from infected preimaginal stages need a blood meal of 24 to 72 h for the development of sporozoites.[24,34] However, unfed ticks have also been found to contain infective sporozoites.[21,35] This means that in some instances cattle may become infected during the first several hours after the attachment of the infective ticks. This makes it difficult to prevent tick transmission by periodic dipping or spraying with acaricides at intervals of several days.

Hyalomma ticks are capable of producing large amounts of *T. annulata* sporozoites. About 50,000 sporozoites per host cell in Type III acini in the tick salivary glands have been estimated.[25] When a frozen suspension of a single macerated infected tick was divided into 100 aliquots, calves could still be infected by the inoculation of a single aliquot.[36]

Examination of salivary glands showed that a high percentage of ticks harvested in barns or in the field were infected with *T. annulata*.[37,38] Consequently, even a few ticks may provoke severe outbreaks. The epizootology of *T. annulata* theileriosis indicates that short of a total eradication of the vector, the only way to prevent losses caused by theileriosis is through vaccination of cattle with effective vaccine.

V. IMMUNOLOGY

A. Immunological Response to *T. annulata* Infection

Infection with *T. annulata* induces a specific immune reponse that may result in clincial recovery followed by resistance to reinfection. It has been thought that the clinical cure was not the result of complete elimination of the parasites;[37] rather it was considered a state of "coinfection relative immunity", or premunition, related to and depending upon the continued presence of parasites.[40] This conclusion was based on the presence of erythrocytic merozoites after recovery and their recrudescence in relatively large numbers following splenectomy. However, the occurrence of erythrocytic parasites is not necessarily evidence for the presence of schizonts, since erythrocytic parasites can multiply independently of the latter.[25,41]

Although schizonts have been reported from chronically infected cattle of uncertain origin,[8,42] and have been isolated in culture from carrier animals,[43] relapse disease with schizonts has never been reported in cattle maintained under experimental conditions.

Infection with *T. annulata* is followed by a marked response of circulating antibody detectable by complement fixation,[44-48] hemagglutination,[49] immunofluorescence,[45,50-52] and recently by the enzyme-linked immunosorbent assay.[53] Antibody titers reach maximal levels shortly after recovery. They decline with time, but positive reactions are still detectable up to 1 to 2 years after initial infection.[47,48,50] Both IgG and IgM classes of antibodies persisted for more than 4 months in cattle inoculated with living or dead *T. annulata* schizonts. No diagnostic application could be assigned to the differentiation of the two classes of antibody.[54]

The passive transfer of circulating antibody to susceptible cattle did not prevent nor mitigate the clinical result of *T. annulata* infection.[55,56] Similar results were reported with *T. parva*.[57] However, field experience shows that cattle born and raised in *T. annulata*-enzootic areas do not suffer clinical theileriosis. It appears that a situation of enzootic stability similar to that observed in babesiosis[58] may occur. This phenomenon may involve transfer of passive immunity from the dam to the offspring through the colostrum.

Recent *in vitro* studies have shown that invasion of peripheral blood leukocytes by *T. annulata* sporozoites is suppressed by hyperimmune sera.[59,60] Similar effects have been obtained with *T. parva* sporozoites by polyclonal and monoclonal antibodies.[61-63] According to authors, resistance to tick-induced infection in immune cattle is mediated initially by serum factors acting both on sporozoites and on the initial stages of the intralymphocytic schizonts. However, the short exposure of the sporozoites to circulating antibody before they penetrate cells,[64] and then the synchronous division of the intracellular schizonts and the infected cells,[65] leaving little opportunity for the antibody to neutralize the parasites.

Evidence has been accumulated that cell-mediated immune mechanisms play an important part in immunity to *T. annulata.*[66] The increased percentage of T lymphocytes in peripheral blood and lymphoid organs in cattle recovered from *T. annulata* suggests a cellular response to the infection.[67] Migration of peripheral leukocytes collected from carrier animals was inhibited in the presence of theilerial antigen.[68] During the acute stage of the disease, appearance of schizonts is accompanied by an enhanced capacity of peripheral blood lymphocytes to form rosettes. Dying animals show a reduced number of rosettes compared to recovering animals.[69-71]

The main pathogenic effects of *T. annulata* infection are observed during the schizont-multiplication phase. Thus, once sporozoites have penetrated the lymphocytes, immunity would need to be mainly directed against the schizont stage. In fact, past and present vaccines against *T. annulata* that contained schizonts induced a fair protection against naturally transmitted theileriosis.[17]

Bovine lymphoblastoid cells transformed by *T. annulata* stimulate proliferation of non-infected autologous lymphocytes when cocultivated *in vitro.*[72] Recovery of calves from *T. annulata* infection was accompanied by the disappearance of macroschizonts from lymph nodes and the appearance of cytotoxic cells in blood and lymph nodes that were capable of lysing schizont-infected cells *in vitro.* The reinfection of recovered cattle was associated with reappearance of these cells even when schizonts could not be detected. The brief response of cytotoxic cells after the reinfection indicated that populations of such cells were established in the immunological memory of the immune cattle. The cytotoxic cell response was found to be genetically restricted to bovine major histocompatibility antigens.[73]

Using monoclonal antibodies, a specific antigen has recently been detected on the surface of infected lyphocytes. This might be the target antigen recognized by the specific cytotoxic cells.[74]

B. Virulence

Tick-induced infections appear to be more virulent than infection caused by inoculation of blood from animals with acute theileriosis. According to Sergent and collaborators,[8] mortality of cattle under experimental conditions was 28.4% when infected by ticks vs. 15% when infected by inoculation of blood. Field isolates vary in virulence. Mortality ranging from 3.2 to 49% was observed in 939 calves inoculated with blood from 5 different Algerian strains.[8] Similar differences in virulence were reported from field isolates of *T. annulata* in Iran,[75] Israel,[76] and the U.S.S.R.[77,78]

Attempts to alter the virulence by successive passages in calves were not successful in Algerian[8] and Israeli strains.[17] An Israeli strain, "Tova", which had been passaged 245 times in cattle, caused a consistent mortality rate of 40%. By the same token, field isolates with low initial virulence did not become more virulent when passaged through cattle.[8] On the other hand, a reduction of virulence during serial passages of strains isolated in Iran was reported,[75] but not to the extent that immunization with such strains became possible. (Attenuation of *T. annulata* schizonts by cultivation *in vitro* and irradiation is discussed in Sections VII and IX.)

C. Immunological Strain Differences

Antigenic differences among isolates became evident when Sergent and collaborators,[79] together with Adler and Ellenbogen,[80,81] investigated the relationship between Algerian and

Israeli *T. annulata*. Out of 25 calves that had recovered from an infection with an Algerian strain and were challenged with an Israeli strain, 5 died and 20 developed clinical theileriosis. Conversely, 12 calves that had recovered from an Israeli strain developed theileriosis when challenged with an Algerian strain of low virulence, but none died; by contrast, calves challenged with homologous Algerian and Israeli strains showed feeble or no clinical reactions to the challenges.

Adler and Ellenbogen[80] obtained similar results when four calves recovered from an Israeli strain died following challenge with a virulent Algerian strain. In another trial, 96 field calves vaccinated with a mild Algerian strain showed a moderate response after inoculation with an Israeli strain. However, when Israeli field isolates were cross-tested, not all of them conferred a reciprocal cross-protection.[76]

In Iran, only a partial cross-immunity was observed when 12 isolates were compared.[75] In a more recent report from India,[82] calves that recovered from tick-induced infection showed complete resistance to homologous challenge, while a lesser degree of protection was observed against heterologous challenge.

Immunization with culture-attenuated schizonts and challenge with virulent schizonts seems to be a highly sensitive method for detecting antigenic differences between *T. annulata* isolates.[83]

In summary, among *T. annulata* field strains, it is relatively rare to encounter a total lack of cross-protection leading to death between heterologous strains. Instead, most isolates confer a high degree of reciprocal immunity.[8,52,76,84] According to Irvin and Boarer,[85] in nature, a constant mixing and crossing of different genetic strains occurs. Most isolates are therefore likely to consist of a mixture of strains and will, as a result, confer a relatively wide range of protection. On the other hand, when theileriosis breaks out in vaccinated cattle,[86] the immunogenic relationship between the vaccine strain and the strain isolated during the outbreak should be assessed.

T. annulata schizonts and erythrocytic merozoites could be clearly differentiated from the corresponding stages of *T. parva* by isoenzyme electrophoresis.[87,88] The same method revealed differences among schizonts from Turkish, Iranian, and Indian isolates, as well as six isolates from Sudan.[88,89] Few differences could be detected when sporozoites from tick salivary glands were compared.[90] However, zymograms do not at present provide a tool for selecting isolates with characteristics appropriate for vaccine production.

D. Stage-Specific Immune Response

The three developmental stages of *T. annulata* that are infective for cattle are: sporozoites derived from ticks, intralymphocytic schizonts, and intraerythrocytic merozoites. At least 5 to 6 d elapse between infection of cattle with sporozoites and the appearance of schizonts in the lymph nodes draining the area of infection.

In an elegant piece of work, Brown and collaborators[91,92] demonstrated that sporozoites of bovine *Theileria* develop into schizonts when incubated *in vitro* with bovine lymphocytes. This line was followed by other investigators[93,94] who observed uninucleated schizonts in cultures about 48 h after introduction of sporozoites, and multinucleated schizonts 24 to 72 h later. It appears, therefore, that there is no additional developmental stage between the sporozoites and the schizonts that might play a role in the immunological response of cattle against *T. annulata* infection.

Infection of cattle with sporozoites results in the appearance of schizonts and erythrocytic stages. Thus, such infection induces immunity in the recovered animals against all stages of *T. annulata* that are infective for cattle.

Recovery from schizont-induced infection engenders a stronger resistance to schizonts than to sporozoites.[8,17] Nevertheless, schizont-induced immunity is usually strong enough to prevent death or severe clinical symptoms following sporozoite-induced infection.[16] This fact provides the basis for the practical approach to vaccination against *T. annulata* using a schizont vaccine.

Immunization with killed schizonts of *T. annulata* plus Freund's adjuvant protected a high percentage of cattle against death and, to a lesser extent, against clinical symptoms caused by challenge with virulent schizonts. All cattle in which schizonts became established as a result of the challenge survived a subsequent tick-induced infection. Cattle that were completely protected by the killed vaccine against the schizont challenge died from the tick infection.[95] It seems, then, that the stage-specific antigens are most strongly expressed when immunization is performed with nonviable schizonts.

The fact that erythrocytic merozoites persist for years, and probably for life, in cattle recovered from theileriosis indicates that the immune response against the schizonts has little or no effect on the erythrocytic merozoites.

Reciprocal experiments showed that erythrocytic merozoites do not confer protection against the pathogenic schizonts. Calves that were inoculated with erythrocytic merozoites carried these forms for 41 to 165 d. After this period, numerous schizonts appeared in the liver of the calves and caused 60% mortality.[96] Although the period during which the calves showed the erythrocytic stages was long enough for an immune response to have been elaborated against the schizonts, the eventual clinical disease and death of the animals following the schizont multiplication indicates a lack of cross-immunity between these two stages in the *T. annulata* life cycle.

The immunology of the erythrocytic merozoites has not been studied intensively, perhaps because this stage has only a weak pathogenic effect, especially in chronic carriers. Presumably, erythrocytic merozoites possess a mechanism for survival in the immunized host similar to that described for other intraerythrocytic parasites (Babesia)[97] of cattle

VI. HISTORY OF VACCINATION AGAINST *T. ANNULATA*

Vaccination against *T. annulata* infection was initiated in North Africa by Sergent and collaborators.[8,24,98] The French investigators isolated a field strain of natural low virulence, "Kouba", and maintained it by needle passage through susceptible cattle. The passages (one every 17 to 23 d) resulted in loss of the erythrocytic stages without modifying the virulence. The strain was used for vaccination from the 52nd to the 220th passage over a period of about 10 years. Vaccination consisted of inoculating 5 to 10 ml of blood from a donor animal in the acute stage of the infection into the animals to be protected. A mortality of 3.2% was caused by this strain among the calves used for passages in the laboratory.

This method was adopted by the Israeli investigators, Adler and Ellenbogen,[14] with the aim of protecting Friesian-type cattle that were imported from Europe in order to improve local breeds. Initially, Algerian strains of low virulence (including a strain "Brunette" that caused a postvaccination mortality in the field of 1.8%) were used for vaccination.[15] Since the protection to field challenge conferred by the Algerian strains alone was not strong enough, a virulent Israeli strain, "Tova", was inoculated 1 to 2 months after the primary inoculation with the Algerian strains in order to reinforce the immunity.[99]

Vaccination proved to be highly beneficial. According to the Chief of the Clinical Veterinary Service, during this period in one settlement with about 3000 cattle, a mortality of 13% caused by theileriosis in nonvaccinated cattle declined to 0.03% following vaccination.[100] To the latter figure must be added the mortality, 1 to 2%, caused by the vaccination itself. However, the benefits were still dramatic.

The strain "Brunette" was maintained in the laboratory of the Israeli Veterinary Institute at Bet Daga, through 420 passages, until 1963 when 2 calves inoculated with the strain did not show schizonts in lymph nodes and liver and their blood did not infect susceptible calves. This event triggered research to develop an alternative method for vaccination.

The successful cultivation *in vitro* of *T. annulata* schizonts by Tsur-Tchermoretz in 1945[101] was the achievement that served as a basis for developing an improved schizonts vaccine against *T. annulata*. However, the plasma-clot cultures of liver and spleen explants used by Tsur[102,103] were not amenable to the large-scale production of cultured schizonts for possible use as vaccine. It was the eminent virologist Abraham Kimron who obtained for the first time (in an unpublished work) a monolayer culture from trypsinized kidney tissues removed from a calf dying of theileriosis. Subsequently, monolayer cultures were also obtained from other internal organs[104] and from buffy coat cells[105] from *T. annulata*-infected calves.

Initial studies showed that supernatant fluid from monolayer cultures contained free schizonts and also lymphoid cells infected with schizonts. This material was invariably infective for cattle, but survival of the schizonts was short, and in some instances they did not provoke infection if stored for 24 to 48 h at 4°C.

Further observations showed that prolonged cultivation led to loss of virulence of the schizonts[106] and that attentuated schizonts conferred protection against homologous virulent parasites.[107] By the end of the 1960s, culture-derived vaccine was being applied in the field against theilerial infection in Israel.

During the last 2 decades, investigators in most of the *T. annulata* enzootic areas have carried out cultivation and immunization trials, and a great deal of experience has been accumulated in this field.[108-115]

VII. CELL CULTURE SCHIZONT VACCINE

The preparation and use of culture-derived schizont vaccine involves four main stages: isolation of *T. annulata* parasites, initiation of cultures and attentuation of virulence by prolonged cultivation, preparation and testing of vaccine, and storage, transport, and application of the vaccine.

A. Isolation of *T. annulata* Parasites
Isolation of *T. annulata* from the field can be done from the following sources.

1. From Cattle Suffering From Theileriosis
Blood from acutely ill cattle is drawn into a vessel containing anticoagulant and is transported to the laboratory in an ice-refrigerated, insulated container. Blood films and smears from lymph node or liver biopsy material should be made from the donor animal in order to confirm the clinical diagnosis. At the laboratory, 20 to 30 ml of blood are inoculated into a susceptible calf. Blood drawn during the acute stage of *T. annulata* theileriosis, when schizonts are present in lymph nodes or liver, will invariably produce infection even in small amounts.

2. From *T. annulata*-Infected Ticks
Engorged preimaginal stages, usually nymphs of *Hyalomma* spp. ticks, can be collected on cattle or from cracks and crevices in the clay or stone walls of barns in *T. annulata* enzootic areas. Field studies have shown that a considerable portion of such ticks carry *T. annulata* infections.[38] After moulting, the ticks are allowed to feed on susceptible cattle, or they are macerated and the suspension is inoculated into candidate cattle. In the feeding method, 10 to 20 female and male ticks are confined in a linen ear bag glued at the base of the ear (See Section VIII). The prepatent period is usually from 1 to 2 weeks after attachment of the ticks. Partially engorged *Hyalomma* adults collected from cattle in theileriosis enzootic areas can also cause infection when allowed to complete a blood meal on susceptible cattle.

Preparation of suspensions of infected macerated ticks is described below in Section VIII.

The same suspensions can be used either for infecting cattle or for inoculating cultures of bovine peripheral leukocytes.

The specific technique of isolation to be used depends upon the facilities available and the working conditions in the laboratory and the field.

B. Cultivation and Attenuation of Virulence

1. Initiation of Cultures

The techniques for establishing cultures of *T. annulata* and *T. parva* schizonts have been thoroughly discussed by Brown in three comprehensive reviews.[116-118]

Cultures of *T. annulata* can be initiated from three sources: from peripheral blood leukocytes (PBL) of *T. annulata*-infected cattle, from internal organs of infected cattle, or by infection *in vitro* of PBL with sporozoites harvested from infected ticks.

a. From PBL of T. annulata-Infected Cattle

The jugular vein area is first clipped, shaved, and disinfected. Blood is drawn in a sterile manner using heparin as anticoagulant. Leukocytes can be separated out by several techniques. Initially, separation was accomplished by centrifuging the blood and harvesting the buffy coat with a pipette. In another method,[105] the plasma was removed after centrifugation and a thin layer of chick embryo extract was placed in the buffy coat. After incubation at 37°C, the clot containing the leukocytes was transferred to an Erlenmeyer flask and dispersed by trypsin. More recently, buffy coat cells were cleansed of red blood cells by lysing the suspension with 0.17 *M* ammonium chloride solution.

A suspension of lymphocytes free of red blood cells can be obtained by a gradient (Ficoll-Paque) technique.[118] Infected heparinized blood is centrifuged in a 50-ml tube and the buffy coat is transferred carefully onto Ficoll-Paque in a 10-ml tube. Smaller amounts of blood (2 to 4 ml) can be directly layered on the Ficoll-Paque[119] without centrifugation. The blood is first diluted with an equal volume of balanced salt solution prepared according to instructions included in the set; 3 ml of Ficoll-Paque solution are introduced into each of four siliconized tubes, and to each tube 4 ml of diluted blood is carefully layered over the solution. After centrifugation, the first layer which contains the plasma is removed and discarded, and the next layer containing the lymphocytes is harvested. The lymphocytes are rinsed with balanced salt solution, diluted with culture medium to yield a concentration of 5×10^5 cells per milliliter and incubated at 37°C.

PBL can also be separated by a simple technique using a plastic syringe. This technique is especially suitable for field veterinary laboratories that have basic equipment for cell culture work.[119] Blood is drawn with an 18-g × 1$^1/_2$-in. disposable needle into a 10-ml conventional disposable plastic syringe containing 0.2 ml of 0.1% heparin in PBS. The needle sheath is replaced, and the plastic rod of the syringe piston is sheared with a scissors at the level of the barrel. The syringe with the sheathed needle still in place is centrifuged, needle-side up, for 20 min at 2000 rpm at 4°C. The syringe now shows a thin layer of buffy coat between an upper plasma layer and a lower red cell layer. With the syringe retained in an upright position, needle-side up, the sheath of the needle is removed and the needle carefully bent to about 90° with a sterile forceps. The piston is now gently driven upward with the help of the forceps in order to discharge the plasma layer, which is discarded. The next few drops containing the buffy coat are delivered into a separate centrifuge tube containing 5 ml of PBS. The syringe with the red cells is discarded. The buffy coat cell suspension is centrifuged, and the sedimented cells resuspended in 5 ml of culture medium. The white cells are counted and then diluted with cell culture medium to obtain a concentration of 5×10^5 cells per milliliter; 5 ml of the cell suspension is introduced into 750-ml plastic culture flasks and incubated at 37°C.

Alternately, the buffy coat can be delivered from the syringe directly into a culture flask containing 5 ml cell culture medium. At this stage, the buffy coat cells are contaminated with red blood cells. However, the latter do not interfere with development and multiplication of the schizont-infected cells. The red blood cells are discarded with the first replacement of the culture medium, while the schizont-infected cells remain attached on the bottom of the culture vessel.

b. From Internal Organs of Cattle with Theileriosis

Cultures of *T. annulata* can be initiated from any organ containing schizont-infected lymphocytes. During the early period of cultivation of *T. annulata,* cultures were initiated from internal organs, mainly liver, spleen, and lymph nodes.[104] Tissues from the above organs can be obtained by dissecting cattle slaughtered *in extremis* and removing pieces of organs under aseptic conditions. The organ pieces are minced with scissors, rinsed with PBS, and digested with 0.25% trypsin solution. After rinsing with PBS, the dispersed cells are resuspended, counted, and incubated with cell culture medium.

Biopsy material from superficial lymph nodes and from the liver can also be used to initiate cultures. The biopsy technique is described in Section III. The material, collected by sterile technique, is ejected into a tube containing 5 ml of PBS. The tube is centrifuged for 10 min at 1000 rpm, and the supernatant is discarded. The lymph node material is diluted in culture medium and incubated at 37°C. The liver material is initially dispersed by trypsin. For this purpose, 5 ml of 0.25% solution of trypsin in PBS is added to the sediment and the tube is shaken for about 10 min at room temperature. The suspension is allowed to settle, and the supernatant is transferred to a tube containing 1 ml calf serum. A second treatment with trypsin is performed and the entire suspension added to the first harvest. After rinsing, the obtained cells are incubated with cell culture medium.

For primary cultures, concentrations of 5×10^5 to 10^6 cells per milliliter yield satisfactory results. It is recommended that the culture medium be replaced after 24 h in order to discard cell debris and dead cells. Except for Hulliger and collaborators, no feeder layer cultures have been used in initiating *T. annulata* cultures. However, feeder cells may improve the chances of obtaining successful cultures. During the first days of cultivation, some of the schizont-infected cells become attached to the bottom of the vessel (Figure 1A), while others remain in suspension. If the cells in suspension are harvested and further passaged by suspension culture techniques, the number of attached cells decreases until only a suspension culture remains. On the other hand, when the attached cells are dispersed with 0.025% EDTA (Versene®) and transferred to other vessels, a monolayer line of schizont-infected cells results after several passages. Some investigators have preferred to grow *T. annulata* in suspension cultures,[109,110,121] while others have preferred monolayers.[104,105]

Intracellular schizonts can be detected in Giemsa-stained preparations of cells harvested from the supernatant fluid of primary and low-passage cultures or of cells grown on a sterile cover slip introduced into a culture flask.

c. By In Vitro Infection of PBL with Sporozoites

This technique was developed by Brown and collaborators[91,92,116,118] for *T. parva* and subsequently for *T. annulata.* To start with, a culture of normal (i.e., noninfected) bovine lymphoid cells is initiated. Buffy coat or lymphocytes obtained by the Ficoll-Paque gradient method or lymphoid cells from internal organs are seeded in a culture vessel. A feeder layer of cells non-susceptible to *T. annulata* may improve establishment of the bovine lymphoid cells and consequently the likelihood of success with sporozoites. The filtrate of a suspension of macerated infected ticks (Section VIII) is added to the bovine cells. The culture is incubated in the presence of 5% CO_2. Growth medium is added after 24 and 48 h. To replace the medium, the supernatant is carefully removed and an equal volume of fresh medium is added. Lymphoid cells that become infected undergo transformation and begin to multiply,

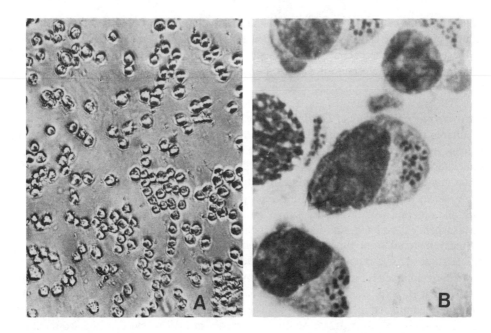

FIGURE 1. (A) Primary culture of *T. annulata*-infected peripheral bovine leukocytes. (Magnification ×
300.) (B) Schizont-infected cells in culture. (Giemsa stain; magnification × 2500.)

a process that is evident from the increased metabolism in the culture. At this stage, passages
to new culture vessels are required.

2. Growth Requirements of T. annulata Cultures

Although the optimal conditions for *in vitro* growth and multiplication of *T. annulata*-
infected cells have not yet been thoroughly studied, it appears that a wide variety of culture
media is suitable for this purpose.[116,118]

In our hands, Eagle's Minimum Essential Medium (MEM) is suitable for growing schizont-
infected cells in monolayer as well as in suspension cultures. An enhanced multiplication
of cells was reported by adding lactalbumin hydrolyzate and yeast extract to the MEM[123]
medium. With regard to media formulations, it is important to consider that once a strain
is attenuated, the cultivation medium is one of the most costly components in vaccine
production. It is thus desirable that cheap formulations for growing *Theileria*-infected cells
be sought. Bovine serum at a concentration of about 20% seems to be optimal.[188] As in
other culture systems, serum may be "toxic" for the cells and delay their multiplication.
It is recommended, therefore, that batches of serum be checked for toxicity in at least two
consecutive passages of cells.

Culture media are supplemented with standard antibiotics (penicillin, streptomycin) and
fungistats (hystatin). *T. annulata* is not sensitive in culture to common antibiotics so that,
if required, other antibiotics can also be added, including tetracyclines.

Plastic culture vessels are preferable for initiation of cultures. However, for long-term,
large-scale propagation for vaccine production, borosilicate glass culture vessels may be
more economical.

3. Cryopreservation of T. annulata Schizonts

Schizonts of *T. annulata* derived from infected blood, internal organs, or cell culture have
been successfully preserved at ultralow temperatures[124-127] with glycerol (10%) or dimethyl-
sulfoxide (DMSO, 7%) as cryoprotectants. Freezing is useful for "banking" theilerial iso-

lates (strains), for storing cultured schizonts at various stages of the *in vitro* passage process (see below), and for storing vaccine needed for vaccination.

The following procedure has been described for cryopreserving schizont-infected cells:[11]

- Mark cryotubes and aluminum canes (sticks) with code identification.
- Measure the volume of the cell suspension. Count the cells and calculate the total number of cells in the suspension.
- Spin the suspension for 10 min at 1500 rpm in a conventional refrigerated centrifuge.
- Discard the supernatant. Resuspend the cells in culture medium containing 7% of DMSO to obtain a final concentration of 5×10^6 cells per milliliter.
- Dispense 1.5 ml of cell suspension into 2-ml cryotubes.
- Attach cryotubes to canes and introduce into a low-temperature cabinet at $-70°C$ or in a dry ice-box for at least 4 h.
- Quickly transfer the canes to canisters and immerse into liquid nitrogen vapor for at least 2 h. (If a $-70°C$ storage cabinet is not available, immerse the cryotube with cell suspension directly into liquid nitrogen vapors.)
- Slowly immerse canister into liquid nitrogen for storage.

Recovery and Subcultivation of Cryopreserved Cells

- Bring the canister into the neck of the liquid nitrogen container and secure it in place by means of an artery forceps clipped on the stem of the canister.
- Remove cryotube, immerse in 40°C water bath, and shake until thawed.
- Dry cryotube with toweling, swab with absolute alcohol, open the cap and transfer contents with a pipette to a centrifuge tube, rinse cryotube with cell culture medium, and add to the centrifuge tube.
- Centrifuge the cell suspension, resuspend the pellet in 5 ml of medium, transfer to a 25-cm² culture flask, and incubate at 37°C.

4. Attenuation of the Virulence of T. annulata Schizonts

The virulence of *T. annulata* schizonts is attenuated by growth and passages in *in vitro* cell cultures. The schizont-infected cells are subcultured every 3 to 4 d. Small culture vessels (for example, 25-cm² plastic flasks) are recommended during the period of attenuation in order to economize on culture medium. At every passage, the cells from one vessel are seeded into two to three new vessels and the vessels from the previous passage are discarded. To reduce the possibility of losing a culture line as a result of contamination or other mishap, cells are frozen at each tenth passage and stored in liquid nitrogen. This insures that if the culture is subsequently lost for any reason, it will not be necessary to start the attenuation process over again from the beginning.

The degree of attenuation is assessed by periodic inoculations of 2 to 5 million infected cells into each of 2 susceptible calves. Body temperature of the inoculated cattle and blood films are checked daily. When a rise in temperature is observed, lymph node and liver biopsies are performed and the material is checked for schizonts. Figure 2 shows a culture schedule protocol for attenuation of a *T. annulata* strain.[119]

During the early period of cultivation, the schizonts produce clinical theileriosis in most inoculated cattle and kill some of them. After several weeks or months, milder clinical reactions, accompanied by rare schizonts in lymph nodes and liver, are observed. Erythrocytic merozoites are detected in the blood films.[106,108,112] Attenuation is complete when inoculated calves show no clinical symptoms, schizonts, or erythrocytic merozoites.[16,107,128-131] Different field isolates have required from 3 months to 3 years of cultivation in order to become completely attenuated.[132] There are no indications that the time needed for attenuation is proportional to the initial virulence of the schizonts.[114,131]

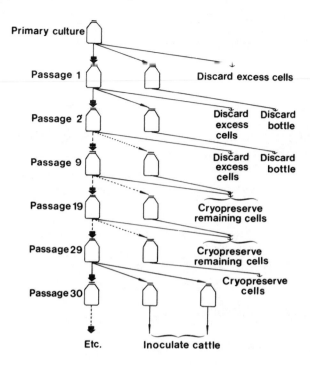

FIGURE 2.   Suggested scheme for attenuation of *T. annulata* schizonts in cell culture.

When the cultured schizonts no longer cause clinical symptoms or parasitemia in inoculated cattle, a final test for safety with four to six susceptible cattle is recommended. The animals are inoculated with the schizonts, and if the attenuation of the cultured schizonts is confirmed by the examinations described above, the same animals can be used for testing the protective capacity of the schizonts. For this purpose, the cattle that received the cultured schizonts and susceptible control cattle are challenged by infestation with ten (five female and five male) infected *Hyalomma* ticks or by inoculation of a suspension equivalent to ten ticks (see Section VIII).

Following the challenge, transient clinical manifestations and a few parasites may appear in the animals that received cultured schizonts, while heavy clinical theileriosis and death should occur in the control animals.[82,127] A field test for potency of the attenuated schizonts is desirable by introduction of vaccinated and susceptible cattle into a *T. annulata*-enzootic area. However, this is a costly procedure if performed under strictly controlled conditions and is not always practical.

After tests for safety and protection have given the expected results, a batch of seed cells to be used for producing vaccine is prepared. The cells from attenuated schizont cultures are frozen in 50 to 60 cryotubes, according to the technique described in the previous section. When the stock of seed cells is exhausted, a new batch is prepared starting from a cryotube from the previous batch.

C. Production of Vaccine

The production of schizont vaccine sufficient to immunize large numbers of cattle requires the mass cultivation of schizonts. The first step in this process is the retrieval of frozen seed cells that have already been tested for safety and potency. For initiating the cultures, 25-cm² plastic flasks are usually used. Further subcultivation of the monolayers requires larger vessels, like Roux culture flasks, Blake bottles, or Roller bottles (for roller cultures). After

3 to 5 d of incubation, Roux or Blake bottles contain from 7×10^7 to 10^8 cells per bottle, while Roller bottles contain about 30 to 40% more. Since the quantity of medium required for all the above vessels is identical, the yield per milliliter of medium is highest with the Roller bottles.

Expanding the number of culture vessels to hundreds or thousands in order to prepare a single large batch of uniform vaccine is not generally practical. A more suitable procedure for parasitological (as opposed to commercial) laboratories is to grow the cells in several tens of culture vessels during any single period and to produce small batches every 3 to 4 d over a period of weeks or months. About two thirds (up to 70%) of the infected cells are used for preparing vaccine, while the remainder goes for further subcultivation. The small batches of vaccine are cryopreserved or used immediately as fresh vaccine.

Obviously, this method involves continuous passaging of the cells during the period of vaccine production (about two passages per week), so that after several months, the vaccine being produced comes from a considerably higher passage than the initial batch. Present experience shows that once schizonts have become attenuated, no reversal to virulence occurs following further passages in culture.[83] On the other hand, it is not yet clear whether immunogenicity of the schizonts is altered during prolonged cultivation. For this reason, it is recommended that when the seed cells have been passaged for 20 to 30 times during the production process, further production should be started anew from another tube of cryopreserved seed cells.

Cells for vaccine production can also be grown in suspension cultures (static or in spinner vessels).[133] However, it should be noted that contamination tends to occur more frequently with this type of culture than with monolayers.

Vaccine may be prepared for immediate use (fresh, chilled vaccine) or may be stored frozen for use over a period of months or years.

The practical shelf life of freshly produced vaccine held at 4°C is 4 to 5 d, although in particular cases, under laboratory conditions, 10-d-old vaccine has induced antibody production in cattle. Viability of the schizont-infected cells in fresh vaccine stored at 4°C was about 60% after 4 d and 30% after 6 d, as determined by dye exclusion counting.

Fresh, chilled vaccine can be used in farms situated near the production center. Field application of this type of vaccine requires relatively cheap, light-weight refrigerating equipment.

1. Preparation of Vaccine for Immediate Use

For preparing fresh vaccine, the following items should be within easy reach of the worker, preferably in the order cited:

- Flasks or bottles of the *T. annulata* culture
- Discard jar for culture supernatants
- Dispenser for PBS
- Discard jar for PBS
- Dispenser for Versene®
- Rack with centrifuge bottles (the largest type available in the laboratory)

Each vessel is processed as follows:

Step 1. Discard supernatant medium.
Step 2. Rinse monolayer with 15 to 20 ml of PBS.
Step 3. Introduce about 10 ml or more Versene®, enough to cover the total cell growing surface of the vessel. Allow to stand at room temperature for 10 to 15 min, then shake vigorously in order to dislodge the cells. If Roller bottles are used they should be returned to the roller apparatus for incubation with Versene®.

Step 4. Transfer the cell suspension to centrifuge bottles, rinse the culture vessel with 10 ml PBS, and add to the centrifuge bottles.

Step 5. When all the cells have been harvested, spin the centrifuge bottles at about 2000 rpm for 20 min at 4°C and decant the supernatant.

Step 6. Resuspend the cells in culture medium without serum and pool the cells from all the centrifuge bottles.

Step 7. Count the cells, measure the volume of the suspension, and compute the total number of cells.

Step 8. Transfer about 30% of the total cells to an Erlenmeyer bottle. These are to be used for seeding culture vessels for the next batch of vaccine.

Step 9. To the remaining 70% of the cells, add culture medium without serum sufficient to yield a final concentration of 5×10^6 cells per 2 ml of suspension. Penicillin (100 μg/ml) and streptomycin (100 μg/ml) are added to the suspension.

Step 10. Dispense in bottles, stop with rubber stoppers, seal with aluminum caps, and label the bottles. Since this material is easily contaminated upon use in the field, small-volume bottles (50 or 100 ml) are recommended for packaging the vaccine. If plastic bottles are used, they should be constructed of high-grade, nontoxic material. Store and transport vaccine at about 4°C.

2. Preparation of Frozen Vaccine

Cells are harvested as described in Steps 1 to 4. The cells should now be pooled and counted before the centrifugation. After centrifugation, the packed cells are resuspended in culture medium containing 7% DMSO to yield a final concentration of 10^8 cells per milliliter.

Aliquots of 1 ml cell suspension each are dispensed in cryoresistant plastic vessels. Appropriate vessels should be of small volume and suitable for filling and withdrawing the vaccine by aseptic techniques. Two types of vessels have been used on a large scale for packing frozen vaccine: screw-cap cryotubes (1.5 ml volume) or plastic disposable pastettes (before use, the pipette portion of the pastette is sheared about 1 cm above the bulb). The vaccine is dispensed with a repeating self-refilling syringe. The screw-cap cryotubes are capped, while the pastettes are heat-sealed. Each vessel provides ten doses of vaccine.

Although 10^3 schizonts can infect a portion of inoculated calves, it appears that at least 10^5 schizonts are required to induce immunity in all inoculated animals.[134,135] The greater number of schizonts recommended here for a single dose of fresh or frozen vaccine (5×10^6 and 10^7, respectively) is intended to counteract possible harmful environmental conditions in the field.

Since DMSO penetrates the cells immediately, the time spent in dispensing the vaccine should also be as short as possible. The vials with the vaccine are introduced into an ultralow freezer at $-70°C$ overnight, then kept in the vapor phase of liquid nitrogen for several hours, and finally immersed into the liquid phase for storage.

For use, the frozen vaccine is thawed and diluted in PBS. For this purpose, bottles containing 19 or 38 ml PBS (for preparing of 10 and 20 doses, respectively) are prepared beforehand, and the bottles are rubber stoppered, sealed with aluminum caps, and autoclaved for 30 min. They can then be stored at room temperature until needed.

3. Testing of Vaccine

There are no official safety standards for *T. annulata* vaccine. In any case, the fresh vaccine has too short a shelf life to be tested adequately before use. Frozen vaccine has been prepared up to now in relatively small batches which do not justify individual testing for safety and potency. As a practical matter, however, the checking for safety occurs when the batch of seed cells from which the vaccine is prepared is tested. This assumption is based on the observation that attenuated schizonts do not revert to virulence during extended subcultivation.

Cell culture vaccine has proved to be safe for all types of cattle, regardless of age or physiological condition. A thorough assessment of the safety of cultured schizont vaccine in hamsters, splenectomized calves, and adult cattle did not detect any potential pathogenic characteristic.[136] In addition, preimaginal stages of *H. excavatum* that were fed on calves vaccinated with cell culture vaccine did not transmit the infection by the subsequent instar.[137]

On the other hand, it is recommended that the fresh and frozen vaccine be tested periodically for contamination with bacteria and mycoplasma.[138] Although the fresh vaccines would already have been inoculated by the time these tests are completed, the testing provides an important control for the quality of the production process. Testing is especially valuable for frozen vaccine since the tests can be completed before the vaccine is put to use and contaminated batches can be discarded in time.

D. Application of Vaccines in the Field

Fresh vaccine is transported to the field in insulated boxes containing cooling bags. Before inoculation the vaccine suspension should be shaken in order to resuspend the cells. Often, clumping of the cells occurs, but this does not affect the viability of the schizonts. Direct sunlight on the vaccine should be avoided.

Although the vaccine contains antibiotics, special attention should be paid to sterility. Vaccine should be withdrawn from the bottle with a different needle from the one used to inoculate the cattle; or alternatively, the vaccine can be inoculated with a repeating syringe. Once a bottle has been started, the entire amount should be used within 1 h.

Application of frozen vaccine proceeds as follows. The frozen vaccine is stored in large liquid nitrogen refrigerators at the producing laboratory. Field centers for storage and supply of vaccine can be set up in theileriosis-enzootic areas. The vaccine is transported to the farms in small liquid nitrogen refrigerators.

The basic equipment required for field application of frozen vaccine includes a wide-mouth vacuum jar for preparing the 40°C water bath, a thermometer for measuring the temperature of water, long forceps, face shield, and asbestos gloves. Hot water is poured into the vacuum jar and adjusted with cold water to obtain 40°C. After donning the face shield and asbestos gloves, the required number of vials is withdrawn with the forceps from the canister of the liquid nitrogen refrigerator. When withdrawing the vials, the canister should be kept as deep as possible in the neck of the refrigerator to avoid warming the remaining vials. Each withdrawn vial should be checked in order to ascertain that liquid nitrogen has not leaked inside. The nitrogen does not alter the vaccine, but may cause the vial to explode when introduced into the water bath. (Such a vial should be set aside for 1 to 2 min to allow the nitrogen to escape. The vial can then be processed in the usual manner.) The vials are gently shaken in the water bath and, after thawing, the vaccine is transferred with a 2-ml syringe to bottles containing PBS. The empty vial is rinsed with 1 ml of PBS withdrawn from the bottle, and the rinse is added back to the bottle. Diluted vaccine should be inoculated within 30 min of thawing.

E. Protection Engendered by Schizont Vaccine

Vaccine from completely attenuated schizonts does not cause a rise in body temperature nor do erythrocytic merozoites appear in blood films of vaccinated cattle.[83] Thus, measuring temperature or examining blood films cannot be used for assessing the response to vaccination. On the other hand, the schizonts induce a marked production of specific antibody,[139] and this provides a means for assessing the results of the vaccination.[136,140,141] Although the antibody detected by a serological test is not necessarily an indication of the immunity that has developed against infection, the very fact of the presence of antibody indicates that a multiplication of schizonts has occurred in the vaccinated animals. Inoculation of the equivalent amount of dead schizonts does not induce a significant level of antibody.

Attenuated schizonts usually engender total protection against virulent homologous schizonts, and a lesser degree of protection against virulent heterologous schizonts.[83,142] However, the protection engendered to tick-transmitted infection is of greater practical interest and importance.

In India,[112] 14 out of 15 calves vaccinated with a schizont vaccine were totally protected against challenge with infected ticks, and one showed mild parasitemia. In the U.S.S.R.,[108] when seven vaccinated calves were challenged by ticks, five exhibited transient fever, but all of them showed a mild rise of parasitemia. In potency tests conducted in Israel,[127] calves that had been vaccinated with completely attenuated schizonts were challenged with ticks infected with field isolates of *T. annulata*. Out of 22 calves, 7 remained asymptomatic and 15 reacted with low parasitemia and mild fever. In all the above trials, control susceptible cattle challenged together with the vaccinated animals showed severe clinical theileriosis, and a considerable portion of the animals succumbed to the infection.

A lesser degree of protection was exhibited when cattle vaccinated with attenuated schizonts were challenged by parasites from ticks originating from a remote geographical area.[114] All of 12 high-grade Friesian calves receiving vaccine prepared from Turkish isolates of *T. annulata* showed clinical symptoms following challenge with sporozoites of Israeli strains; 11 animals survived and 1 died from mixed theilerial and bacterial infection. The response of these calves was considerably more severe than the response to challenge of calves vaccinated with Israeli isolates. It seems, therefore, that it is desirable to prepare schizont vaccine from local isolates of *T. annulata*.

Field observations have confirmed the efficacy of schizont vaccine in protecting cattle against theileriosis. In a wide field trial during 3 consecutive years, Stepanova and collaborators[143] introduced 410 vaccinated and 66 susceptible bull calves into theileriosis-enzootic pastures. None of the vaccinated animals showed clinical theileriosis, while all control calves contracted the disease and 22 (33%) of them died.

During 2 decades, more than 3000 young bulls from European breeds (Charolais, Simental, Brown Swiss, and others) were imported to Israel, and, after vaccination, most of them were introduced into theileria-infected pastures. Except for rare sporadic cases, theileriosis has not been reported among these animals. Local or imported cattle that were accidentally not vaccinated contracted theileriosis on the same pastures. In most of the vaccinated cattle erythrocytic stages of *T. annulata* were detected in blood films after various periods on pasture. Since schizonts used for vaccination do not yield erythrocytic merozoites, the presence of these stages testified to exposure of the animals to tick-transmitted infection.

In the past few years, a few vaccinated adult high-grade Friesian cows have exhibited clinical symptoms when exposed to tick infection. Although these animals recovered from the infection, reduced milk production and abortions were noted. Reinforcement of immunity by a two-step vaccination procedure was considered for such animals,[127] but the low incidence of cases did not justify a modification of the vaccination regimen at the time.

Controversial data about the length of protection engendered by vaccination have been reported. Periods of more than $3^1/_2$ years[144] to less than 1 year[145] have been reported. The longevity of immunity against theileriosis after a preliminary exposure is still an open question (see Section V), but the high cost of this type of study handicaps investigations along these lines.

As far as the epizootological implications of vaccination are concerned, vaccinated cattle do not provide a source of infection for other cattle.[137] On the other hand, even when complete protection against clinical theileriosis is obtained, schizont vaccine does not prevent the appearance of erythrocytic stages resulting from tick infections of vaccinated cattle. It follows, then, that vaccination with schizont vaccine cannot lead to eradication of theileriosis in enzootic areas.

VIII. SPOROZOITE VACCINE

A technique that involves infecting cattle with sporozoites and then mitigating the subsequent symptoms of theileriosis by chemotherapy was proposed for immunization against *T. parva*.[146] The considerable work performed on this subject was summarized by Cunningham,[147] Purnell,[148] and Radley.[149] The technique has been evaluated also for immunization against *T. annulata* theileriosis,[150-153] with promising results under laboratory conditions. However, the various parameters involved in preparing and using sporozoites for immunization against *T. annulata* have not been studied as thoroughly as in the case of *T. parva*.

The main stages in the sporozoite vaccine technique are breeding large batches of ticks infected with *T. annulata,* preparing and storing suspensions of infected tick tissues, and inoculating cattle with the suspensions and administering specific chemoprophylaxis.

A. Breeding *T. annulata*-Infected Ticks

Livestock and laboratory animals used for breeding ticks should be kept in tick-proof accommodations from considerations of safety. The barns and laboratory should be surrounded by a tick-proof moat. In tropical and subtropical areas, where high ambient temperatures occur, low-temperature incubators might be needed for storage of ticks. On the other hand, heating of the premises may be required in the cold season during the period of engorgement of ticks on large and small animals.

For rearing noninfected ticks, small rodents (e.g., gerbils) can be used as hosts for larval stages.[154] The rodent is exposed to the ticks while confined in an iron mesh cylinder set in a tray on clean wood shavings. The entire rim of the tray is smeared with tree tanglefoot (castor oil 70%, natural gum resins 25%, vegetable wax 5%) mixed with dimethyl-phthalate. After attachment of the larvae, the tick-infested rodent is transferred to a small iron mesh cage set on white filter paper in a small tray. The latter, in its turn, is set in a bigger tray, the rim of which is also smeared with tanglefoot. Engorged ticks that drop off onto the filter paper are aspirated from the trays into an Erlenmeyer flask by means of a vacuum pump.

Rabbits may be used for engorgement of all stages, but adults of some *Hyalomma* spp. may not fully engorge on rabbits. In such cases, calves or goats may be necessary. For feeding on rabbits, the ticks are usually applied in ear bags glued or fixed by adhesive tape at the base of the ear. For feeding on cattle, ear bags or cloth patches stuck with adhesive tape (or glued) onto the neck, tail or other parts of the body are used. Since rabbits and cattle often attempt to rub or scratch the ear bags or cloth patches, the animals should be properly restrained to avoid partial or complete detachment of the tick containers. If nymphs have to be infected with *T. annulata,* they can also be fed on the whole body of young calves maintained in a special device that allows the harvesting of the engorged ticks.[155]

Repeated feedings of ticks on the same animal often results in extravasation of body fluid in the feeding area, which interferes with the normal attachment of the ticks. Animals should, therefore, be used only once for feeding the ticks.

Any potential vector of *T. annulata* can be used for preparing suspensions of sporozoites. In selecting a particular *Hyalomma* species from those occurring in a specific geographical area, two main criteria should be considered: the potential rate of infection of the salivary glands with *T. annulata,* and the ability of the tick to develop and multiply under laboratory conditions.

The behavior of a *Hyalomma* spp. as a two- or three-host tick may vary according to the breeding conditions (ambient temperature, humidity) and the species of host used for engorgement. Details given here should be complemented by practical experience with tick populations in different ecological areas.

A colony of noninfested ticks should be maintained continuously. Preimaginal stages from this colony can then be released onto cattle carrying *T. annulata* in order to pick up the infection. If ticks are collected in the field in order to initiate colonies for producing sporozoite vaccines, the ticks will first need to be tested for other protozoan and rickettsial infections by feeding them on susceptible hosts.

In all stages, the moulted ticks require a prefeeding rest period before being ready to feed again. Various minimal prefeeding periods have been reported for *Hyalomma* species,[156] but in general, longer periods are required for the adults than for the preimaginal stages. For most species the prefeeding period is 10 to 20 d for larvae and nymphs and 25 to 35 d for adults. During moulting and prefeeding, the ticks of all stages require specific conditions of temperature and humidity,[156] but generally a temperature of 28°C and 75% relative humidity (RH) are adequate. For storage of nonfeeding stages, larvae and nymphs are maintained at 18 to 20°C and 85% RH and adults at 12 to 13°C and 50% RH.

To obtain *T. annulata*-infected, two-host *Hyalomma* ticks (e.g., *H. detritum*), unfed larvae are released on infected cattle and the engorged nymphs are harvested and allowed to moult to adults. The adults are then fed for 3 to 5 d as described below in order to induce maturation of sporozoites.

The technique of infecting three-host ticks (*H. excavatum*) has been described by Samish et al.[157] Larvae of *H. excavatum* are fed on gerbils and, after moulting, the nymphs are allowed to engorge on cattle infected with erythrocytic stages of *T. annulata*. The adult ticks are then treated like those of the two-host *Hyalomma*.

Most three-host *Hyalomma* ticks transmit theileriosis from larva to nymph and from nymph to adult. Therefore, suspensions of sporozoites can be prepared also from infected nymphs. However, it appears that adult *Hyalomma* produce more *T. annulata* sporozoites than nymphs.[158] A similar relationship was found with *T. parva*. In point of fact, most vaccination trials with *T. annulata* have been made with sporozoite suspensions prepared from adult ticks.[150,152,161]

According to some authors,[162] there is little correlation between the level of *T. parva* parasitemia in donor cattle and the rate of infection in *R. appendiculatus* ticks which feed upon them. However, others[163] have reported that ticks fed as nymphs on animals with rising or peak parasitemias were more likely to become heavily infected than those engorged on animals with falling parasitemias. On the other hand, *Hyalomma* ticks failed to transfer *T. annulata* when fed on cattle with low parasitemias.[137,164] It appears, therefore, that a higher parasitemis in the donor calves is more likely to ensure a higher yield of *T. annulata* sporozoites in the engorged ticks.

Since considerably higher parasitemias can be reached in splenectomized cattle, such animals can be used for infecting preimaginal stages. Particularly high parasitemias occur in acutely infected animals. Thus, if preimaginal stages are released on infected calves during their prepatent period, the ticks will complete their meal during the acute stage of the disease when the cattle are heavily infected. According to the experience in our laboratory, young calves inoculated with sporozoite suspensions from five female and five male *H. excavatum* showed schizonts in the lymph nodes on day 7 after inoculation, and erythrocytic merozoites on day 9 or 10, succumbing on days 11 to 16 with parasitemias of 21 to 99%. Most nymphs of three-host *Hyalomma* spp. engorge within 5 to 9 d after being put onto cattle. Thus, nymphs that are released on cattle 3 to 6 d after infection will generally complete their engorgement around day 12 when most of the calves are still alive and suffering high parasitemias, thus ensuring a high level of infection in the ticks.

On the other hand, when acutely infected calves are used as just described, a portion of the ticks may be lost when some of the calves die before engorgement is completed. For this reason, more engorged ticks can be obtained by using splenectomized calves that have survived a previous infection. So, with reference to *T. annulata,* more precise information is needed concerning the relationship of cattle parasitemia to tick infection levels before a specific program can be drawn up for the optimal infection of ticks.

Nymphs engorged on animals that are not susceptible to theileriosis (e.g., rabbits) can be infected by inoculating the ticks with *Theileria*-infected red blood cells from cattle.[165-167] However, this method has not yet been developed to the stage of being available for mass production of infected ticks.

Although ticks derived from infected preimaginal stages are also infected, maturation of the parasite is arrested until the ticks begin feeding in their next instar. Recently, it was found that incubation at 37°C and high RH[21] may stimulate maturation of sporozoites in the tick salivary glands even without a blood meal. However, parasites in blood-fed ticks mature more completely and rapidly than in incubated ticks[168,169] and provide a considerably higher number of sporozoites.

Infective stages of *T. annulata* are detected in ticks 24 to 72 h after attachment[170-172] and persist in some *Hyalomma* species during the entire feeding period.[171] A period of 3 to 5 d of feeding appears to be optimal for obtaining a high rate of mature sporozoites in the tick salivary glands.[20,157,172]

Since nymphs are infected by feeding them on cattle with variable degrees of parasitemia, and since environmental conditions are practically never identical, it is recommended that the infection rate in a sample batch of ticks be checked before feeding the whole batch. For this purpose, 20 to 30 each of female and male ticks are introduced into the ear bags of a rabbit and fed for 4 d. Alternatively, the ticks can be maintained at 37°C for 4 to 6 d, but as already indicated, maturation of sporozoites is less effective by this method than by feeding.

The ticks are examined as described by Walker et al.[173] and Walker and McKellar. The salivary glands are removed and mounted on microscopic slides. Upon drying, the glands are fixed for 5 min in a solution of 60% ethanol, 30% chloroform, and 10% acetic acid. The slides are rinsed with distilled water and immersed in methylgreen-pyronin for 7 min. A successfully infected batch may show an infection rate of nearly 100% of the ticks.[37,38] Minimum rates of infection, below which it is not worthwhile processing the entire batch, have not yet been determined, and these may vary depending upon the *Hyalomma* species used. Titration of different batches of vaccine for infectivity in cattle will give the relationship between rate of infection of the ticks and the efficacy of the vaccine.

After checking the sample, the whole batch of ticks is fed for 4 d. For this purpose, rabbits are usually used, but sheep and especially goats with long ears have also proved to be suitable.[157,161] The ticks are confined in ear bags or cloth patches as described previously. Nonattached ticks are discarded during the first 2 d after applying the bags. After 4 d, the feeding ticks are transferred with forceps into large test tubes, the openings of which are covered by rubber diaphragms in which a 1-cm-long slit has been cut. This arrangement allows easy introduction of ticks while, at the same time, preventing escape of those already in the tube. The process of obtaining infected adult ticks, starting from ova deposited by noninfected females, requires about 140 to 160 d.

B. Preparation and Testing of Sporozoite Vaccine

Sporozoite vaccine is essentially the supernatant extract obtained after crushing and triturating infected ticks in a suitable liquid carrier.

The partially engorged ticks are divided into several batches of 500 to 1000 ticks depending upon the capacity of the equipment available for preparing the suspension of sporozoites. It has been shown that female *H. excavatum* ticks carry considerably more sporozoites than males.[36,174] Similar results have been observed also with *R. appendiculatus* and *T. parva*. Nevertheless, the harvest of sporozoites from the males cannot be dismissed as negligible, and both sexes are used in preparing the vaccine.

For triturating the ticks a tissue homogenizer[162] or Ultra Turrax®[157] can be employed. In the latter instance, the apparatus should be set at low speed to avoid damaging the sporozoites. Ticks can also be ground manually in a mortar containing sterile glass powder[157] or sand.[116]

The devices and solutions used for triturating the ticks should be precooled, and the recovered supernatants should be kept under refrigeration until all the ticks have been processed.

The ticks are first washed free of host tissue exudate and associated debris. The procedure proposed by Brown[118] for obtaining sterile ground-up tick supernatant for *in vitro* infection of cells is also applicable for preparing sporozoite vaccine. The ticks are washed first with 1% benzalkonium chloride, followed by three washes with Eagle's Minimum Essential Medium (MEM) containing antibiotics (200 IU/ml penicillin, 200 mg/ml streptomycin, and 100 kg/ml nystatin). If necessary, the ticks may be left in the last wash solution until proceeding.

Before crushing the ticks the last wash solution is decanted and 3.5% bovine plasma albumin in Eagle's MEM is added. A solution of 5 parts Eagle's MEM and 4.25 parts fetal calf serum also appears satisfactory (0.75 parts of glycerol is added before cryopreservation).[157,175] Antibiotics at the same concentration as in the wash solution are recommended for the grinding solution. A ratio of 10 ml solution for each 100 ticks is used initially. It is preferable to divide the grinding solution into two to three portions — crushing the ticks and removing the supernatant with each portion until all the solution is used up. A homogenous suspension will considerably improve the yield of sporozoites and the quality of the vaccine. After triturating all the ticks, the supernatants are pooled, and gross sediments that have settled from the refrigerated supernatants are discarded. The pool is centrifuged in 50- to 100-ml centrifuge tubes for about 5 min at about 100 to 200 × g in order to separate the lighter sporozoites from the heavier debris. The yield of sporozoites can be improved by decanting the supernatant after centrifugation, resuspending the pellet in 20 to 30 ml of the grinding medium, and centrifuging again. Two or three such cycles can be performed. Finally, all the supernatants are pooled and the total volume determined. At this stage, each milliliter contains the product of about 8 to 10 ground-up ticks when starting with a batch of 5000 ticks.

Although the shelf life of sporozoite suspension stored at 4°C has not been evaluated, it is probably too short to allow the use of fresh preparations for vaccination of field cattle. On the other hand, sporozoites can be preserved successfully at very low temperature. Glycerol has been used as the cryoprotectant of choice for *T. annulata* and *T. parva* sporozoites.[175,176] A solution of 25% glycerol in MEM/BSA solution is added with constant mixing to the sporozoite suspension at a ratio of three parts glycerol solution to seven parts of suspension. This yields a final concentration of 7.5% glycerol in a sporozoite suspension that now contains the equivalent of six to seven ticks per milliliter. While stirring constantly, the material is distributed in aliquots of 1 to 1.5 ml into the same type of vessels used for cryopreserving the schizont vaccine (see Section VII.C). The equilibrium time, i.e., the time needed for glycerol to penetrate the theilerial organisms, is about 30 min. This includes the time needed for dispensing the material into the cryovessels. For this reason, if the volume of original sporozoite suspension is greater than what can be processed within 30 min, it should be divided into several portions, and glycerol should be added to each portion about 30 min before freezing.

The cryovessels containing the vaccine are introduced into an ultralow temperature cabinet (−80°C) overnight. The optimal cooling rate for *T. annulata* sporozoites has not been evaluated, but it is likely to be similar to that recommended for *T. parva*. Vials packed in cardboard cooling racks in a −80°C freezer have a cooling rate of about 1°/min, which is claimed to be optimal for *T. parva* sporozoites.[162] The next day, the cryotubes are transferred into liquid nitrogen vapors for 4 to 6 h and then immersed in the liquid nitrogen for long-term storage. Although developed on a rather empirical basis, this freezing procedure ensures a high rate of survival for *T. annulata* sporozoites.[36] In the absence of an electric ultralow freezer, an alternative procedure that works well is to introduce the vials containing the sporozoites directly into liquid nitrogen vapors overnight, before immersing them into the liquid phase.

As in the case of the schizont vaccine, it is desirable to test sporozoite vaccine for purity and efficacy (ability to cause infection in cattle). Testing for safety is not relevant since, if effective at all, the vaccine provokes heavy clinical theileriosis accompanied by relatively high mortality unless specific medication is administered. Unlike the case of *T. parva* Lawrencei,[149] *T. annulata* strains resistant to chemoprophylaxis have not been reported to date, so that testing for that purpose does not appear necessary.

Testing for purity can be performed in any bacteriological laboratory according to the standard procedures used for this purpose.[138] It is difficult to obtain a tick-derived suspension of sporozoites that is sterile, and a low-level contamination with bacteria not pathogenic for cattle should be accepted.

The efficacy of the vaccine is tested by inoculating susceptible cattle. Four to six cryotubes are thawed in a water batch at 40°C and pooled. Fivefold dilutions are prepared with the cryopreservative medium (including glycerol), and these are inoculated subcutaneously into at least two animals for each dilution. Response of the animals is monitored as described in Section III.

Since it is difficult to obtain a completely even suspension of sporozoites, it is recommended that the working dose be set about tenfold higher than the minimal dose causing infection in all inoculated cattle. With a good suspension, the working dose will be about one tick equivalent per animal.[36]

C. Vaccination with Sporozoite Vaccine

The rationale for vaccination with live sporozoites followed by medication is based on the observation of Neitz[177] that aureomycin administered in repeated large doses during the incubation period of tick-induced *T. parva* theileriosis mitigates the severity of the infection. Other investigators[178] have proposed replacing the extended treatment regime by a single inoculation of long-acting oxytetracycline formulation.

Initial trials conducted with tick-induced infection of *T. annulata* showed that oral medication with chlortetracycline for 4 to 16 d at a daily dose of 16 mg/kg prevented death in crossbred calves.[150] Similar results were obtained with 4 to 12 daily treatments with water-soluble oxytetracycline formulation or 1 to 2 treatments with long-acting oxytetracycline.[151-153,179,180]

It appears from these results that the administration of 20 mg/kg long-acting oxytetracycline simultaneously with the sporozoite vaccine can prevent severe clinical theileriosis in crossbred calves. Vaccination of more sensitive breeds may require repeated postvaccination medication.[151,153]

Recently, the antitheilerial drug buparvaquone[181] has been evaluated in vaccination against theileriosis. Given simultaneously with *T. annulata* sporozoites at the rate of 2.5 mg/kg, i.m., the drug prevented clinical symptoms of theileriosis, but did not interfere with development of immunity.[182] However, these preliminary results need further confirmation.

Buparvaquone administered with supportive therapy also shows therapeutic activity during the acute stage of theileriosis. For this reason, it can be used to treat vaccinated animals when the chemoprophylactic treatment is not successful.

Sporozoite vaccine has yet not been used for routine vaccination of field cattle against *T. annulata* theileriosis, so that no wide practical experience has been accumulated. The information that follows, although based on experimental trials only, nevertheless has a strong foundation of factual data accumulated during those trials.

The first step in using a live, nonattenuated vaccine is to decide which type of cattle can be vaccinated safely. Work published to date describes vaccination only of young cattle of various breeds[151,152,179,180] including high-grade Friesian calves.[153] When such highly susceptible animals were inoculated with the sporozoite vaccine, most showed fever for several days despite chemoprophylactic treatment. On the other hand, no mortality has yet been reported from the vaccination.

At present, the sporozoite vaccine is not recommended for mass vaccination of high grade, adult dairy cattle, especially pregnant cows. If such animals are to be vaccinated, small experimental groups should be tried first in order to elaborate an appropriate treatment regime and to assess the results of vaccination.

When doses of frozen-thawed sporozoite suspension equivalent to one to ten ground ticks were inoculated simultaneously with 20 mg/kg oxytetracycline, the prepatent period was between 10 to 12 d.[151-153,179] Severe reactions may occur, if at all, within a week after this period. It follows that vaccinated cattle should be confined in barns or corrals where they can be observed regularly for about 3 weeks. It is important that facilities for easy restriction of the animals be available in order to allow for treating animals that develop postvaccination theileriosis despite the chemoprophylactic regime. Cattle showing clinical manifestations should have their body temperature recorded, and lymph node biopsy material should be examined for schizonts. When severe theileriosis occurs, chemotherapy with buparvaquone is recommended.[121]

IX. OTHER METHODS OF VACCINATION

A. Nonviable Antigens

Cattle vaccinated with dead, culture-derived schizonts plus incomplete Freund's adjuvant exhibited various degrees of resistance to challenge with virulent schizonts.[95] A high degree of protection was obtained with ultrasonic-disrupted and lyophilized vaccine. Only four to ten animals immunized with this preparation developed fever upon challenge, and schizonts were seen in two of them. However, all animals survived, while 6 of 11 nonvaccinated controls died.

The killed schizont vaccine did not protect against challenge with sporozoites. Furthermore, this type of dead vaccine cannot be used for priming cattle before vaccination with live virulent schizonts since animals that were completely protected against the virulent schizonts died from the sporozoite infection. Vaccination with lyophilized, inactivated sporozoites plus *Corynebacterium parvum* adjuvant prevented death in calves upon challenge with tick-derived parasites.[183] However, as mentioned below, this type of adjuvant alone is capable of inducing a nonspecific resistance to theileria infection.

B. Irradiated Parasites

Tick suspensions containing sporozoites of *T. annulata* were exposed to various doses of gamma irradiation in an attempt to influence the severity of response caused by the parasites in cattle. Irradiated parasites caused generally milder infections than nonirradiated organisms.[184-186] It was suggested[185] that irradiation might have altered the virulence of the sporozoites rather than have reduced the number of infective (viable) organisms in the inoculum. Cattle vaccinated with irradiated parasites were in most instances resistant to challenge infections. Irradiation of 6 to 7 krad appeared to have an optimal effect, decreasing the severity of the response to vaccination, while inducing a high level of immunity. However, some paradoxical effects were observed where immunization with parasites irradiated with 5 krad gave lower protection than 10-krad-irradiated parasites,[184] thus suggesting variability in the results of irradiation.

C. Nonspecific Vaccination

Corynebacterium parvum induced a nonspecific protection in cattle against tick-induced theileriosis. In 2 separate experiments,[183,187] 15 calves were each inoculated with 200 mg of killed, freeze-dried preparations of *C. parvum*. The animals were challenged with infected ticks 45 to 90 d after vaccination. Few schizonts (below 1%) were detected in lymph node biopsy smears, and rare erythrocytic parasites were seen in blood films. Out of 15 vaccinated calves, 14 recovered from the challenge infection, while all 11 nonvaccinated control calves died. Despite these encouraging results, this approach has not yet been further investigated for possible practical application in field immunization against theileriosis.

APPENDIX

Selected Equipment and Supplies Cited in the Text:

Identification and Sources

The items and supplies listed here are provided as a convenient point of departure for materials cited in the chapter. However, they should not be considered as exclusive or necessarily superior to similar items available from other suppliers.

1. Aspirating hypodermic needle (size 1.4×110 mm), The Holborn Surgical Instrument Co., Ltd., London, E.C.1.
2. Blake bottle (Blake culture bottle), Cat. No. 5634-01000, Bellco Glass, Inc., Vineland, NJ 08360.
3. Cryotube 48×12.5, Cat. No. 3-63495., Nunc, Postbox 280-Kamstrup-DK, 4000 Roskilde, Denmark.
4. Ficoll-Pague, Pharmacia, Uppsala.
5. Pastettes = Beral "graduated pipet". Low density polyethylene, one-piece disposable plastic transfer pipette, Cat. No. B 80-800, Beral Enterprises, Inc., 8907 Woodman Ave., Arleta, CA 91331.
6. Roller bottle, Cat. No. 7730-38285, Bellco Glass, Inc., Vineland, NJ 08360.
7. Roux culture flask, Cat. No. 3501-M10, Thomas Scientific, 99 High Hill Rd., P.O. Box 99, Swedesboro, NJ 08085-0099.
8. Spinner vessels (Spinner flasks), Cat. No. 1969, Bellco Glass, Inc., Vineland, NJ 08360.
9. Ultra Turrax, Janke & Kunkel GMBH & Co., KG, Ika-Werk Staufen, D-7813 Staufen, West Germany.

ACKNOWLEDGMENTS

I would like to thank Dr. M. Goldman, emeritus of this Institute, for his valuable editorial advice, and Ms. Meira Frank for general editorial assistance including, especially, the compilation of the bibliography.

REFERENCES

1. **Dschunkowsky, E. and Luhs, J.**, Die piroplasmosen der Rinder, *Zentralbl. Bakteriol. Parasitenkd. Infektionskr.*, 35, 486, 1904.
2. **Koch, R.**, Weiterer Bericht uber das Texasfieber, *Zentralbl. Bakteriol. Parasitenkd. Infektionskr. Hyg. Abt. 1: Orig.*, 24, 202, 1898.

3. **Gonder, R.,** On the development of *Piroplasma parvum* (Protozoa) in the various organs of cattle, *Trans. R. Soc. S. Afr.,* 2, 63, 1910.

4. **Wenyon, C. M.,** Theileria of cattle, in *Protozoology,* Vol. II, Bailliere, Tindall and Cox, London, 1926, 1029.

5. **Carpano, M.,** Piroplasmosi tipo "parvum" nei bovini del basso bacino del Mediterraneo (Febbre della costa mediterranea), *Clin. Vet.,* 38, 479, 1915.

6. **Mason, F. E.,** Egyptian fever in cattle and buffaloes, *J. Comp. Pathol. Ther.,* 35, 33, 1922.

7. **Gilbert, S. J.,** A case of *Theileria mutans* infection (Egyptian fever) in Palestine, *J. Comp. Pathol. Ther.,* 37, 158, 1924.

8. **Sergent, E., Donatien, A., Parrot, L., and Lestoquard, F.,** Theileriose a *Theileria dispar.* in *Etudes sur les Piroplasmoses bovines,* Institut Pasteur d'Algerie, Alger, 1945, 241.

9. **Witenberg, G.,** On the status of the name commonly cited as *Piroplasma annulatum* Dschunkowsky and Luhs, 1904 (class Sporozoa, order Coccidiida), *Bull. Zool. Nomen.* February, 233, 1947.

10. **Levine, N. D., Corliss, J. O., Cox, F. E. G., Deroux, G., Grain, J., Honigberg, B. M., Leedale, G. F., Loeblich, A. R., Lom, J., Lynn, D., Merinfeld, E. G., Page, F. C., Poljansky G., Sprague, V., Varva, J., and Wallace, F. G.,** A newly revised classification of the Protozoa, *J. Protozool.,* 27, 37, 1980.

11. **Purnell, R. E.,** Tick-borne diseases of cattle - a case for pragmatism?, *Vet. Rev.,* 25, 56, 1979.

12. **McHardy, N., Wekesa, L. S., Hudson, A. T., and Randall, A. W.,** Antitheilerial activity of BW720C (buparvaquone): a comparison with parvaquone, *Res. Vet. Sci.,* 39, 29, 1985.

13. **Donatien, A.,** Les vaccinations contre les piroplasmoses bovines, *Arch. Inst. Pasteur Alger.,* XVI, 37, 1938.

14. **Adler, S. and Ellenbogen, V.,** A note on the pre-immunisation of calves against *Theileria annulata, Vet. Rec.,* 14, 91, 1934.

15. **Zur (Tchernomoretz), I.,** Immunization against *Theileria annulata* (Hebrew-English summary), *Refu. Vet.,* 5, 69, 1949.

16. **Pipano, E.,** Immunization against intracellular blood protozoans of cattle, in *New Developments With Human and Veterinary Vaccines,* Mizrahi, A., Hertman, I., Klinberg, M. A., and Kohn A., Eds., Alan R. Liss, New York, 1980, 301.

17. **Pipano, E.,** Immunological aspects of *Theileria annulata* infection, *Bull. Off. Int. Epizool.,* 81, 139, 1974.

18. **Schein, E., Buscher, G., and Friedhoff, K. T.,** Lichtmikroskopische Untersuchungen uber die Entwicklung von *Theileria annulata* (Dschunkowsky und Luhs, 1904) in *Hyalomma anatolicum excavatum* (Koch, 1844), *Z. Parasitenkd.* 48, 123, 1975.

19. **Mehlhorn, H. and Schein, E.,** Electron microscopic studies of the development of kinetes in *Theileria annulata* Dschunkowsky & Luhs, 1904 (Sporozoa, Piroplasmea), *Protozoology,* 24, 249, 1977.

20. **Samish, M. and Pipano, E.,** Transmission of *Theileria annulata* by two and three host ticks of the genus *Hyalomma* (Ixodidae), in *Tick-Borne Diseases and Their Vectors,* Wilde, J. K. H., Ed., University of Edinburgh Press, Edinburgh, 1976, 371.

21. **Samish, M.,** Infective *Theileria annulata* in the tick without a blood meal stimulus, *Nature (London),* 270, 51, 1977.

22. **Schein, E. and Friedhoff, K. T.,** Lichtmikroskopische Untersuchunger uber die Entwicklung von *Theileria annulata* (Dschunkowsky und Luhs, 1904) in *Hyalomma anatolicum excavatum* (Koch, 1988), *Z. Parasitenkd.,* 56, 287, 1978.

23. **Fawcett, D. W., Doxsey, S., Stagg, D. A., and Young, A. S.,** The entry of sporozoites of *Theileria parva* into bovine lymphocytes *in vitro.* Electron microscopic observation, *Eur. J. Cell. Biol.,* 27, 10, 1982.

24. **Levine, N. D.,** *Veterinary Protozoology,* Iowa State University Press, Ames, 1985.

25. **Mehlhorn, H. and Schein, E.,** The piroplasms: life cycle and sexual stages, in *Advances in Parasitology,* Vol. 23, Baker, J. R. and Muller, R., Eds., Academic Press, London, 1984, 37.

26. **Sergent, E., Donatien, A., Parrot, L., and Lestoquard, F.,** Suppression experimentale de la reproduction sexuelle chez un Hematozoaire, *Theileria dispar, C. R. Acad. Sci.* 195, 1054, 1932.

27. **Pipano, E.,** Piroplasmosis — *Theileria* — a review, *Refu. Vet.,* 22, 181, 1966.

28. **Yakimoff, W. L. and Gousseff, W. F.,** Zur Frage der Modifikation der Virulenz der *Theileria annulata, Z. Infektionskr. Haust.,* 50, 65, 1936.

29. **Pipano, E. and Hadani, A.,** Epidemiological and immunological aspects of *Theileria annulata* infection, *Proc. Third Int. Congr. Parasitol.,* Vol. 1, Facta, Munich, 1974, 140.

30. **Pipano, E.,** Control of bovine theileriosis and anaplasmosis in Israel, *Bull. Off. Int. Epizool.,* 86, 55, 1976.

31. **Sergent, E., Donatien, A., Parrot, L., and Lestoquard, F.,** La transmission naturelle de la theileriose bovine dans l'Afrique du Nord, *Arch. Inst. Pasteur Alger.,* 9, 527, 1931.

32. **Sharma, R. D.**, Some epidemological observations on tropical theileriosis in India, in *Haemoprotozoan Diseases of Domestic Animals*, Gautam, O. P., Sharma, R. D., and Dhar, S., Eds., Department of Veterinary Medicine, Haryana Agricultural University, Hissar, India, 1980, 30.

33. **Robinson, P. M.**, Theileriosis annulata and its transmission — a review, *Trop. Anim. Health Prod.*, 14, 3, 1982.

34. **Bhattacharyulu, Y., Chaudhri, R. P., and Gill, B. S.**, Transstadial transmission of *Theileria annulata* through common ixodid ticks infesting Indian cattle, *Parasitology*, 71, 1, 1975.

35. **Mazlum, Z.**, Transmission of *Theileria annulata* by the crushed infected unfed *Hyalomma dromedarii*, *Parasitology*, 59, 597, 1969.

36. **Pipano, E., Samish, M., and Krigel, Y.**, Relative infectivity of *Theileria annulata* (Dschunkowsky and Luhs, 1904) stabilates derived from female and male *Hyalomma excavatum* (Koch, 1844) ticks, *Vet. Parasitol.*, 10, 21, 1982.

37. **Jongejan, F., Morzaria, S. P., Om El Hassan Mustafa, and Latif, A. A.**, Infection rates of *Theileria annulata* in the salivary glands of the tick *Hyalomma marginatum rufipes*, *Vet. Parasitol.*, 13, 121, 1983.

38. **Walker, A. R., Latif, A. A., Morzaria, S. P., and Jongejan, F.**, Natural infection rates of *Hyalomma anatolicum anatolicum* with theileria in Sudan, *Res. Vet. Sci.*, 35, 87, 1983.

39. **Sergent, E.**, Infections latentes et infections actives, *Arch. Inst. Pasteur Alger.*, 16, 3, 1938.

40. **Donatien, A. and Lestoquard, F.**, La premunition et les vaccinations premunitives dans la pathologie Veterinaire, *Bull. Acad. Vet. France*, 8, (2) 1, 1935.

41. **Conrad, P. A., Kelly, B. G., and Brown, C. G. D.**, Intraerythrocytic schizogony of *Theileria annulata*, *Parasitology*, 91, 67, 1985.

42. **Velu, H.**, Existe-t-il rechutes dans la theilerose Nord-Africaine, *Rev. Vet. Milit.*, 17, 779, 1933.

43. **Sharma, R. D. and Brown, C. G. D.**, *In Vitro* studies on two strains of *Theileria annulata*, in *Advances in the Control of Theileriosis*, Irvin, A. D., Cunningham, M. P., and Young, A. S., Eds., Martinus Nijhoff, Dordrecht, Netherlands, 1981, 140.

44. **Schindler, R. and Wokatsch, R.**, Versuche zur Differenzierung der Theilerienspezies des Rindes durch serologische Untersuchungen, *Z. Tropenmed. Parasitol.*, 16, 17, 1965.

45. **Markov, A. A., Stepanova, N. I., Laptev, V. I., Dubovy, S. Z., and Storozhev, U. I.**, Application of xenodiagnostic and complement fixation test for cattle *Theileria* differentiation, in *Proc. 1st Int. Congr. of Parasitology*, Corradetti, A., Eds., Pergamon Press, Oxford, 1964, 275.

46. **Tutushin, M. I.**, Some problems of immunity to *Theileria annulata* infection in cattle, *Veterinariya (Moscow)*, 43 (6), 46, 1966.

47. **Konyukhov, M. P. and Poluboyarova, G. V.**, Complement fixation tests for detecting immunity in cattle recovering from *Theileria annulata* infection, *Veterinariya (Moscow)*, 44 (10), 56, 1967.

48. **Stepanova, N. I.**, The complement fixation test in diagnosis and differentiation of blood parasites, *Veterinariya (Moscow)*, 44 (1), 55, 1968.

49. **Tutushin, M. I.**, Haemagglutination and HI tests for *Theileria annulata* infection, *Veterinariya (Moscow)*, 39 (6), 47, 1962.

50. **Pipano, E. and Cahana, M.**, Measurement of the immune response to vaccine from tissue culture of *Theileria annulata* by fluorescent antibody test, *J. Protozool.*, 15 (Suppl.), 45, 1968.

51. **Pipano, E. and Cahana, M.**, Fluorescent antibody test for the serodiagnosis of *Theileria annulata*, *J. Parasitol.*, 55, 765, 1969.

52. **Askarov, E. M.**, Immunofluorescent reaction for studying the antigenic and immunogenic properties of *Theileria annulata*, *Veterinariya (Moscow)*, (6), 74, 1975.

53. **Gray, M. A., Luckins, A. G., Rae, P. F., and Brown, C. G. D.**, Evaluation of an enzyme immunoassay for serodiagnosis of infections with *Theileria parva* and *T. annulata*, *Res. Vet. Sci.*, 29, 360, 1980.

54. **Goldman, M. and Pipano, E.**, Specific IgM and IgG antibodies in cattle immunized or infected with *Theileria annulata*, *Tropenmed. Parasitol.*, 29, 85, 1978.

55. **Dhar, S. and Gautam, O. P.**, A note on the use of hyperimmune serum in bovine tropical theileriasis, *Indian Vet. J.*, 55, 738, 1978.

56. **Samad, M. A., Dhar, S., and Gautam, O. P.**, Effect of humoral antibodies on *Theileria annulata* infection of cattle, *Haryana Agric. Univ. J. Res.*, 14, 441, 1984.

57. **Muhammed, S. I., Lauermann, L. H., and Johnson, L. W.**, Effect of humoral antibodies on the course of *Theileria parva* infection (East Coast fever) of cattle, *Am. J. Vet. Res.*, 36, 399, 1975.

58. **Mahoney, D. F. and Ross, D. R.**, Epizootiological factors in the control of bovine babesiosis, *Aust. Vet. J.*, 48, 292, 1972.

59. **Gray, M. A. and Brown, C. G. D.**, *In vitro* neutralization of theilerial sporozoite infectivity with immune serum, in *Advances in the Control of Theileriosis*, Irvin, A. D., Cunningham, M. P., and Young, A. S., Eds., Martinus Nijhoff, Dordrecht, Netherlands, 1981, 127.

60. **Preston, P. M. and Brown, C. G. D.**, Inhibition of lymphocyte invasion by sporozoites and the transformation of trophozoite infected lymphocytes *in vitro* by serum from *Theileria annulata* immune cattle, *Parasite Immunol.*, 7, 301, 1985.

61. **Musoke, A. J., Nantulya, V. M., Buscher, G., Masake, R. A., and Otim, B.,** Bovine immune response to *Theileria parva:* neutralizing antibodies to sporozoites, *Immunology,* 45, 663, 1982.
62. **Musoke, A. J., Nantulya, V. M., Rurangirwa, F. R., and Buscher, G.,** Evidence for a common protective antigenic determinant on sporozoites of several *Theileria parva* strains, *Immunology,* 52, 231, 1984.
63. **Dobblelaere, D. A. E., Spooner, P. R., Barry, W. C., and Irvin, A. D.,** Monoclonal antibody neutralizes the sporozoite stage of different *Theileria parva* stocks, *Parasite Immunol.,* 6, 361, 1984.
64. **Irvine, A. D. and Morrison, W. I.,** Immunopathology, immunology, and immunoprophylaxis of *Theileria* infections, in *Immune Responses in Parasitic Infections: Immunology, Immunopathology, and Immunoprophylaxis,* Vol. 3, Soulsby, E. J. L., Ed., CRC Press, Boca Raton, FL, 1987, 223.
65. **Hulliger, L., Wilde, J. K. H., Brown, C. G. D., and Turner, L.,** Mode of replication of *Theileria* in cultures of bovine lymphocytic cells, *Nature (London),* 203, 728, 1964.
66. **Cox, F. E. G.,** Cell mediated immunity in theileriosis, *Nature (London),* 295, 14, 1982.
67. **Samad, M. A., Dhar, S., Gautam, O. P., and Kaura, Y. K.,** T- and B-lymphoid cell population in calves immunized against *Theileria annulata, Vet. Parasitol.,* 13, 109, 1983.
68. **Singh, D. K., Jagdish, S., and Gautam, O. P.,** Cell-mediated immunity in tropical theileriasis (*Theileria annulata* infection), *Res. Vet. Sci.,* 23, 391, 1977.
69. **Rehbein, G., Ahmed, J. S., Schein, E., Horchner, F., and Zweygarth, E.,** Immunological aspects of *Theileria annulata* infection in calves. II. Production of macrophage migration inhibition factor (MIF) by sensitized lymphocytes from *Theileria annulata* infected calves, *Tropenmed. Parasitol.,* 32, 154, 1981.
70. **Rehbein, G., Ahmed, J. S., Zweygarth, E., Schein, E., and Horchner, F.,** Immunological aspects of *Theileria annulata* infection in calves. I. E, EA, and EAC rosette forming cells in calves infected with *T. annulata, Tropenmed. Parasitol.,* 32, 101, 1981.
71. **Samad, M. A., Gautam, O. P., Dhar, S., and Kaura, Y. K.,** Lymphocytic response in calves immunized against *Theileria annulata, Indian J. Anim. Sci.,* 54, 757, 1984.
72. **Preston, P. M. and Brown, C. G. D.,** Transformation of bovine lymphocytes in co-cultivation with autologous *Theileria annulata*-transformed cell lines, *Trans. R. Soc. Trop. Med. Hyg.,* 75, 328, 1981.
73. **Preston, P. M., Brown, C. G. D., and Spooner, R. L.,** Cell-mediated cytotoxicity in *Theileria annulata* infection of cattle with evidence for BoLA restriction, *Clin. Exp. Immunol.,* 53, 88, 1983.
74. **Shiels, B. R., McDougall, C., Tait, A., and Brown, C. G. D.,** Identification of infected-associated antigens in *Theileria annulata* transformed cells, *Parasite Immunol.,* 8, 69, 1986.
75. **Rafyi, A., Maghami, G., and Hooshmand-Rad, P.,** Sur la virulence de *Theileria annulata* (Dschunkowsky et Luhs, 1904) et la premunition contre la theileriose bovine en Iran, *Bull. Off. Int. Epizool.,* 64, 431, 1965.
76. **Pipano, E., Weisman, Y., and Benado, A.,** The virulence of four local strains of *Theileria annulata, Refu. Vet.,* 31, 59, 1974.
77. **Askarov, E. M.,** Immunofluorescent reaction for studying the antigenic and immunogenic properties of *Theileria annulata, Veterinariya (Moscow),* (6), 74, 1975.
78. **Shirinov, N., Agaev, A. A., Mirzabekov, D. A., Mamedov, A. M., and Godzhaev, A. N.,** Control of *Theileria annulata* infection of cattle (in Azerbaijan) *Veterinariya (Moscow),* (3), 62, 1975.
79. **Sergent, E., Donatien, A., Parrot, L., and Lestoquard, F.,** Theilerioses bovines de l'Afrique du Nord et du Proche-Orient, *Arch. Inst. Pasteur Alger.,* 13, 472, 1935.
80. **Adler, S. and Ellenbogen, V.,** Observation on theileriosis in Palestine, *Arch. Inst. Pasteur Alger.,* 13, 451, 1935.
81. **Adler, S. and Ellenbogen, V.,** Remarks on the relationship between the Palestinian and Algerian pathogenic *Theileria, Arch. Inst. Pasteur Alger.,* 14, 66, 1936.
82. **Gill, B. S., Bansai, G. C., Bhattacharyulu, Y., Kaur, D., and Singh, A.,** Immunological relationship between strains of *Theileria annulata,* Dschunkowsky and Luhs 1904, *Res. Vet. Sci.,* 29, 93, 1980.
83. **Pipano, E.,** Basic Principles of *Theileria annulata* control, in *Theileriosis,* Henson, J. B. and Campbell, M., Eds., International Development Research Center, Ottawa, Canada, 1977, 55.
84. **Rasulov, I. K.,** Immunological properties of strains of *Theileria annulata, Veterinariya (Moscow),* 40, (6), 52, 1963.
85. **Irvin, A. D. and Boarer, C. D. H.,** Some implications of a sexual cycle in *Theileria, Parasitology,* 80, 571, 1980.
86. **Tsur, I., Hadani, A., Pipano, E., Cwilich, R., Senft, Z., and Cohen, R.,** Outbreaks of theileriasis in premunized cattle, *Refu. Vet.,* 21, 167, 1964.
87. **Musisi, F. L., Kilgour, V., Brown, C. G. D., and Morzaria, S. P.,** Preliminary investigations on isoenzyme variants of lymphoblastoid cell lines infected with *Theileria* species, *Res. Vet. Sci.,* 30, 38, 1981.
88. **Melrose, T. R., Brown, C. G. D., and Sharma, R. D.,** Glucose phosphate isomerase isoenzyme patterns in bovine lymphoblastoid cell lines infected with *Theileria annulata* and *T. parva,* with an improved enzyme visualisation method using meldola blue, *Res. Vet. Sci.,* 29, 298, 1980.

89. **Melrose, T. R., Brown, C. G. D., Morzaria, S. P., Ocama, J. G. R., and Irvin, A. D.**, Glucose phosphate isomerase polymorphism in *Theileria annulata* and *T. parva*, *Trop. Anim. Health Prod.*, 16, 239, 1984.

90. **Melrose, T. R., Walker, A. R., and Brown, C. G. D.**, Identification of *Theileria* infections in the salivary glands of *Hyalomma anatolicum anatolicum* and *Rhipicephalus appendiculatus* using isoenzyme electrophoresis, *Trop. Anim. Health Prod.*, 13, 70, 1981.

91. **Brown, C. G. D., Stagg, D. A., Purnell, R. E., Kanhai, G. K., and Payne, R. C.**, Infection and transformation of bovine lymphoid cells *in vitro* by infective particles of *Theileria parva*, *Nature (London)*, 245, 101, 1973.

92. **Brown, C. G. D.**, *In vitro* infection and transformation of lymphoid cells by sporozoites of *Theileria parva* and *T. annulata*, *J. S. Afr. Vet. Assoc.*, 50, 345, 1979.

93. **Kurtti, T. J. and Munderloh, U. G.**, Factors influencing the *in vitro* establishment of bovine lymphoblastoid cells infected with *Theileria parva*, in *Proc. 11th Int. Congr. Disease of Cattle*, Mayer, E., Ed., Bregman Press, Tel Aviv, 1980, 684.

94. **Kurtti, T. J., Munderloh, U. G., Irvin, A. D., and Buscher, G.**, *Theileria parva*: early events in the development of bovine lymphoblastoid cell lines persistently infected with macroschizonts, *Exp. Parasitol.*, 52, 280, 1981.

95. **Pipano, E., Goldman, M., Samish, M., and Friedhoff, K. T.**, Immunization of cattle against *Theileria annulata* using killed schizont vaccine, *Vet. Parasitol.*, 3, 11, 1977.

96. **Pipano, E.**, Development of schizonts in calves inoculated with red blood cell forms of *Theileria annulata*, *J. Protozool.*, 19(Suppl.), 54, 1972.

97. **Curnow, J. A.**, Studies on antigenic changes and strain differences in *Babesia argentina* infection, *Aust. Vet. J.*, 49, 279, 1973.

98. **Sergent, E., Donatien, A., Parrot, L., and Lestoquard, F.**, *Etudes sur les Piroplasmoses bovines*, Institut Pasteur d'Algerie, 1945, 723.

99. **Adler, S.**, Les piroplasmoses ches les bovides en Israel, *Bull. Off. Int. Epizool.*, 38, 570, 1952.

100. **Sturman, M.**, Tick fever in Israel, *Symp. for Veterinarian, Rehovoth*, Israel, 1959.

101. **Tsur, I.**, Multiplication *in vitro* of Koch bodies of *Theileria annulata*, *Nature (London)*, 156, 391, 1945.

102. **Tsur, I.**, Multiplication *in vitro* of Koch bodies of *Theileria annulata*, *Refu. Vet.*, 4, 86, 1947.

103. **Tsur, I.**, *Theileria annulata* et *Leishmania* en culture de tissue, in *15th Proc. Veterinary Congr.*, Stockholm, 1, 26, 1953.

103. **Tsur, I.**, *Theileria annulata* et *Leishmania* en culture de tissue, *15th Veterinary Congr.*, Stockholm, 1, 25, 1953.

104. **Tsur, I., and Adler, S.**, Cultivation of *Theileria annulata* schizonts in monolayer tissue cultures, *Refu. Vet.*, 19, 225, 1962.

105. **Tsur, I. and Adler, S.**, The cultivation of lymphoid cells and *Theileria annulata* schizonts from infected bovine blood, *Refu. Vet.*, 22, 62, 1965.

106. **Tsur, I., Adler, S., Pipano, E., and Senft, Z.**, Continuous growth of *Theileria annulata* schizonts in monolayer tissue culture, in *Proc. 1st Int. Congr. Parasitology*, Vol. 1, Corradetti, A., Ed., Pergamon Press, Oxford, 1964, 266.

107. **Pipano, E. and Tsur, I.**, Experimental immunization against *Theileria annulata* with a tissue culture vaccine, Laboratory trails, *Refu. Vet.*, 23, 186, 1966.

108. **Zablotskii, V. T.**, Use of tissue culture in the study of *Theileria annulata*, *Veterinariya (Moscow)*, (9), 66, 1967.

109. **Hooshmand Rad, P. and Hashemi Fesharki, R.**, The effect of virulence on cultivation of *Theileria annulata* strains in lymphoid cells which have been cultured in suspension, *Arch. Inst. Razi*, 20, 85, 1968.

110. **Van Den Ende, M. and Edlinger, E.**, Culture de lignees lymphocytaires bovines infectees par *Theileria annulata*, *Arch. Inst. Pasteur Tunis*, N. (1-2) 45, 1971.

111. **Al-Hammawi, H. M.**, The sanitary position and methods of control used in Iraq, *Bull. Off. Int. Epizool.*, 84, 563, 1975.

112. **Gill, B. S., Bhattacharyulu, Y., Kaur, D., and Singh, A.**, Vaccination against bovine tropical theileriasis (*Theileria annulata*), *Nature (London)*, 264, 355, 1976.

113. **Ozkoc, U., Vural, A., Onar, E., and Pipano, E.**, Isolation and cultivation of three Turkish strains of *Theileria annulata*, *J. Pendik Vet. Bakteriol. Serol. Inst.*, 10, 31, 1978.

114. **Ozkoc, U. and Pipano, E.**, Trials with cell culture vaccine against theileriosis in Turkey, in *Advances in the Control of Theileriosis*, Irvin, A., Cunningham, M. P., and Young, A. S., Eds., Martinus Nijhoff, Dordrecht, Netherlands, 1981, 256.

115. **Xia, W. J., Hou, Y. Z., Dai, R. L., and Liu, D.**, Development of Koch bodies of *Theileria annulata* from infected cattle in tissue culture, *Acta Vet. Zootech. Sin.*, 13 (2), 125, 1982.

116. **Brown, C. G. D.**, Propagation of *Theileria*, in *Practical Tissue Culture Applications*, Maramorosch, K. and Hirumi, H., Eds., Academic Press, New York, 1979, 223.

117. **Brown, C. G. D.,** *In vitro* cultivation of *Theileria,* in *The In Vitro Cultivation of the Pathogens of Tropical Diseases,* Rowe, D. S. and Hirumi, H., Eds., Schwabe, Basel, 1980, 127.

118. **Brown, C. G. D.,** *Theileria, In vtiro* cultivation of protozoan parasites, Jensen, J. B., Ed., CRC Press, Boca Raton, 1983, 243.

119. **Pipano, E.,** Immunization against *Theileria annulata* infection, in *Control of Ticks and Tick-Borne Diseases of Cattle, A Practical Field Manual,* Food and Agriculture Organization, Rome, 1982, 36.

120. **Hulliger, L.,** Cultivation of three species of *Theileria* in lymphoid cells *in vitro, J. Protozool.,* 12, 649, 1965.

121. **Sharma, R. D. and Brown, C. G. D.,** *In vitro* studies on two strains of *Theileria annulata,* in *Advances in the Control of Theileriosis,* Irvin, A. D., Cunningham, M. P., and Young, A. S., Eds., Martinus Nijhoff, Dordrecht, Netherlands, 1981, 140.

122. **Mutuzkina, Z. P.,** Cultivation of *Theileria annulata* in tissue cultures, (Russian), *Veterinariya, (Moscow),* (4), 56, 1975.

123. **Hooshmand-Rad, P.,** The growth of *Theileria annulata* infected cells in suspension culture, *Trop. Anim. Health Prod.,* 7, 23, 1975.

124. **Tsur, I. and Pipano, E.,** The successful infection of cattle with *Theileria annulata* parasites preserved in the frozen state, *Refu. Vet.,* 19, 110, 1962.

125. **Tsur, I. and Pipano, E.,** Survival of *Theileria annulata* schizonts at low temperatures, *J. Protozool.,* 10 (Suppl.), 35, 1963.

126. **Stepanova, N. I., Zablotskii, V. T., and Mutuzkina, Z. P.,** Storage in liquid nitrogen of cell cultures infected with *Theileria annulata, Veterinariya (Moscow),* (3), 61, 1975.

127. **Pipano, E.,** Schizonts and tick stages in immunization against *Theileria annulata* infection, in *Advance in the Control of Theileriosis,* Irvin, A. D., Cunningham, M. P., and Young, A. S., Eds., Martinus Nijhoff, Dordrecht, Netherlands, 1981, 242.

128. **Pipano, E.,** Immunization against *Theileria annulata* Infection, in *Proc. 20th World Veterinary Congr.,* Vol. 1, Papageorgiou Publishing, Thessaloniki, Greece, 1975, 487.

129. **Pipano, E.,** Sur quelques caracteristiques pathogeniques et immunologiques des schizontes de *Theileria annulata,* in *Proc. 18th World Veterinary Congr.,* Vol. 2, 1967, 749.

130. **Pipano, E. and Israel, V.,** Absence of erythrocyte forms of *Theileria annulata* in calves inoculated with schizonts from a virulent field strain grown in tissue culture, *J. Protozool.,* 18 (Suppl.), 37, 1971.

131. **Pipano, E.,** Immunization of calves with attenuated wild strains of *Theileria annulata,* in *Comptes-Rendus 1er Multicolloque Europeen de Parasitologie,* 1971, 202.

132. **Pipano, E. and Shkap, V.,** Attenuation of two Turkish strains of *Theileria annulata, J. Protozool.,* 3(Suppl. 26), 80A, 1979.

133. **Hashemi-Fesharki, R.,** Seven years study on immunization of cattle against *Theileria annulata* infection with lymphoid cell suspension culture vaccine in Iran, in *Fourth Int. Congr. Parasitol.,* 1978, 111.

134. **Pipano, E.,** Immune response of calves to varying numbers of attenuated schizonts of *Theileria annulata, J. Protozool.,* 17 (Suppl.), 31, 1970.

135. **Pipano, E.,** Virulence and immunogenicity of cultured *Theileria annulata* schizonts, *J. S. Afr. Vet. Assoc.,* 50, 332, 1979.

136. **Pipano, E., Klopfer, U., and Cohen, R.,** Inoculation of cattle with bovine lymphoid cell lines infected with *Theileria annulata, Res. Vet. Sci.,* 15, 388, 1973.

137. **Samish, M., Krigel, Y., Frank, M., Bin, C., Pipano, E.,** The transmission of *Theileria annulata* to cattle by *Hyalomma excavatum* infected with *Theileria* cultivated *in vitro, Refu. Vet.,* 41, 62, 1984.

138. U.S. Code of Federal Regulations. Animals and Animal Products, U.S. Government Printing Office, Washington, D.C., 1987, 354 and 384; Published by the Office Register National Archives and Records Administration.

139. **Pipano, E. and Cahana, M.,** Measurement of the immune response to vaccine from tissue culture of *Theileria annulata* by the fluorescent antibody test, *J. Protozool.,* 15 (Suppl.), 45, 1968.

140. **Pipano, E., Cahana, M., Feller, B., Shabat, Y., and David, E.,** A serological method for assessing the response to *Theileria annulata* immunization, *Refu. Vet.,* 26, 145, 1969.

141. **Frank, M., Pipano, E., and Rosenberg, A.,** The diagnosis of *Theileria annulata* infection by the indirect fluorescent antibody method using small quantities of blood dried on absorbent paper, *Refu. Vet.,* 28, 78, 1971.

142. **Hashemi-Fesharki, R. and Shad-Del, F.,** Vaccination of calves and milking cows with different strains of *Theileria annulata, Am. J. Vet. Res.,* 34, 1465, 1973.

143. **Stepanova, N. I., Zablotskii, V. T., Mutuzkina, Z. P., Rasulov, I. Kh., Umarov, I. S., and Tukhtaev, B. T.,** Live cell culture vaccine against *Theileria annulata* infection, *Veterinariya (Moscow),* (3), 69, 1977.

144. **Zablotskii, V. T.,** Duration of postvaccinal immunity in bovine theileriosis, *Tr. Vses. Inst. Eksp. Vet.,* 57, 56, 1983.

145. **Arshadi, M.,** Control of tick-borne diseases in Iran (anaplasmosis, piroplasmosis and theileriosis) XLIV General Session of the O.I.E. Committee, Paris, May 17 to 22, 1976.

146. **Brocklesby, D. W. and Bailey, K. P.,** The immunization of cattle against East Coast Fever (*Theileria parva* infection) using tetracyclines: a review of the literature and a reappraisal of the method, *Bull. Epizoot. Dis. Afr.,* 13, 161, 1965.

147. **Cunningham, M. P.,** Immunization of cattle against *Theileria parva,* in *Immunity to Blood Parasites of Animals and Man,* Miller, L. H., Pino, J. A., and McKelvey, J. J., Jr., Eds., Plenum Press, New York, 1977, 189.

148. **Purnell, R. E.,** East Coast Fever: some recent research in East Africa, in *Adv. Parasitol.,* 15, 83, 1977.

149. **Radley, D. E.,** Infection and treatment method of immunization against Theileriosis, in *Advances in the Control of Theileriosis,* Irvin, A. D., Cunningham, M. P., and Young, A. S., Eds., Martinus Nijhoff, Dordrecht, Netherlands, 1981, 227.

150. **Gill, B. S., Bhattacharyulu, Y., and Kaur, D.,** Immunization against bovine tropical theieriosis *(Theileria annulata* infection), *Res. Vet. Sci.,* 21, 146. 1976.

151. **Gill, B. S., Bhattacharyulu, Y., Kaur, D., and Singh, A.,** Chemoprophylaxis with tetracycline drugs in the immunization of cattle against *Theileria annulata* infection, *Int. J. Parasitol.,* 8, 467, 1978.

152. **Jagdish, S., Singh, D. K., Gautam, O. P., and Dhar, S.,** Chemoprophylactic immunization against bovine tropical theileriosis, *Vet. Rec.,* 104, 140, 1979.

153. **Pipano, E., Samish, M., Kriegel, Y., and Yeruham, I.,** Immunization of Friesian cattle against *Theileria annulata* by the infection - treatment method, *Br. Vet. J.,* 137, 416, 1981.

154. **Hadani, A., Cwilich, R., Rechav, Y., and Dinur, Y.,** Some methods for the breeding of ticks in the laboratory, *Refu. Vet.,* 26, 87, 1969.

155. **Samish, M.,** Mass-rearing devices for *Hyalomma anatolicum excavatum* (Acari: Ixodidae) on calves and rodents exposed to whole-body infestation, *J. Med. Entomol.,* 19, 6, 1982.

156. **Hoogstraal, H.,** Hyalomma, in *African Ixodoidea,* Vol. 1, U. S. Government Printing Office, 1956, 388.

157. **Samish, M., Ziv, M., and Pipano, E.,** Preparation of suspensions of *Hyalomma excavatum* ticks infected with *Theileria annulata, Vet. Parasitol.,* 13, 267, 1983.

158. **Srivastava, P. S. and Sharma, N. N.,** Studies on the infectivity of *Theileria annulata* infected nymphs, adults and ground tissues of the tick *Hyalomma anatolicum, Vet. Parasitol.,* 4, 83, 1978.

159. **Purnell, R. E., Boarer, C. D. H., and Peirce, M. A.,** *Theileria parva:* comparative infection rates of adult and nymphal *Rhipicephalus appendiculatus, Parasitology,* 62, 349, 1971.

160. **Purnell, R. E., Young, A. S., Brown, C. G. D., Burridge, M. J., and Payne, R. C.,** Comparative infectivity for cattle of stabilates of *Theileria lawrencei* (Sergenti) derived from adult and nymphal ticks, *J. Comp. Pathol.,* 84, 533, 1974.

161. **Samish, M. and Pipano, E.,** Preparation and application of *Theileria annulata* infected stabilate, in *Advances in the Control of Theileriosis,* Irvin, A. D., Cunningham, M. P., and Young, A. S., Eds., Martinus Nijhoff, Dordrecht, Netherlands, 1981, 253.

162. Immunization against *Theileria parva,* in *Ticks and Tick-Borne Disease Control, A Practical Field Manual,* Vol. 2, Food and Agriculture Organization, Rome, 1984, 457.

163. **Purnell, R. E., Ledger, M. A., Omwoyo, P. L., Payne, R. C., and Pierce, M. A.,** *Theileria parva* variation in the infection rate of the vector tick *Rhipicephalus appendiculatus, Int. J. Parasitol.,* 4, 513, 1974.

164. **Hadani, A., Tsur, I., Pipano, E., and Zenft, Z.,** Studies on the transmission of *Theileria annulata* by ticks (Ixodidea, Ixodidae). I. *Hyalomma excavatum, J. Protozool.,* 10, 35, 1963.

165. **Schreuder, B. E. C. and Uilenberg, G.,** Studies on Theileriidae (Sporozoa) in Tanzania. V. Preliminary experiments on a new method for infecting ticks with *Theileria parva* and *Theileria mutans, Tropenmed. Parasitol.,* 27, 422, 1976.

166. **Walker, A. R., Brown, C. G. D., Bell, L. J., and McKellar, S. B.,** Artificial infection of the tick *Rhipicephalus appendiculatus* with *Theileria parva, Res. Vet. Sci.,* 26, 264, 1979.

167. **Jongejan, F., Perie, N. M., Franssen, F. F. J., and Uilenberg, G.,** Artificial infection of ticks by percutaneous injection using deep-frozen blood, in *Advances in the Control of Theileriosis,* Irvin, A. D., Cunningham, P. M., and Young, A. S., Eds., Martinus Nijhoff, Dordrecht, Netherlands, 1981, 136.

168. **Walker, A. R. and McKellar, S. B.,** The maturation of *Theileria* in *Hyalomma anatolicum anatolicum* stimulated by incubation or feeding to produce sporozoites, *Vet. Parasitol.,* 13, 13, 1983.

169. **Reid, G. D. F. and Bell, L. J.,** The development of *Theileria annulata* in the salivary glands of the vector tick *Hyalomma anatolicum anatolicum, Ann. Trop. Med. Parasitol.,* 78, 409, 1984.

170. **Gautam, O. P.,** *Theileria annulata* infection in India, in *Tick-Borne Diseases and Their Vectors,* Wilde, J. K. D., Ed., University of Edinburgh Press, Edinburgh, 1976, 374.

171. **Samish, M. and Pipano, E.,** Development of infectivity in *Hyalomma detritum* (Schulze, 1919) ticks infected with *Theileria annulata* (Dschunkowsky and Luhs, 1904), *Parasitology,* 77, 375, 1978.

172. **Singh, D. K., Jagdish, S., Gautam, O. P., and Dhar, S.,** Infectivity of ground-up tick supernates prepared from *Theileria annulata* infected *Hyalomma anatolicum anatolicum, Trop. Anim. Health Prod.,* 11, 87, 1979.

173. **Walker, A. R., McKellar, S. B., Bell, L. J., and Brown, C. G. D.,** Rapid quantitative assessment of *Theileria* infection in ticks, *Trop. Anim. Health Prod.,* 11, 21, 1979.

174. **Cunningham, M. P., Joyner, L. P., Brown, C. G. D., Purnell, R. E., and Bailey, K. P.,** Infection of cattle with East Coast fever by inoculation of the infective stage of *Theileria parva* harvested from the tick vector *Rhipicephalus appendiculatus, Bull. Epizoot. Dis. Afr.,* 21, 235, 1973.

175. **Cunningham, M. P., Brown, C. G. D., Purnell, R. E., and Branagan, D.,** The preservation at low temperature of infective particles of *Theileria parva,* in *Proc. Second Int. Congr. Parasitology,* Washington, D.C., September 6 to 12, 1970, 60.

176. **Samish, M. and Pipano, E.,** Low temperature preservation of *Theileria annulata* infective particles derived from ticks, *J. Protozool.,* 24 (Suppl.), 19A, 1977.

177. **Neitz, W. O.,** Aureomycin in *Theileria parva* infection, *Nature (London),* 171, 34, 1953.

178. **Radley, D. E., Brown, C. G. D., Cunningham, M. D., Kimber, C. D., Musisi, F. L., Payne, R. C., Purnell, R. E., Stagg, S. M., and Young, A. S.,** East Coast fever. III. Chemoprophylactic immunization of cattle using oxytetracycline and a combination of theilerial strains, *Vet. Parasitol.,* 1, 51, 1975.

179. **Gill, B. S., Bhattacharyulu, Y., Kaur, D., and Singh, A.,** Immunisation of cattle against tropical theileriosis (*Theileria annulata* infection) by infection-treatment method, *Ann. Rech. Vet.,* 8, 285, 1977.

180. **Khanna, B. M., Dhar, S., and Gautam, O. P.,** Immunisation against bovine tropical theileriosis by using the infection and treatment method, in *Haemoprotozoan Diseases of Domestic Animals,* Gautam, O. P., Sharma, R. D., and Dhar, S., Eds., Department of Veterinary Medicine, Haryana Agricultural University, Hissar, India, 1980, 91.

181. **Morgan, D. W. T. and McHardy, N.,** The therapy and prophylaxis of theileriosis with a new naphtho-quinone buparvaquone (BW720C), *Proc. 14th World Congr. on Diseases of Cattle,* Coopers Annual Health, Dublin, 1986, 1271.

182. **Dhar, S., Malhotra, D. V., Bhushan, C., and Gautam, O. P.,** Chemoimmunoprophylaxis with bupar-vaquone against theileriosis in calves, *Vet. Rec.,* 120, 375, 1987.

183. **Manickam, R., Dhar, S., and Singh, R. P.,** Protection of cattle against *Theileria annulata* injection using *Corynebacterium parvum, Trop. Anim. Health Prod.,* 15, 209, 1983.

184. **Srivastava, P. S. and Sharma, N. N.,** Studies on the potential of immunoprophylaxis using *Theileria annulata* attenuated by Cobalt-60 irradiation in bovine lymphocytes, *Vet. Parasitol.,* 3, 23, 1977.

185. **Samantaray, S. N., Bhattacharyulu, Y., and Gill, B. S.,** Immunization of calves against bovine tropical theileriosis *(Theileria annulata)* with graded doses of sporozoites and irradiated sporozoites, *Int. J. Parasitol.,* 10, 355, 1980.

186. **Gautam, O. P., Sastry, K. N. U., Dhar, S., and Singh, R. P.,** Effect of irradiation on *Theileria annulata* particles derived from ticks, *Indian Vet. J.,* 59, 581, 1982.

187. **Manickam, R., Dhar, S., and Singh, R. P.,** Non-specific immunization against bovine tropical theileriosis. *(Theileria annulata)* using killed *Corynebacterium parvum, Vet. Parasitol.,* 13, 115, 1983.

188. **Pipano, E.,** Unpublished data.

INDEX